T0299717

DISCRETE MATHEMATICS AND ITS APPLICATIONS

Series Editor KENNETH H. ROSEN

INTRODUCTION TO COMBINATORIAL DESIGNS

SECOND EDITION

DISCRETE MATHEMATICS AND ITS APPLICATIONS

Series Editor
Kenneth H. Rosen, Ph.D.

Juergen Bierbrauer, Introduction to Coding Theory

Kun-Mao Chao and Bang Ye Wu, Spanning Trees and Optimization Problems

Charalambos A. Charalambides, Enumerative Combinatorics

Henri Cohen, Gerhard Frey, et al., Handbook of Elliptic and Hyperelliptic Curve Cryptography

Charles J. Colbourn and Jeffrey H. Dinitz, Handbook of Combinatorial Designs, Second Edition

Steven Furino, Ying Miao, and Jianxing Yin, Frames and Resolvable Designs: Uses, Constructions, and Existence

Randy Goldberg and Lance Riek, A Practical Handbook of Speech Coders

Jacob E. Goodman and Joseph O'Rourke, Handbook of Discrete and Computational Geometry, Second Edition

Jonathan L. Gross and Jay Yellen, Graph Theory and Its Applications, Second Edition

Jonathan L. Gross and Jay Yellen, Handbook of Graph Theory

Darrel R. Hankerson, Greg A. Harris, and Peter D. Johnson, Introduction to Information Theory and Data Compression, Second Edition

Daryl D. Harms, Miroslav Kraetzl, Charles J. Colbourn, and John S. Devitt, Network Reliability: Experiments with a Symbolic Algebra Environment

Leslie Hogben, Handbook of Linear Algebra

Derek F. Holt with Bettina Eick and Eamonn A. O'Brien, Handbook of Computational Group Theory

David M. Jackson and Terry I. Visentin, An Atlas of Smaller Maps in Orientable and Nonorientable Surfaces

Richard E. Klima, Neil P . Sigmon, and Ernest L. Stitzinger, Applications of Abstract Algebra with Maple™ and MATLAB® , Second Edition

Patrick Knupp and Kambiz Salari, Verification of Computer Codes in Computational Science and Engineering

William Kocay and Donald L. Kreher, Graphs, Algorithms, and Optimization

Donald L. Kreher and Douglas R. Stinson, Combinatorial Algorithms: Generation Enumeration and Search

Continued Titles

Charles C. Lindner and Christopher A. Rodgers, Design Theory

Hang T. Lau, A Java Library of Graph Algorithms and Optimization

Alfred J. Menezes, Paul C. van Oorschot, and Scott A. Vanstone, Handbook of Applied Cryptography

Richard A. Mollin, Algebraic Number Theory

Richard A. Mollin, Codes: The Guide to Secrecy from Ancient to Modern Times

Richard A. Mollin, Fundamental Number Theory with Applications

Richard A. Mollin, An Introduction to Cryptography, Second Edition

Richard A. Mollin, Quadratics

Richard A. Mollin, RSA and Public-Key Cryptography

Carlos J. Moreno and Samuel S. Wagstaff, Jr., Sums of Squares of Integers

Dingyi Pei, Authentication Codes and Combinatorial Designs

Kenneth H. Rosen, Handbook of Discrete and Combinatorial Mathematics

Douglas R. Shier and K.T. Wallenius, Applied Mathematical Modeling: A Multidisciplinary Approach

Jörn Steuding, Diophantine Analysis

Douglas R. Stinson, Cryptography: Theory and Practice, Third Edition

Roberto Togneri and Christopher J. deSilva, Fundamentals of Information Theory and Coding Design

W. D. Wallis, Introduction to Combinatorial Designs, Second Edition

Lawrence C. Washington, Elliptic Curves: Number Theory and Cryptography

DISCRETE MATHEMATICS AND ITS APPLICATIONS
Series Editor KENNETH H. ROSEN

INTRODUCTION TO COMBINATORIAL DESIGNS

SECOND EDITION

W.D. WALLIS

Boca Raton London New York

Chapman & Hall/CRC is an imprint of the
Taylor & Francis Group, an informa business

Front cover art: Shallow Waters by Burghilde Gruber

Chapman & Hall/CRC
Taylor & Francis Group
6000 Broken Sound Parkway NW, Suite 300
Boca Raton, FL 33487-2742

International Standard Book Number-13: 978-1-58488-838-3 (Hardcover)

Library of Congress Cataloging-in-Publication Data

Wallis, W. D.
Introduction to combinatorial designs / W.D. Wallis. -- 2nd ed.
p. cm. -- (Discrete mathematics and its applications)
Rev. ed. of: Combinatorial designs. c1988.
Includes bibliographical references and index.
ISBN-13: 978-1-58488-838-3 (alk. paper)
ISBN-10: 1-58488-838-5 (alk. paper)
1. Combinatorial designs and configurations. I. Wallis, W. D. Combinatorial designs. II. Title. III. Series.

QA166.25.W34 2007
511'.6--dc22 2007003650

Visit the Taylor & Francis Web site at
http://www.taylorandfrancis.com

and the CRC Press Web site at
http://www.crcpress.com

This book is dedicated to the coffee growers of the world, without whom many of its theorems would not exist.

(Paul Erdös said, "A mathematician is a machine for turning coffee into theorems.")

This book is dedicated to the coffee growers of the world,
without whom many of the theorems would not exist.

(Paul Erdős said, "A mathematician is a machine
for turning coffee into theorems.")

Contents

List of Figures

Preface

This book is a revised edition of my 1988 *Combinatorial Designs,* after eighteen years.

I have deleted some things from the first edition that some readers (myself included) thought were a little too specialized. In their place, I have added some material that reflects recent results, and a very broad outline of some applications. I have also significantly reorganized the development, with the necessary algebra and number theory background being incorporated into the chapters where they are first used.

It has been suggested that combinatorial theory is the fastest-growing area of modern mathematics. The largest part of the subject is graph theory, as evidenced by the huge number of research publications in that part of the discipline. Accordingly, there have appeared a good many textbooks devoted to graph theory, suitable for use at junior, senior, and graduate levels. A second major part of combinatorial mathematics is enumeration. The third—the one we treat here—is the theory of combinatorial designs.

Design theory began with Euler's invention of the Latin square 200 years ago. The next area of particular interest was the balanced incomplete block design and the special case of finite projective and affine planes. This interest was fanned by the statistical interpretation of designs and by the geometrical nature of planes. Many textbooks are written as though these, what we might call "classical designs," are the only combinatorial designs. However, over the last 25 years, combinatorial researchers have discussed a wider range of designs: one-factorizations, Room squares, and other designs based on unordered pairs, various tournament designs, nested designs, and so on. This interest has been spurred by the discovery of numerous applications. These include coding theory and cryptography, both theoretical and applied computer science, design of enterprise networks and other aspects of communications theory, industrial engineering, chemistry and biology.

In writing the book my aims were:

- to enable instructors to present combinatorial designs to undergraduate and beginning graduate students in more than a cursory fashion;
- to give a good groundwork in the classical areas of design theory: block designs, finite geometries, and Latin squares; and then to introduce some modern extensions of design theory;

- to lead students toward the current boundaries of the subject—I hope they will be capable of reading current research papers after studying the book;
- to present a well-rounded textbook, providing motivation, instructive examples, theorems, and exercises for all its topics.

After an overview we introduce balanced designs. There are two chapters on finite geometries; Chapters 5 and 6 go further into balanced incomplete block designs (difference methods, residual and derived designs, resolvability), and Chapter 7 is on the existence theorem of Bruck, Ryser and Chowla. There are two chapters each on Latin squares, one-factorizations and triple systems, and chapters on Hadamard matrices and Room squares. We finish with a rough sketch of some statistical and other applications of designs. The applications are outside our scope, but a number of references for further reading are given in that chapter.

I have included exercises with each section, and selected answers are given. Exercises with answers or (complete or partial) solutions are indicated by an asterisk after the question number. More complete solutions will appear on the web at

`http://www.math.siu.edu/Wallis/designs`

and errata, comments, updates and extensions will appear on that site.

I would like to thank Ebad Mahmoodian and Katherine Heinrich for their extensive comments on the first edition, and Elizabeth Billington for her careful reading and helpful suggestions on this edition.

W. D. Wallis
Department of Mathematics
Southern Illinois University
Carbondale, IL 62901-4408

Chapter 1

Basic Concepts

1.1 Combinatorial Designs

Informally, one may define a *combinatorial design* to be a way of selecting subsets from a finite set in such a way that some specified conditions are satisfied. Although we shall not define the type of condition precisely, they generally involve *incidence*: set membership, set intersection, and so on.

As an example, say it is required to select 3-sets from the seven objects $\{0, 1, 2, 3, 4, 5, 6\}$ in such a way that each object occurs in three of the 3-sets and every intersection of two 3-sets has precisely one member. The solution to this problem—the way of selecting the 3-sets—is a combinatorial design. One possible solution is

$$\{012, 034, 056, 135, 146, 236, 245\}$$

where 012 represents $\{0, 1, 2\}$ and so on. (We shall often omit set brackets, when no confusion can arise.)

The example above does not use the arithmetical properties of the numbers $0, 1, 2, 3, 4, 5, 6$. In fact, we do not even know what these symbols stand for. They could be the corresponding integers; they could be the integers modulo 7, or modulo some larger base; or they could be labels attached to any seven objects. The elements of the set are not viewed from an arithmetical viewpoint (although arithmetic will sometimes be very useful in showing that a design exists or in solving other problems).

Our example involved showing that a certain design exists. Other problems can arise. For example, in our problem say there are n 3-sets. Then they have between them $3n$ entries. But there are seven objects, and each occurs in three of the subsets. So when they are listed, there are 21 entries. Thus $3n = 21$ and $n = 7$. In other words, the number of subsets is determined; we can prove that, if a solution exists, it must contain precisely seven subsets. So one can establish *properties* of possible solutions.

We now prove the uniqueness of the given solution of our example, in the following sense. If we have any one solution, we can obviously obtain another by relabeling the objects in some way. For example, the relabeling of "1" as "2" and "2" as "1" in the given solution leads to the different-looking answer

$$\{012, 034, 056, 235, 246, 136, 145\}.$$

But this is essentially the same as the original. We shall consider two solutions to be *isomorphic* if the list of 3-sets in one can be obtained from the list of 3-sets of the other by carrying out some permutation of the symbols consistently. What we now show is that the solution of the problem is unique, up to isomorphism.

Assume that we have a solution to the problem. There must be three 3-sets that contain the object 0. Suppose they are $0ab$, $0cd$, and $0ef$, where a, b, c, d, e and f are chosen from the numbers 1 through 6. Since the intersection of any two sets is to have one entry, and since 0 is common to $0ab$ and $0cd$, we must have

$$a \neq c, a \neq d, b \neq c, b \neq d.$$

Moreover, since $0ab$ is to have three members, we know that

$$a \neq b, a \neq 0, b \neq 0.$$

By similar observations we see that a, b, c, d, e, and f must all be different from each other and must all be nonzero. So

$$\{a, b, c, d, e, f\} = \{1, 2, 3, 4, 5, 6\}.$$

Whichever ordering is used, we may permute the symbols at will, so we may take

$$a = 1, b = 2, c = 3, d = 4, e = 5, f = 6$$

up to isomorphism, and the design contains the sets

$$012, 034, 056.$$

Now 1 belongs to three sets. One of them is 012. Suppose $1gh$ is another. Since two sets have one common member, we have

$$g \neq 0, g \neq 2, h \neq 0, h \neq 2$$

(consider the two sets 012 and $1gh$). Also,

$$|\{g, h\} \cap \{3, 4\}| = 1.$$

If h is the common member, we can swap the labels g and h; if 4 is the common member, we can swap the labels 3 and 4. So up to isomorphism, we can assume that

$$g = 3, h \neq 3, h \neq 4.$$

If $h = 6$, we can exchange labels 5 and 6. So we can assume that $h = 5$. Therefore, two of the sets containing 1 are 012 and 135. It is easy to see that the third set cannot contain 0, 2, 3, or 5, so it is 146.

We next consider the sets containing 2. In the same way we see that they must be

$$012, 236, 245.$$

So the design is uniquely determined, up to isomorphism.

We could also show that our problem has no solution unless the number of objects is precisely seven. More generally, suppose we replaced "3" with "k": Each object is to belong to precisely k of the k-sets. Then we could prove that exactly $k^2 - k + 1$ objects are needed. But this necessary condition is not sufficient: Even with the correct number of elements, there is no solution for $k = 7$. This will be proven in Section 7.2.

Finally, our example design has the following property. If you select any two of the seven objects, there is exactly one 3-set that contains both of them. We say that the design is *balanced*; this important property will be explored in Section 1.3.

Exercises 1.1

1.1.1* Show that there exists a combinatorial design on the nine symbols $\{0, 1, 2, 3, 4, 5, 6, 7, 8\}$, that consists of 12 3-sets with the property that any two symbols occur together in exactly one 3-set. Show that the design is uniquely defined up to isomorphism.

1.1.2* A combinatorial design consists of the selection of 2-sets and 3-sets from the set {0, 1, 2, 3, 4, 5}, such that:

(a) Every pair of symbols occur together in exactly one subset;

(b) There are at least three 3-sets.

Prove that no two 3-sets can be disjoint. Show that the number of 3-sets can be either three or four, but no larger; and show that there are precisely two nonisomorphic solutions, one with three and one with four 3-sets.

1.1.3 A tennis club has 20 members. One day the members schedule exactly 14 singles (that is, two-player) matches among themselves, with each member playing in at least one game. Prove that the schedule includes a set of six games that are disjoint—that is, no two of them contain a common player—but no set of seven disjoint matches.

1.1.4 Can problem 1.1.3 be generalized to n players and a schedule of m matches?

1.1.5 It is required to select 3-sets from the seven objects $\{0, 1, 2, 3, 4, 5, 6\}$ in such a way that each object occurs in three of the 3-sets and every pair of objects belongs to precisely one 3-set. Show that every intersection of two 3-sets has precisely one member.

1.2 Some Examples of Designs

Many examples of combinatorial designs have been discussed in the literature. We shall define some of these, to provide examples of the concept. Some of these ideas will be restated when we consider the concepts again in later chapters.

Linked Design. A *linked design* is a way of selecting subsets from a set in such a way that any two subsets have intersection size, μ, where μ is a constant for the design. Such a constant is called a *parameter* of the design; one might say that a linked design on v objects, in which every pair of blocks has intersection size μ, is a "linked design with parameters (v, μ)." The designs discussed in Section 1.1 were linked designs with parameters $(7, 1)$, but with certain other conditions (constant subset size, and every element belonging to the same number of subsets).

Balanced Design. A way of selecting subsets from a set is called *balanced* if any two elements occur together in precisely λ of the subsets, where λ is a constant. Balanced designs are similar to linked designs, and we shall see that they are in a sense equivalent classes of objects. However, balanced designs have significant applications, particularly in statistical theory, and will be studied extensively in this book. Of special interest are *balanced incomplete block designs*, which have constant subset size; see below, Section 1.3.

One-Factorization. A *one-factorization* on a set S of n symbols is a way of organizing the $n(n-1)/2$ unordered pairs of members of S into classes called *one-factors*, or simply *factors*, such that every pair belongs to precisely one factors and such that every symbol belongs to precisely one member of each class. A one-factorization is a combinatorial design; one may interpret the phrase "way of selecting subsets" in the definition to refer to the selection of (all $n(n-1)/2$) unordered pairs from the set S, and interpret the arrangement of these pairs into classes as a "specified condition"; alternatively, one could say that the universal "finite set" is the set of unordered pairs on S. A one-factorization has only one parameter, the number n, the size of the symbol set S.

Latin Square. A *Latin square* L of side n is an $n \times n$ array whose entries all come from some n-set S, with the property that each row and each column is a permutation of S. One can either interpret the rows of L as subsets (of size n) that are ordered, or interpret the entries of L as subsets (of size 1) that are subjected to a two-dimensional positioning rule. Examples of Latin squares are

$$\boxed{\begin{array}{cc} 1 & 2 \\ 2 & 1 \end{array}} \quad \boxed{\begin{array}{ccc} 1 & 3 & 2 \\ 3 & 2 & 1 \\ 2 & 1 & 3 \end{array}} \quad \boxed{\begin{array}{ccc} 1 & 3 & 2 \\ 2 & 1 & 3 \\ 3 & 2 & 1 \end{array}}$$

Graph. A *graph* $G = G(V, E)$ consists of a finite set V of objects called *vertices* (or *points*), and a set E of unordered pairs of members of V called *edges.* This definition is consistent with the conventions that the edges of a graph are not directed, that the two vertices constituting an edge (its "*endpoints*") are distinct, and that two vertices belong together to no edge or to one edge ("*are adjacent*") but to no more. However, the reader should be aware that the word "graph" is sometimes used for objects with directed edges, with loops on vertices, with two or more edges with the same endpoints ("multiple edges") or with infinite vertex sets. If this more general terminology is used, "graphs" without loops or multiple edges are called *simple.* For our part, we use the word *multigraph* for the case where multiple edges are allowed.

Graph theory is a large area of study in combinatorial mathematics, and the reader has probably already encountered it. It is not our intention to claim that graph theory is a part of design theory, and the methods of approach of a graph theorist and a design theorist are often (but not always) quite different. However, it should be noted that a graph is a form of design. Moreover, the terminology of graph theory is occasionally useful to us; see Chapter 10 in particular.

Exercises 1.2

1.2.1 Prove that the parameter of a one-factorization must be even.

1.2.2* Construct a one-factorization on the sets $\{1, 2, 3, 4\}$.

1.2.3 Construct a one-factorization on the set $\{1, 2, 3, 4, 5, 6\}$.
Hint: You might as well start with factor $\{12 \ \ 34 \ \ 56\}$. *(Factorizations starting* $\{12 \ \ 35 \ \ 46\}$, *for example, can be transformed by exchanging all 4's and 5's throughout.)*

1.2.4 Prove that there are precisely two Latin squares of side 3 with first row

$$\boxed{\begin{array}{ccc} 1 & 3 & 2 \end{array}}$$

namely, those shown in this section.

1.2.5 Exhibit a Latin square of side 4.

1.2.6 Graphs G and H are isomorphic if there is a bijection ϕ from the vertex set of G to the vertex set of H such that, when $\{x, y\}$ is an edge of G, then $\{x\phi y\phi\}$ is an edge of H.

1.2.7* How many different graphs are there with vertex set $\{1, 2, 3\}$? How many isomorphism classes of these graphs are there?

1.2.8 What is the smallest value of v such that there exist nonisomorphic graphs on v vertices that have the same number of edges?

1.3 Block Designs

The members of the universal set S in a combinatorial design are usually called *treatments*, or *varieties*, and the subsets chosen are called *blocks*. S is called the *support* of the design. We shall say that a design of a certain type is a *block design* if in the definition of the type of design, the blocks are simply unordered sets of treatments and there is no structural ordering or pattern in the blocks. This vague definition (containing the undefined phrase "structural ordering or pattern") is really a statement of intention: If the blocks are ordered internally (as in a Latin square) or arranged in a pattern (as in a one-factorization), we shall usually study them in a different way from a linked design, for example.

There is no requirement that all the blocks in a design should be different. If two blocks have the same set of elements, we say there is a "repeated block." A design that has no repeated block is called *simple*.

We now define an important class of block design. We say a design is *regular* if every treatment occurs equally often in the design. Formally, a *regular design* based on a v-set S is a collection of k-sets from S such that every member of S belongs to r of the k-sets. It is usual to write b for the number of blocks in a design. So a regular design has four parameters: v, b, r, and k. However, they are not independent.

THEOREM 1.1

In any regular design,

$$bk = vr. \tag{1.1}$$

Proof. One counts, in two different ways, all the ordered pairs (x, y) such that treatment x belongs to block y. Since every treatment belongs to r blocks, there are r ordered pairs for each treatment, so the number is vr. Similarly, each block contributes k ordered pairs, so the summation yields bk. Therefore $bk = vr$. □

The number of blocks that contain a given treatment is called the *replication number* or *frequency* of that treatment. So the defining characteristics of a regular design are that all elements have the same replication number and that all blocks have the same size.

If all v treatments occur in a block of a design, that block is called *complete*. If a regular design has that property, then $k = v$ and obviously $r = b$. Such a design is a complete design, and it has very little interest unless some further structure is imposed (as, for example, in a Latin square). We say that a design is *incomplete* if at least one block is incomplete. If $v = b$, the design is called *symmetric*.

If x and y are any two different treatments in an incomplete design, we shall refer to the number of blocks that contain both x and y as the *covalency* of x and y, and write it as λ_{xy}. Many important properties of block designs are concerned with this covalency function. The one that has most frequently been studied is the property of *balance:* a *balanced incomplete block design*, or BIBD, is a regular incomplete design in which λ_{xy} is a constant, independent of the choice of x and y. These designs were defined (essentially as a puzzle) by Woolhouse [181], in the annual *Lady's and Gentleman's Diary*, a "Collection of mathematical puzzles and aenigmas" that he edited; Yates [188] first studied them from a statistical viewpoint.

It is usual to write λ for the constant covalency value, in a balanced incomplete block design, λ is sometimes called the *index* of the design. One often refers to a balanced incomplete block design by using the five parameters (v, b, r, k, λ); for example, we say "a $(9, 12, 4, 3, 1)$-design" or "a $(9, 12, 4, 3, 1)$-BIBD" to mean a balanced incomplete block design with 12 blocks of size 3, based on 9 treatments, with replication number 4 and index 1.

A balanced design with $\lambda = 0$, or a null design, is often called "trivial," and so is a complete design. We shall demand that a balanced incomplete block design be not trivial; since completeness is already outlawed, the added restriction is that $\lambda > 0$.

THEOREM 1.2

In a (v, b, r, k, λ)-BIBD,

$$r(k - 1) = \lambda(v - 1). \tag{1.2}$$

Proof. Consider the blocks of the design that contain a given treatment, x say. There are r such blocks. Because of the balance property, every treatment other than x must occur in λ of them. So if we list all entries of the blocks, we list x r times and we list every other treatment λ times.

The list contains rk entries. So

$$rk = r + \lambda(v-1),$$
$$r(k-1) = \lambda(v-1).$$

□

The word *incidence* is used to describe the relationship between blocks and treatments in a design. One can say block B is incident with treatment t, or treatment t is incident with block B, to mean that t is a member of B. A block design may be specified by its *incidence matrix:* If the design has b blocks $B_1, B_2, \ldots, B_b$ and v treatments $t_1, t_2, \ldots, t_v$, define a $v \times b$ matrix A with (i, j) entry a_{ij} as follows:

$$a_{ij} = \begin{cases} 1 & \text{if } t_i \in B_j; \\ 0 & \text{otherwise.} \end{cases}$$

This matrix A is called the *incidence matrix of the design.* The definition means that each block corresponds to a column of the incidence matrix, and each treatment corresponds to a row. For example, the regular design with treatments t_1, t_2, t_3, t_4 and blocks $B_1 = \{t_1, t_2\}$, $B_2 = \{t_3, t_4\}$, $B_3 = \{t_1, t_4\}$, $B_4 = \{t_2, t_3\}$, has incidence matrix

$$A = \begin{bmatrix} 1\,0\,1\,0 \\ 1\,0\,0\,1 \\ 0\,1\,0\,1 \\ 0\,1\,1\,0 \end{bmatrix}.$$

As another example, observe that the following blocks form a $(6, 10, 5, 3, 2)$-design:

123, 124, 135, 146, 156,
236, 245, 256, 345, 346.

This design has incidence matrix

$$\begin{bmatrix} 1\,1\,1\,1\,1\,0\,0\,0\,0\,0 \\ 1\,1\,0\,0\,0\,1\,1\,1\,0\,0 \\ 1\,0\,1\,0\,0\,1\,0\,0\,1\,1 \\ 0\,1\,0\,1\,0\,0\,1\,0\,1\,1 \\ 0\,0\,1\,0\,1\,0\,1\,1\,1\,0 \\ 0\,0\,0\,1\,1\,1\,0\,1\,0\,1 \end{bmatrix}.$$

It is often useful to discuss designs that lie within other designs. The generic term "subdesign" is used. However, in some cases it is not clear what the appropriate definition should be: For example, does one require that each block be a subset of a block in the original design, or that the set of blocks be a subset of the original set? Usually, we take the latter view, that a subdesign of a block design is the block design formed from a subset of the block set and a (sufficiently large) subset of the treatment set. However, different definitions may be used for different types of designs. Observe that, for example, a subdesign of a balanced incomplete block design will not necessarily be a balanced incomplete block design.

As was pointed out in Section 1.1, we do not wish to consider two designs to be different if it is possible to relabel the treatments and blocks of one in such a way that the other design is obtained. To formalize the earlier discussion, block designs $\mathcal{D}$ and $\mathcal{E}$ are defined to be *isomorphic* if there is a one-to-one map from the set of treatments of $\mathcal{D}$ onto the set of treatments of $\mathcal{E}$ such that when the mapping is applied to a block of $\mathcal{D}$, the result is a block of $\mathcal{E}$, and such that every block of $\mathcal{E}$ can be derived from some block of $\mathcal{D}$ in this way. A map with this property is called an *isomorphism.*

Alternatively, one could define an isomorphism to consist of two one-to-one maps, one mapping the set of treatments of $\mathcal{D}$ onto the set of treatments of $\mathcal{E}$, and the other mapping the set of blocks of $\mathcal{D}$ onto the set of blocks of $\mathcal{E}$, with the property that if treatment t maps to treatment $t\phi$ and block B maps to block $B\phi$, then t belongs to B if and only if $t\phi$ belongs to $B\phi$:

$$t \in B \Leftrightarrow t\phi \in B\phi.$$

This property is described by saying that "isomorphisms preserve incidence."

In terms of incidence matrices, two block designs are isomorphic if the incidence matrix of one can be transformed into the incidence matrix of the other by row and column permutations. As an example of this, consider the design that was discussed in Section 1.1. It is easy to verify that it is a balanced incomplete block design with parameters $(7, 7, 3, 3, 1)$. The first representation of the design had blocks

$$\{012, 034, 056, 135, 146, 236, 245\}.$$

The corresponding incidence matrix is

$$A = \begin{bmatrix} 1\,1\,1\,0\,0\,0\,0 \\ 1\,0\,0\,1\,1\,0\,0 \\ 1\,0\,0\,0\,0\,1\,1 \\ 0\,1\,0\,1\,0\,1\,0 \\ 0\,1\,0\,0\,1\,0\,1 \\ 0\,0\,1\,1\,0\,0\,1 \\ 0\,0\,1\,0\,1\,1\,0 \end{bmatrix}.$$

When we interchanged treatments 1 and 2 we obtained blocks

$$\{012, 034, 056, 235, 246, 136, 145\},$$

with incidence matrix

$$B = \begin{bmatrix} 1\,1\,1\,0\,0\,0\,0 \\ 1\,0\,0\,0\,0\,1\,1 \\ 1\,0\,0\,1\,1\,0\,0 \\ 0\,1\,0\,1\,0\,1\,0 \\ 0\,1\,0\,0\,1\,0\,1 \\ 0\,0\,1\,1\,0\,0\,1 \\ 0\,0\,1\,0\,1\,1\,0 \end{bmatrix}.$$

Obviously, B can be converted into the original matrix A by interchanging the second and third rows (which correspond to treatments 1 and 2, respectively). There are, in fact, other row and column permutations that have the same effect: If the last two rows of B are interchanged and then the permutation (46)(57) is carried out on the columns of B, we obtain A.

Exercises 1.3

1.3.1* Prove that there are exactly three nonisomorphic regular designs with $v = 4, b = 6, r = 3$ and $k = 2$. (Some will have repeated blocks.)
Hint: Subdivide the problem by asking, "How many solutions are there in which some block occurs n times but none occurs more than n times?"

1.3.2 Prove that there is a (7, 7, 4, 4, 2)-design and that it is unique up to isomorphism.

1.3.3 Prove that every (6, 10, 5, 3, 2)-design is simple.

1.3.4* Find a (7, 14, 6, 3, 2)-design that is not simple.

1.3.5 Show that there is a (7, 7, 3, 3, 1)-design whose blocks include 124, 235 and 347; find its other blocks. What is its incidence matrix?

1.3.6 A balanced incomplete block design has eight treatments and blocks of size four. Prove that the design has at least 14 blocks.

1.3.7 Write down the incidence matrices of the three designs you found in Exercise 1.1.1.

1.3.8* A design is constructed in the following way. The names of 16 treatments are written in a 4×4 array: for example,

$$\begin{array}{cccc} 1 & 2 & 3 & 4 \\ 5 & 6 & 7 & 8 \\ 9 & 10 & 11 & 12 \\ 13 & 14 & 15 & 16 \end{array}.$$

There are 16 blocks: To form a block, one chooses a row and a column of the array, deletes the common treatment, and takes the other six elements. (For example, row 1 and column 2 yield the block $\{1, 3, 4, 6, 10, 14\}$; treatment 2 is deleted.)

(i) Prove that the design is a balanced incomplete block design. What are its parameters?

(ii) Does the construction above work if the array is of a size different from 4×4?

1.3.9* In each row of the following table, fill in the blanks so that the parameters are possible parameters for a balanced incomplete block design, or else show that this is impossible.

v	b	r	k	λ
	35		3	1
		6	4	2
14	7	4		
		13	6	1
21	28		3	
17		8	5	
21	30		7	
17			7	1

1.3.10 Suppose a balanced incomplete block design satisfies $v = b$, and λ is odd. Show that v cannot be even.
Hint: Use (1.2).

1.4 Systems of Distinct Representatives

Suppose $\mathcal{D}$ is a block design based on $S = \{1, 2, \ldots, v\}$ with blocks $B_1, B_2, \ldots, B_b$. We define a *system of distinct representatives* (SDR) for $\mathcal{D}$ to be a way of selecting a member x_i from each block B_i such that $x_1, x_2, \ldots$ are all different.

As an example, consider the blocks

$$124, 124, 134, 235, 246, 1256.$$

One system of distinct representatives for them is

$$1, 2, 3, 5, 4, 6$$

(where the representatives are listed in the same order as the blocks). There are several others. On the other hand, the blocks

$$124, 124, 134, 23, 24, 1256$$

have no SDR.

If the design $\mathcal{D}$ is to have an SDR, it is clearly necessary that $\mathcal{D}$ have at least as many treatments as blocks. The example above shows that this is not sufficient: If we consider the first five blocks of the design, they constitute a design with $b = 5$ and $v = 4$. If there were an SDR for the six-block design, then the first five elements would be an SDR for the five-block subdesign, which is impossible. The implied necessary condition, that $\mathcal{D}$ contain no subdesign with $v < b$, is in fact sufficient, as we now prove.

THEOREM 1.3 [64]

A block design $\mathcal{D}$ has a system of distinct representatives if and only if it never occurs that some n blocks contain between them fewer than n treatments.

Proof. We proceed by induction on the number of blocks. If $\mathcal{D}$ has one block, the result is obvious. Assume the theorem to be true for all designs with fewer than b blocks. Suppose $\mathcal{D}$ has b blocks $B_1, B_2, \dots, B_b$ and the v treatments $\{1, 2, \dots, v\}$, and suppose $\mathcal{D}$ satisfies the hypothesis that the union of any n blocks has size at least n, for $1 \le n \le b$. By induction, any set of n blocks has an SDR, provided that $n < b$. We distinguish two cases.

(i) Suppose no set of n blocks contains less than $n + 1$ treatments in its union, for $n < b$. Select any element $x_1 \in B_1$, and write $B_i^* = B_i \backslash \{x_i\}$, for $i \in \{1, 2, \dots, b\}$. Then the union of any n of the B_i^* has at least n elements, for $n = 1, 2, \dots, b-1$. By the induction hypothesis, there is an SDR $x_2, x_3, \dots, x_b$ for $B_2, B_3, \dots, B_b$, so $x_1, x_2, x_3, \dots, x_b$ is an SDR for the original design $\mathcal{D}$.

(ii) Suppose there is a set of n blocks whose union has precisely n elements, for some n less than b. Without loss of generality, take these blocks as $B_1, B_2, \dots, B_n$. For $i > n$, write B_i^* to mean B_i with all members of $B_1 \cup B_2 \cup \dots \cup B_n$ deleted. It is easy to see that the design with blocks $B_{n+1}^*, B_{n+2}^*, \dots, B_b^*$ satisfies the conditions of the theorem (if $B_{n+i_1}^*, B_{n+i_2}^*, \dots, B_{n+i_k}^*$ were k blocks whose union has less than k elements, then $B_1, B_2, \dots, B_n, B_{n+i_1}^*, B_{n+i_2}^*, \dots, B_{n+i_k}^*$ would be $n + k$ blocks of $\mathcal{D}$ whose union has less than $n + k$ elements, which is impossible). From the induction hypothesis both sets have SDRs, and clearly they are disjoint, so together they comprise an SDR for $\mathcal{D}$. $\square$

We give two important applications of this result. Another important application occurs later, as Theorem 8.1.

THEOREM 1.4

A set of tasks are to be performed by a set of workers; no task may be assigned to a worker who is not qualified to perform it, and no worker may be required to perform more than one task. A necessary and sufficient condition that all tasks can be assigned is that for any set of n tasks, there are at least n workers capable of performing at least one of them each.

Proof. Number the tasks; for task i, write B_i for the set of workers qualified to perform it. Then what is required is an SDR for the B_i, and the necessary and sufficient condition is precisely that of Theorem 1.3. □

THEOREM 1.5

If M is a square matrix of zeros and ones with the property that every row and every column sums to r, then there are permutation matrices $P_1, P_2, \ldots, P_r$ such that

$$M = P_1 + P_2 + \ldots + P_r.$$

Proof. We proceed by induction on r. If $r = 1$, the theorem is clearly true. Assume it holds for $r < k$; let M be a $v \times v$ matrix of zeros and ones with constant row and column sums k. Define

$$B_i = \{j : m_{ij} = 1\}.$$

These B_i form a design with blocks of constant size k.

We show that the union of any n blocks contains at least n elements. Without loss of generality, say they are blocks $B_1, B_2, \ldots, B_n$. Consider

$$\{(i, j) : j \in B_i, 1 \leq i \leq n\}.$$

This set has kn elements. But no integer j can occur as right-hand element in more than k of the pairs (since column j has only k nonzero entries), so at least n different values of j must be represented.

Now the blocks B_i have a system of distinct representatives. If x_i is the representative of B_i, write P_k for the permutation matrix with (i, j) entry 1 if and only if $j = x_i$. Then $M - P_k$ is a zero-one matrix. Moreover, it has constant row-sum and column-sum $k - 1$. So by the induction hypothesis there are permutation matrices $P_1, P_2, \ldots, P_{k-1}$ such that

$$M - P_k = P_1 + P_2 + \ldots + P_{k-1},$$

giving the result. □

Exercises 1.4

1.4.1 Do there exist SDRs for the following blocks?
(i) 12, 145, 12, 123
(ii) 12, 145, 12, 13, 23.

1.4.2* Show that the blocks

$$123, 124, 134, 235, 246, 1256$$

have precisely 17 SDRs.

1.4.3* Count the SDRs of the following sets of blocks:
(i) 123, 234, 345, 451, 512;
(ii) 123, 124, 125, 345, 345;
(iii) 123, 124, 135, 245, 345.

1.4.4 Suppose the design $\mathcal{D}$ has an SDR, and suppose each of the b blocks has at least t elements. Prove that if $t \leq b$, then $\mathcal{D}$ has at least $t!$ SDRs.

1.4.5 The set S_i has $i + 1$ members. Show that the system $\{S_1, S_2, \ldots, S_n\}$ has at least 2^n SDRs. Construct such a system with precisely 2^n SDRs.

1.4.6 S and T are sets of v elements each. P is a set of ordered pairs(s, t), where s is in S and t is in T. Define

$$A(s) = \{t : t \in T, (s, t) \in P\}$$
$$B(t) = \{s : s \in S, (s, t) \in P\}.$$

Prove that the following two properties are equivalent:
(a) For every k-subset X of S, $1 \leq k \leq v$, $\cup_{s \in X} A(s)$ has at least k elements;
(b) For every k-subset Y of T, $1 \leq k \leq v$, $\cup_{s \in Y} B(t)$ has at least k elements.

Chapter 2

Balanced Designs

2.1 Pairwise Balanced Designs

A *pairwise balanced design* of index λ is a way of selecting blocks from a set of treatments (support set) such that any two treatments have covalency λ. If there are v treatments and if every block size is a member of some set K of positive integers, the design is designated a $PB(v; K; \lambda)$. So the parameters consist of two positive integers and one set of positive integers. To avoid trivial differences between designs, block size 1 is not allowed.

The number of blocks is *not* normally treated as a parameter; one can have two pairwise balanced designs with the same parameters but with different numbers of blocks. Both

$$123, 145, 24, 25, 34, 35$$

and

$$123, 14, 15, 24, 25, 34, 35, 45$$

are $PB(5; \{3, 2\}; 1)$s, but they have six and eight blocks, respectively.

It should be observed that we do not require every member of K to be a block size. For instance, the two examples are also PB$(5; \{4, 3, 2\}; 1)$s. More generally, any $PB(v; K; \lambda)$ is also a $PB(v; L; \lambda)$ whenever K is a subset of L.

Various easy results may be proven about pairwise balanced designs. For example, two copies of a $PB(v; K; \lambda)$, taken together, constitute a $PB(v; K; 2\lambda)$. Just as obviously, there is a $PB(v; \{v\}; \lambda)$ for all positive integers v and λ (it is the complete design with λ sets, each containing all v elements). The following theorem is more interesting.

THEOREM 2.1

Suppose there exists a $PB(v; K; \lambda)$, and for every element k of K there exists a $PB(k; L; \mu)$. Then there exists a $PB(v; L; \lambda\mu)$.

Proof. Suppose we have a $PB(v;K;\lambda)$ based on a v-set V. Suppose its blocks are $B_1, B_2, \ldots, B_n$, where B_i has k_i elements. We replace each block by a new collection of blocks. Given B_i, form a $PB(k_i;L;\mu)$, but instead of taking the numbers $1, 2, \ldots, k_i$ as treatments, use the elements of B_i. This is done for every i. If x and y are any two treatments, then $\{x, y\}$ was contained in λ of the B_i, and in each case B_i has been replaced by a collection of blocks, μ of which contain $\{x, y\}$. So x and y occur together $\lambda\mu$ times in total. Therefore, the total collection of new blocks is a pairwise balanced design of index $\lambda\mu$, based on V. The size of any block of the new design is a member of L, so the design is a $PB(v;L;\lambda\mu)$. □

As an example, we construct a $PB(15;\{4,3\};3)$ from the $PB(15;\{5,3\};1)$ with blocks

01234		56789		$ABCDE$
$05A$	$06C$	$07E$	$08B$	$09D$
$16B$	$17D$	$18A$	$19C$	$15E$
$27C$	$2SE$	$29B$	$25D$	$26A$
$38D$	$39A$	$35C$	$36E$	$37B$
$49E$	$45B$	$46D$	$47A$	$48C$.

We need a $PB(5;\{4,3\};3)$ and a $PB(3;\{4,3\};3)$. The former has blocks

$$1234 \quad 1235 \quad 1245 \quad 1345 \quad 2345 \tag{2.1}$$

and the latter is the complete design

$$123 \qquad 123 \qquad 123.$$

The block 01234 is replaced by a copy of the set of blocks (2.1) in which the treatments have been relabeled using the mapping

$$(1, 2, 3, 4, 5) \mapsto (0, 1, 2, 3, 4).$$

The new blocks are

$$0123 \quad 0124 \quad 0134 \quad 0234 \quad 1234. \tag{2.2}$$

Similarly, 56789 and ABCDE are replaced by

$$5678 \quad 5679 \quad 5689 \quad 5789 \quad 6789 \tag{2.3}$$

and

$$ABCD \quad ABCE \quad ABDE \quad ACDE \quad BCDE, \tag{2.4}$$

respectively. The block 05A is replaced by

$$05A \quad 05A \quad 05A$$

and similarly for every other 3-block. So the required design consists of the blocks listed in (2.2), (2.3) and (2.4) and three copies of every 3-set in the original design.

The $PB(15; \{5, 3\}; 1)$ that we used above is an example of a useful class of designs. We now give a more general construction.

THEOREM 2.2

There exists a $PB(3k; \{3, k\}; 1)$ whenever k is odd.

Proof. We construct a design with treatment set $\{a_1, a_2, \ldots, a_k, b_1, b_2, \ldots, b_k, c_1, c_2, \ldots, c_k\}$. There are three blocks of size k (which we call "big blocks"), namely

$$\{a_1 a_2 \ldots a_k\}, \quad \{b_1 b_2 \ldots b_k\}, \quad \{c_1 c_2 \ldots c_k\}.$$

The blocks of size 3 are the blocks

$$\{a_i b_{i+x} c_{i-2x} : 1 \le i \le k, 1 \le x \le k\}$$

where subscripts that exceed k are reduced modulo k.

The pairs $\{a_i, a_j\}$, $\{b_i, b_j\}$, and $\{c_i, c_j\}$ occur once each, in the big blocks. The pair $\{a_i, b_j\}$ occurs in $\{a_i b_{i+x} c_{i+2x}\}$ if and only if $j \equiv i + x \pmod{k}$; this will happen for only one value of x, either $x = j - i$ (if $j > i$) or $x = k - j - i$ (if $j \le i$). So only one block contains $\{a_i, b_i\}$. A similar remark applies to $\{b_i, c_j\}$. If $\{a_i, c_j\}$ occurs in $\{a_i b_{i+x} c_{i+2x}\}$ then $i - 2x \equiv j \pmod{k}$, and this also uniquely defines x; since k is odd, the solution is $x \equiv \frac{1}{2} \cdot (k - 1)(j - 1) \pmod{k}$. □

The construction above does not generalize to even values of k, because $\frac{1}{2}(k - 1)(j - i)$ is not necessarily an integer in that case. However, we shall see in Chapter 8 that another construction is available for even k.

We now prove an interesting theorem concerning the number of blocks in a pairwise balanced design. We start with an easy remark.

LEMMA 2.3

In a pairwise balanced design with $\lambda = 1$, no two blocks have two common elements.

Proof. Suppose $B_1 = \{x, y, \ldots\}$ and $B_2 = \{x, y, \ldots\}$. Then$\{x, y\}$ is a subset of B_1, and also of B_2. So $\lambda_{xy} \ge 2$. But this contradicts the property that $\lambda = 1$. □

THEOREM 2.4 [27]

Suppose there is a $PB(v; K; 1)$ with b blocks, where $b > 1$. Then $b \geq v$. If $b = v$, then either the $PB(v; K; 1)$ has one block of size $v - 1$ and the rest of size 2, or else $b = v = k^2 - k + 1$ for some integer k and all the blocks have size k.

Proof. Suppose a $PB(v; K; 1)$ has treatments $t_1, t_2, \ldots, t_v$ and b blocks $B_1, B_2, \ldots, B_b$. Say k_i is the number of elements in B_i, and that t_j belongs to r_j blocks. (We call r_j the *frequency* or *replication number* of t_j.) Then, counting all elements of all blocks,

$$\sum_{j=1}^{v} r_j = \sum_{i=1}^{b} k_i. \tag{2.5}$$

If t_j does not belong to B_i, then t_j must belong to at least k_i blocks: For every element x of B_i there is a block that contains t_j and x, and these blocks are all disjoint because of Lemma 2.3. So

$$t_j \notin B_i \Rightarrow k_i \leq r_j. \tag{2.6}$$

The blocks are incomplete, and there are no blocks of size 1. So

$$1 < k_i < v \text{ for } 1 \leq i \leq b. \tag{2.7}$$

There must be some treatment whose replication number is minimal. Say it is t_v, and write $r_v = m$. Relabel the blocks so that those containing t_v are $B_1, B_2, \ldots, B_m$. We can select an element of each block, other than t_v, and clearly all these elements are different. Suppose (after relabeling) that $t_i \in B_i, t_i \neq t_v$. Then if $1 \leq i \leq m$, $1 \leq j \leq m$, and $i \neq j, t_j \notin B_i$; in particular,

$$t_1 \notin B_2, t_2 \notin B_3, \ldots, t_{m-1} \notin B_m, t_m \notin B_1$$

so from (2.6)

$$k_2 \leq r_1, k_3 \leq r_2, \ldots, k_m \leq r_{m-1}, k_1 \leq r_m, \tag{2.8}$$

whence

$$\sum_{i=1}^{m} k_i \leq \sum_{j=1}^{m} r_j \tag{2.9}$$

Also, $t_v \notin B_i$ for $i > m$, so $k_i \leq r_v$ for $i > m$, so

$$\sum_{i=m+1}^{b} k_i \leq \sum_{j=m+1}^{b} r_v. \tag{2.10}$$

Adding (2.9) and (2.10) and comparing with (2.5), we obtain

$$\begin{aligned}\sum_{j=1}^{v} r_j &= \sum_{i=1}^{b} k_i \\ &= \sum_{i=1}^{m} k_i + \sum_{i=m+1}^{b} k_i \\ &\le \sum_{j=1}^{m} r_j + \sum_{j=m+1}^{b} r_v \le \sum_{j=1}^{b} r_j\end{aligned}$$

(since $r_v \le r_j$ for all j), and this is impossible if $b < v$ because the r_i are all positive.

In particular, suppose $b = v$. Then each of the inequalities in (2.8) must be an equality, and also $k_i = r - i$ for all $i > m$. If we relabel the treatments $t_1, t_2, \ldots, t_m$, we obtain $r_i = k_i$, all $i \in \{1 \ldots v\}$ for some ordering of treatments and blocks. Moreover, t_v is unchanged. Let us further relabel the treatments (and simultaneously, the blocks) so that

$$r_1 \ge r_2 \ge \ldots \ge r_v.$$

(Since t_v had minimum frequency, it has still not been disturbed.)

We consider the various possibilities.

(i) Suppose $r_1 > r_2$. Then $r_1 > r_j$ for all $j \ge 2$. So $k_1 = r_1 > r_j (j \ge 2)$. From (2.6), $t_j \in B_1$ for all $j > 1$. Of course, $t_r \notin B_1$. So

$$B_1 = \{t_2, t_3, \ldots, t_v\},$$

and the other blocks must be

$$\{t_l, t_2\}, \{t_1, t_3\}, \ldots, \{t_2, t_v\}.$$

(ii) Suppose $r_1 = r_2 = \ldots = r_{j-1} > r_j$, where $j > 2$. Then $t_j \in B_1$ and $t_j \in B_2$ (from (2.6)); since $t_v \in B_1 \cap B_2$, the only possibility (according to Lemma 2.3) is $t_j = t_v$ and $j = v$. So we consider that case. Since $r_v < r_{v-1} = k_{v-1} < v$ (from (2.7)) there are at least two blocks not containing t_v. One might be B_v, but suppose the other is B_x, where $x \ne v$. Then from (2.6) we have $r_x = k_x \le r_v$, a contradiction.

(iii) Finally, suppose $r_1 = r_2 = \ldots = r_v$. We have constant block size and constant frequency, a balanced incomplete block design. From (1.1) and (1.2) we immediately deduce that $b = v = k^2 - k + 1$, where k is the common block size. □

We could generalize pairwise balanced designs by requiring that every set of t treatments occurs in a fixed number of blocks. This is called a *t-wise balanced design.*

Exercises 2.1

2.1.1 Assume there exists a $PB(v; \{k, 3\}; 1)$ with $v \equiv 2 \pmod 3$. Prove that $k \equiv 2 \pmod 3$.

2.1.2* Assume a $PB(7; \{5, 4, 3\}; 1)$ exists. Prove that the number of blocks of size 5 is divisible by 3, and that consequently no such blocks exist. Can there be any blocks of size 4?

2.1.3* Prove that no $PB(8; \{4, 3\}; 1)$ can exist.

2.1.4 Prove[1] that there is no $PB(v; \{4, 3\}; 1)$ when $v \equiv 2 \pmod 3$.

2.1.5 Suppose there is a $PB(7; \{5, 4, 3\}; 1)$. Prove that the number of blocks of size 5 is divisible by 3, and that consequently no such blocks exist. Can there be any blocks of size 4?

2.1.6 Does there exist any $PB(5; \{3, 2\}; 1)$ not isomorphic to the one given at the beginning of this section?

2.1.7 Use Theorems 2.1 and 2.2 and the design exhibited in Section 1.1 to prove that there is a balanced incomplete block design on 7×3^t treatments with $k = 3$ and $\lambda = 1$ for every integer $t \geq 0$.

2.2 Balanced Incomplete Block Designs

One can consider a balanced incomplete block design as a kind of pairwise balanced design, so all the results of the preceding section apply to them.

In particular, Theorem 2.1 can be applied. If the set L in that theorem has only one element, the resulting design is regular (as defined in Section 1.3). So we have the following corollary to Theorem 2.1.

THEOREM 2.5

Suppose there exists a $PB(v; K; \lambda)$, and for every member k of K there is a balanced incomplete block design on k treatments with block size l and balance parameter μ. Then there exists a balanced incomplete block design with parameters

$$\left(v, \frac{\lambda\mu v(v-1)}{l(l-1)}, \frac{\lambda\mu(v-1)}{l(l-1)}, l, \lambda\mu\right).$$

[1] See also exercises 12.1.8 and 13.1.2.

In a balanced incomplete block design all the sets chosen are the same size, every treatment occurs equally often and the design is pairwise balanced. It is possible to relax the condition of pairwise balance but keep the other two conditions, and the result is a regular block design, which we discussed in Chapter 1. It is also possible to retain balance and retain the constant number of replications but have more than one block size; such designs are called (r, λ)-designs, where r and λ are the appropriate parameters. However, we now prove that there are no examples where the restriction on the number of replications is the only one that is dropped.

THEOREM 2.6

A pairwise balanced design in which every set has the same size is a balanced incomplete block design.

Proof. If treatment x has frequency r_x, then

$$(k-1)r_x = \lambda(v-1)$$

as in Theorem 1.2. But this means that all treatments have the same frequency r, where

$$r = \lambda\frac{v-1}{k-1}.$$

□

If one treatment is deleted from a balanced incomplete block design, the result is a pairwise balanced design with two block sizes, k and $k-1$. These pairwise balanced designs will have constant replication number (unless $k = 2$, and the blocks of size 1 that arise are then deleted). This gives rise to many examples of (r, λ)-designs, but there are other examples D that are not constructed in this way.

We shall now discuss the incidence matrices of balanced incomplete block designs. Incidence matrices belong to the class of $(0, 1)$-matrices, matrices each of whose entries is 0 or 1. It is interesting to observe that two equations characterize the incidence matrices of balanced incomplete block designs among the wider class of (0, 1)-matrices. To express these equations, we use two notations. An $n \times n$ identity matrix is denoted I_n, and an $m \times n$ matrix with every entry 1 is denoted $J_{m\times n}$. (If $m = n$, we simply write J_m for $J_{m\times n}$; and in any event, we usually omit the subscripts whenever possible.)

THEOREM 2.7

If A is the incidence matrix of a (v, b, r, k, λ) design, then

$$AA^T = (r-\lambda)I_v + \lambda J_v \tag{2.11}$$

and

$$J_v A = kJ_{v\times b}. \tag{2.12}$$

Conversely, if there is a $v \times b$ $(0,1)$*-matrix* A *that satisfies* (2.11) *and* (2.12), *then*

$$v = \frac{r(k-1)}{\lambda} + 1,$$

$$b = \frac{vr}{k},$$

and provided that $k < v$, A *is the incidence matrix of a* (v, b, r, k, λ)*-design.*

Proof. First, suppose A is the incidence matrix of a (v, b, r, k, λ)-design. As $a_{ij} = 1$ if and only if treatment i belongs to block j, the number of entries 1 in column j equals the number of members of block j, that is, k. All other entries of A are zero, so the number of entries 1 in a column equals the sum of the entries in the column. But each entry of column j of JA equals the sum of the entries in column j of A. Therefore, (2.12) holds.

The (i, j) entry of AA^T is

$$\sum_{n=1}^{b} a_{in}a_{j}n. \tag{2.13}$$

Now $a_{in}a_{jn}$ equals 1 when both treatments i and j belong to block n, and 0 otherwise. So the sum (2.13) equals the number of blocks that contain both t_i and t_j; this is r when $i = j$ and λ when $i \neq j$. Therefore,

$$\begin{aligned} AA^T &= rI + \lambda(J - I) \\ &= (r - \lambda)I + \lambda J. \end{aligned} \tag{2.11}$$

Conversely, suppose A is a $(0, 1)$ matrix satisfying (2.11) and (2.12). Define a block design with treatments $t_1, t_2, \ldots, t_v$ and blocks $B_1, B_2, \ldots, B_b$ by $t_i \in B_j$ if and only if $a_{ij} = 1$.

One sees easily that this is a balanced incomplete block design (provided that $k < v$). Consequently, v and b must satisfy the given equations. □

It is useful to know the determinant of the matrix AA^T. We prove a slightly more general result, in that we do not assume the parameters r and λ to be positive integers.

LEMMA 2.8

The determinant of the $v \times v$ *matrix*

$$M = (r - \lambda)I + \lambda J$$

is $(r-\lambda)^{v-1}[r+(v-1)\lambda]$.

Proof. The matrix M is transformed as follows. First, subtract column 1 from every other column; second, add each of rows $2, 3, \ldots, n$ to row 1.

The process is

$$M = \begin{bmatrix} r & \lambda & \lambda & \ldots & \lambda \\ \lambda & r & \lambda & \ldots & \lambda \\ \lambda & \lambda & r & \ldots & \lambda \\ \cdot & \cdot & \cdot & \ldots & \cdot \\ \lambda & \lambda & \lambda & \ldots & r \end{bmatrix} \mapsto \begin{bmatrix} r & \lambda-r & \lambda-r & \ldots & \lambda-r \\ \lambda & r-\lambda & 0 & \ldots & 0 \\ \lambda & & r-\lambda & \ldots & 0 \\ \cdot & \cdot & \cdot & \ldots & \cdot \\ \lambda & 0 & 0 & \ldots & r-\lambda \end{bmatrix}$$

$$\mapsto \begin{bmatrix} r+(v-1)\lambda & 0 & 0 & \ldots & 0 \\ \lambda & r-\lambda & 0 & \ldots & 0 \\ \lambda & 0 & r-\lambda & \ldots & 0 \\ \cdot & \cdot & \cdot & \ldots & \cdot \\ \lambda & 0 & 0 & \ldots & r-\lambda \end{bmatrix}.$$

These transformations do not change the determinants; since the final matrix is zero above the diagonal, the determinant equals the product of the diagonal elements, which is $(r-\lambda)^{v-1}[r+(v-1)\lambda]$. □

THEOREM 2.9 **[53]**

In a balanced incomplete block design, $b \geq v$.

Proof. Let A be the incidence matrix of a (v, b, r, k, λ)-BIBD. Since $k < v$, the equation $\lambda(v-1) = r(k-1)$ implies that $r > \lambda$. So the determinant of AA^T, which equals $(r-\lambda)^{v-1}[r+(v-1)\lambda]$, is nonzero. Therefore, AA^T has rank v, and $v = \text{rank}\,(AA^T) \leq \text{rank}\,(A) \leq \min(v, b)$. So $v \leq b$. □

Theorem 2.9 is called *Fisher's inequality.* The original proof did not involve matrices, but rather used an ingenious method that takes its inspiration from statistics. (That proof is reproduced in the next section.)

If A is the incidence matrix of a regular design with parameters v, b, r and k, then A^T is the incidence matrix of a regular design with parameters b, v, k and r. The design is easy to construct. If the original design had blocks $B_1, B_2, \ldots, B_b$ and treatments $t_1, t_2, \ldots, t_v$, the new design has blocks $C_1, C_2, \ldots, C_v$ and treatments $u_1, u_2, \ldots, u_b$, and u_i belongs to C_j if and only if t_j belongs to B_i. The new design is called the *dual* of the original.

It is clear that the dual of an incomplete design is an incomplete design. We shall, however, prove that balance is preserved if and only if $v = b$. We would need to show that a $v \times v$ matrix of zeros and ones that satisfies

$$AA^T = (k-\lambda)I + \lambda J \tag{2.14}$$
$$JA = kJ \tag{2.15}$$

[the results of substituting $r = k$ in (2.11) and (2.12)] will also satisfy

$$A^TA = (k-\lambda)I + \lambda J \tag{2.16}$$
$$JA^T = kJ. \tag{2.17}$$

In fact, we shall prove the following, stronger theorem.

THEOREM 2.10 [122]

If A is a nonsingular matrix of side v, which satisfies one of (2.14), (2.16) *and one of* (2.15), (2.17), *then it satisfies all four equations, and*

$$k(k-1) = \lambda(v-1).$$

Proof. All matrices in the proof are $v \times v$. Since A is nonsingular, both AA^T and A^TA have nonzero determinant. So either (2.14) or (2.16) implies that $(k-\lambda)I + \lambda J$ has nonzero determinant; from Lemma 2.8, we have

$$k - \lambda \neq 0, \lambda(v-1) + k \neq 0.$$

First suppose (2.14) and (2.15) hold and A is nonsingular. From (2.14) we deduce that

$$JAA^T = (k-\lambda)J + \lambda J^2$$

and substituting kJ for JA (from (2.15)) and vJ for J^2 yields

$$kJA^T = (k-\lambda)J + \lambda vJ.$$

So

$$\begin{aligned} kJA^TJ &= k - \lambda + \lambda v)J^2 \\ KJ(JA)^T &= (k - \lambda + \lambda v)J^2 \\ kJ(kJ)^T &= (k - \lambda + \lambda v)J^2 \\ k^2J^2 &= (k - \lambda + \lambda v)J^2 \end{aligned}$$

whence $k^2 = k - \lambda + \lambda v$ and $k(k-1) = \lambda(v-1)$. Also, from

$$kJA^T = (k - \lambda + \lambda v)J$$

we can now deduce

$$kJA^T = k^2J;$$

since $k^2 = k - \lambda + \lambda v = \lambda(v-1) + k$, which we know to be nonzero, k is nonzero, so we can divide by k to get $JA^T = kJ$ and

$$A^TA = A^{-1}(AA^T)A = (k-\lambda)I + \lambda A^{-1}JA; \tag{2.18}$$

but we have $AJ = kJ = JA$, so $A^{-1}JA = A$, and (2.18) becomes (2.16).

Next, suppose (2.14) and (2.17) hold for the nonsingular matrix A. Since $JA^T = kJ$ we have $AJ = kJ$, so $J = kA^{-1}J$, whence k must be nonzero and $A^{-1}Jk^{-1}J$. Now

$$\begin{aligned} A^T &= A^{-1}(AA^T) \\ &= (k-\lambda)A^{-1} + \lambda A^{-1}J \\ &= (k-\lambda)A^{-1} + \lambda k^{-1}J; \qquad (2.19) \\ kJ &= JA^T \\ &= (k-\lambda)JA^{-1} + \lambda k^{-1}J^2 \\ &= (k-\lambda)JA^{-1} + \lambda k^{-1}vJ. \end{aligned}$$

Therefore (division being possible because we know $k - \lambda \neq 0$),

$$\begin{aligned} JA^{-1} &= \frac{k - \lambda k^{-1}v}{k-\lambda}J, \qquad (2.20) \\ JA^{-1}J &= \frac{k - \lambda k^{-1}v}{k-\lambda}(vJ). \end{aligned}$$

But $vk^{-1}J = k^{-1}J^2 = k^{-1}J(kA^{-1}J) = JA^{-1}J$. So we have

$$\begin{aligned} vk^{-1}J &= \frac{k - \lambda k^{-1}v}{k-\lambda}v, \\ k - \lambda &= k^2 - \lambda v, \end{aligned}$$

which is $k(k-1) = \lambda(v-1)$. We can substitute k^{-1} for the right-hand coefficient in (2.21):

$$JA^{-1} = k^{-1}J,$$

whence $JA = kJ$, which is (2.15). Returning to (2.20) and multiplying on the right by A, we get

$$\begin{aligned} A^TA &= (k-\lambda)A^{-1}A + \lambda k^{-1}J \\ &= (k-\lambda)I + \lambda k^{-1}JA \\ &= (k-\lambda)I + \lambda k^{-1}(kJ) \\ &= (k-\lambda)I + \lambda J. \qquad (2.16) \end{aligned}$$

Finally, suppose we start with (2.16) and (2.15). We see that these are just (2.14) and (2.17) with A replaced by A^T; so A^T satisfies (2.15) and (2.16) and $k(k-1) = \lambda(v-1)$, which means that A satisfies (2.14) and (2.17) and $k(k-1) = \lambda(v-1)$. A similar remark applies if we start with (2.16) and (2.17). □

COROLLARY 2.10.1

The dual of a balanced incomplete block design is a balanced incomplete block design if and only if the design is symmetric: $v = b$.

Proof. (i) If $b > v$ in a balanced incomplete block design, its dual design will have more treatments than blocks, and by Theorem 2.9 it cannot be a balanced incomplete block design.

(ii) If $b = v$, Theorem 2.10 applies. □

COROLLARY 2.10.2

The intersection of two distinct blocks of a symmetric balanced incomplete block design always contains λ *elements.*

Proof. Let A be the incidence matrix of a symmetric balanced incomplete block design. The number of treatments common to block i and block j equals the number of places where column i and column j both have entry 1. As A is a $(0,1)$-matrix, this number must equal the scalar product of column i with column j, which in turn is the (i,j) entry of A^TA; this always equals λ, by Corollary 2.10.1. □

Suppose there is a balanced incomplete block design with parameters (v, b, r, k, λ), having blocks $B_1, B_2, \ldots, B_b$. Write S for the v-set of all treatments in the design. Then the sets

$$S\backslash B_1, S\backslash B_2, \ldots, S\backslash B_b$$

form a design with treatment set S, called the *complementary design* or *complement* of the original.

THEOREM 2.11

The complementary design of a (v, b, r, k, λ)*-design is a balanced incomplete block design with parameters*

$$(v, b, b-r, v-k, b-2r+\lambda),$$

provided $b - 2r + \lambda$ *is nonzero.*

1	2	3	4	5	6	7	8	9
4	5	6	1	2	3	7	8	9
7	8	9	1	2	3	4	5	6
1	4	7	2	3	5	6	8	9
2	5	8	1	3	4	6	7	9
3	6	9	1	2	4	5	7	8
1	6	8	2	3	4	5	7	9
2	4	9	1	3	5	6	7	8
3	5	7	1	2	4	6	8	9
1	5	9	2	3	4	6	7	8
2	6	7	1	3	4	5	8	9
3	4	8	1	2	5	6	7	9
	$\mathcal{A}$				$\mathcal{C}$			
	(original)				(complement)			

FIGURE 2.1: A design and its complement.

Proof. There are v treatments and b blocks in each design. Since each treatment belonged to r blocks in the original design, it will belong to the other $b - r$ blocks in the complement; since B_i has k elements, S has v elements, and B_i is a subset of S, $S\backslash B_i$ has $v - k$ elements. So we have a regular design and the first four parameters are verified. Finally, consider two distinct treatments t_1 and t_2 in the original design. They occur together in λ blocks. As t_1 belongs to r blocks, there will be $r - \lambda$ that contain t_1 but not t_2. Similarly, $r - \lambda$ blocks contain t_2 but not t_1. So the number of blocks that contain neither t_1 nor t_2 is

$$b - (r - \lambda) - (r - \lambda) - \lambda = b - 2r + \lambda,$$

and this will be the number of complementary blocks that contain both of them. The constancy of this number implies that the design is balanced. □

As an example, Figure 2.1 shows the blocks of a $(9, 12, 4, 3, 1)$-BIBD $\mathcal{A}$ and its complement, a $(9, 12, 8, 6, 5)$-BIBD $\mathcal{C}$.

Exercises 2.2

2.2.1 A is a (0,1)-matrix. Prove that A is the incidence matrix of a regular block design if and only if A satisfies the equations

$$\begin{aligned} J_v A &= k J_{v \times b} \\ A J_b &= r J_{v \times b} \end{aligned}$$

for some integers v, b, r, k.

2.2.2* S is a set and $\mathcal{B}$ is the set of all k-subsets of S. Prove that if $\mathcal{B}$ is considered as a set of blocks on the treatment set S, the result is a balanced incomplete block design. If S has v elements, what are the parameters of the design?

2.2.3* Are there any balanced incomplete block designs with $k = 2, \lambda = 1$, other than those that arise as part of Exercise 2.2.2?

2.2.4* Suppose $\mathcal{D}$ is a BIBD with parameters (v, b, r, k, λ) whose blocks are all different. Write S for the support of $\mathcal{D}$; and write $\mathcal{T}$ for the collection of all k-sets on S other than the blocks of $\mathcal{D}$. Prove that if $b < \binom{v}{k}$, the set $\mathcal{T}$ of blocks forms a balanced incomplete block design based on S, and find its parameters.

2.2.5 Suppose B is a block of a BIBD. Write x_i for the number of blocks other than B that intersect B in precisely i elements. Prove:

$$\sum_i ix_i = k(r-1) \quad \text{and} \quad \sum_i i(i-1)x_i = k(k-1)(\lambda-1).$$

2.2.6 In each case, what are the parameters of the complement of a design with the parameters shown?

(i) $(7, 7, 3, 3, 1)$
(ii) $(16, 20, 5, 4, 1)$
(iii) $(13, 26, 6, 3, 1)$
(iv) $(7, 14, 8, 4, 4)$
(v) $(16, 24, 9, 6, 3)$
(vi) $(10, 30, 9, 3, 2)$
(vii) $(12, 44, 33, 9, 24)$
(viii) $(8, 14, 7, 4, 3)$

2.2.7 The balanced incomplete block design $\mathcal{D}$ has parameters $(6, 10, 5, 3, 2)$ and blocks

$$\begin{array}{lll} B_0 = \{0,1,2\} & B_5 = \{1,2,5\} \\ B_1 = \{0,1,3\} & B_6 = \{1,3,4\} \\ B_2 = \{0,2,4\} & B_7 = \{1,4,5\} \\ B_3 = \{0,3,5\} & B_8 = \{2,3,4\} \\ B_4 = \{0,4,5\} & B_9 = \{2,3,5\}. \end{array}$$

Construct the blocks of the dual of $\mathcal{D}$, and verify that it is not balanced.

2.2.8 $\mathcal{D}$ is the $(15, 15, 7, 7, 3)$-design with blocks

$$\begin{array}{ll} B_0 = \{0,1,2,3,4,5,6\} & B_1 = \{0,3,4,9,10,13,14\} \\ B_2 = \{0,1,2,7,8,9,10\} & B_3 = \{0,5,6,7,8,13,14\} \\ B_4 = \{0,1,2,11,12,13,14\} & B_5 = \{0,5,6,9,10,11,12\} \\ B_6 = \{0,3,4,7,8,11,12\} & B_7 = \{1,3,5,7,9,11,13\} \\ B_8 = \{1,3,6,7,10,12,14\} & B_9 = \{2,3,6,8,9,11,14\} \\ B_{10} = \{1,4,5,8,10,11,14\} & B_{11} = \{2,4,5,7,9,12,14\} \\ B_{12} = \{1,4,6,8,9,12,13\} & B_{13} = \{2,4,6,7,10,11,13\} \\ B_{14} = \{2,3,5,8,10,12,13\} & \end{array}$$

(i) Write down the blocks of the dual design $\mathcal{D}^*$ of $\mathcal{D}$.

(ii) A *triplet* is a set of three blocks whose mutual intersection has size 3. (For example, $\{B_0, B_1, B_6\}$ is a triplet in D.) By considering triplets, prove that $\mathcal{D}$ and $\mathcal{D}^*$ are not isomorphic.

2.2.9* Suppose $\mathcal{D}$ is a balanced incomplete block design with parameters (v, b, r, k, λ). Form a new design by replacing each block of $\mathcal{D}$ by n identical copies of itself. Prove that this new structure is a balanced incomplete block design (called the *n-multiple* of $\mathcal{D}$). What are its parameters?

2.2.10 In this question we restrict ourselves to blocks of size 3. Suppose we relax the requirement that the elements in a block of a design should be distinct. Blocks could be like $x\,x\,x$ or $x\,x\,y$ or $x\,y\,z$. In any block, there are three *pairs* of elements:

- $x\,x\,x$ contains the pair $x\,x$ three times (x is a *triple* element);
- $x\,x\,y$ contains the pair $x\,x$ once and $x\,y$ twice (x is a *double* element);
- $x\,y\,z$ contains the pairs $x\,y$, $x\,z$, $y\,z$ once each.

We require each pair of elements (distinct or not) to occur λ times. Such a design is called a *balanced multiset design.*

(i) Show that in a balanced multiset design the smallest possible λ is 2.

(ii) Show that if $\lambda = 2$, then:

(a) there are no triple elements;

(b) $v \geq 5$;

(c) $v \equiv 0$ or $2 \pmod 3$;

(d) each element appears $v+1$ times ($v-3$ times singly in a block, and in two blocks as a double element).

(iii) Give examples of balanced multiset designs in the cases $v = 5$ and $v = 6$.

(iv) When $v = 6$ is it possible to have a balanced multiset design without a repeated block?

2.3 Another Proof of Fisher's Inequality

The original proof of Fisher's inequality did not involve matrices; instead it followed some statistical ideas. This proof is called the *variance method*. Our version is based on [64]. We assume we are given a (v, b, r, k, λ)-design with blocks $B_1, B_2, \ldots, B_n$.

Given n readings $f_1, f_2, \ldots, f_n$, their mean $\overline{f}$ is defined as $n^{-1} \sum f_i$, and their variance is $v = n^{-1} \sum (f_i - \overline{f})^2$. Obviously $\sum (f_i - \overline{f})^2$ is nonnegative. In order to reduce the size of numbers when calculating variances, the identity

$$\sum (f_i - \overline{f})^2 = \sum f_i^2 - n\overline{f}^2 \tag{2.21}$$

is used.

Now write $n = b - 1$. Define f_i to be the size of the intersection $B_i \cap B_b$. We count the occurrences of pairs of members of B_b in the other blocks. Each pair has $\lambda - 1$ further occurrences, so

$$\sum \tfrac{1}{2} f_i(f_i - 1) = \tfrac{1}{2} k(k-1)(\lambda - 1).$$

On the other hand,

$$\sum f_i = k(r-1)$$

whence we have

$$\overline{f} = (b-1)^{-1} k(r-1)$$

and also

$$\sum f_i^2 = k(k-1)(\lambda - 1) + k(r-1).$$

Therefore

$$\sum f_i^2 - (b-1)\overline{f}^2 = k(k-1)(\lambda-1) + k(r-1) - (b-1)^{-1}k^2(r-1)^2.$$

Several identities are now used. The least obvious is the observation that

$$\begin{aligned} k^2(r-1)^2 - r^2(k-1)^2 &= (k(r-1) - r(k-1))\,(k(r-1) + r(k-1)) \\ &= (r-k)(2kr - k - r), \end{aligned}$$

so that

$$\begin{aligned} k^2(r-1)^2 &= r^2(k-1)^2 + (r-k)(2kr-k-r) \\ &= \lambda(v-1)r(k-1) + (r-k)(2kr-k-r) \end{aligned}$$

(using (1.2)). Others needed are

$$\begin{aligned} k(b-1) &= vr - k = r(v-1) + (r-k), \\ \lambda v &= \lambda + rk - r, \end{aligned}$$

which follow from (1.1) and (1.2) respectively. We then proceed:

$$
\begin{aligned}
& (b-1)\left(\sum f_i^2 - (b-1)\overline{f}^2\right) \\
&= (b-1)k(k-1)(\lambda-1) + (b-1)k(r-1) - k^2(r-1)^2 \\
&= (b-1)k\left((k-1)(\lambda-1) + r - 1\right) \\
&\quad - (\lambda(v-1)r(k-1) + (r-k)(k+r-2kr)) \\
&= r(v-1)(k\lambda - k - \lambda + r) + (r-k)(k\lambda - k - \lambda + r) \\
&\quad - \lambda(v-1)r(k-1) + (r-k)(k+r-2kr) \\
&= r(v-1)(k\lambda - k - \lambda + r - k\lambda + \lambda) \\
&\quad + (r-k)(k\lambda - k - \lambda + r + k + r - 2kr) \\
&= r(v-1)(r-k) + (r-k)(k\lambda - \lambda + 2r - 2kr) \\
&= (r-k)(rv + k\lambda - 2kr + r - \lambda) \\
&= (r-k)(rv + k\lambda - kr - r(k-1) - \lambda) \\
&= (r-k)(rv + k\lambda - kr - \lambda(v-1) - \lambda) \\
&= (r-k)(rv + k\lambda - kr - \lambda v) \\
&= (r-k)(v-k)(r-\lambda).
\end{aligned}
$$

Now $v - k$ and $r - \lambda$ are nonnegative, so

$$r - k \geq 0,$$

and it follows from (1.1) that $b \geq v$.

Exercises 2.3

2.3.1 Prove the inequality (2.21).

2.3.2 (Yet another proof of Fisher's inequality.) Suppose $V_1, V_2, \ldots, V_v$ are the rows of the incidence matrix of a (v, b, r, k, λ)-BIBD. Define

$$
\begin{aligned}
K_1 &= V_1; \\
K_i &= V_i - \frac{\lambda}{r + (i-2)\lambda}(V_1 + V_2 + \ldots + V_{i-1}),\ 2 \leq i \leq v.
\end{aligned}
$$

Prove that the vectors $K_1, K_2, \ldots, K_v$ are orthogonal, and therefore independent, vectors of length b, whence $v \leq b$. [136]

2.4 t-Designs

Suppose t, v, k, and λ are positive integers with $1 < k < \lambda$. We define a *t-design* with parameters (v, k, λ), or t-(v, k, λ) design, to be a way of selecting blocks of size k from a v-set so that any set of t treatments appears as a subset of exactly λ blocks. The relationship between t-designs and t-wise balanced designs is analogous to the relationship between balanced incomplete block designs and pairwise balanced designs; in fact, a balanced incomplete block design is a 2-design. We define b and r in the usual way. As an example, we exhibit a 3-$(8,4,1)$ design that was first found by Barrau [10]:

$$\begin{array}{ccccccc} 1248 & 2358 & 3468 & 4578 & 1568 & 2678 & 1378 \\ 3567 & 1467 & 1257 & 1236 & 2347 & 1345 & 2456. \end{array}$$

LEMMA 2.12

In a t-(v,k,λ) design,

$$b = \lambda\frac{\binom{v}{t}}{\binom{k}{t}}.$$

Proof. We count all the ordered pairs whose first element is a block and whose second element is a t-set contained in that block. Since there are b blocks and each contains $\binom{k}{t}$ t-sets, these are $b\binom{k}{t}$ pairs. On the other hand, there are $\binom{v}{t}$ t-sets in all, and each must appear in λ blocks, so the number is $\lambda\binom{v}{t}$. Equating these we get the result. □

Instead of computing r, we shall prove a more general theorem. Suppose S is any set of s treatments of a t-design $\mathcal{D}$. We define the *derivative* $\mathcal{D}_S$ of $\mathcal{D}$ with regard to S to be the block design whose treatments are the treatments of $\mathcal{D}$ other than S, and whose blocks are the sets $B\backslash S$ where B was a block that contained S. If S is singleton, $\mathcal{D}_S$ is called the *contraction* of $\mathcal{D}$ at S.

LEMMA 2.13

The derivative of a t-design with regard to a subset S of size s, $s \le t$, is a $(t-s)$-design with parameters $(v-s, k-s, \lambda)$.

Proof. Let X be any set of $t-s$ treatments of $\mathcal{D}_S$. The blocks of $\mathcal{D}_S$ containing X are precisely the blocks $B\backslash S$ where B is a block of $\mathcal{D}$ that contains the t-set $S \cup X$. There are λ such blocks B. So the design is a $(t-s)$-$(v-s, k-s, \lambda)$ design. □

THEOREM 2.14

There is a constant r_s such that every s-set of treatments of a t-(v,k,λ)

design belongs to exactly r_s blocks, when $0 \leq s \leq t$; and

$$r_s = \frac{\binom{v-s}{t-s}}{\binom{k-s}{t-s}}.$$

Proof. The number of blocks of $\mathcal{D}$ containing all elements of a set S equals the number of blocks of $\mathcal{D}_S$. From Lemma 2.13, if S is an s-set, this number equals the number of blocks in a $(t-s)$-$(v-s, k-s, \lambda)$-design. The result now follows from Lemma 2.12. □

COROLLARY 2.14.1

A t-design is an s-design for $s \leq t$.

It follows from Theorem 2.14 that if there is a t-(v, k, λ)-design, the expression for r_s must be an integer for $0 \leq s \leq t$. In other words,

$$\binom{k-s}{t-s} \text{ divides } \lambda\binom{v-s}{t-s}, 0 \leq s \leq t. \tag{2.22}$$

If the numbers (v, k, λ, t) satisfy (2.22), we refer to them as an *admissible quadruple*. Admissibility is clearly a necessary condition for a t-(v, k, λ) design, but it may be shown that it is not sufficient.

Steiner [140] asked about t-designs with $k = t+1$ and $\lambda = 1$, and these are called *Steiner systems*. The quadruple $(v, 3, 1, 2)$ is admissible if and only if $v \equiv 1$ or $3 \pmod 6$, and we show in Chapter 12 that these designs (Steiner triple systems) always exist. For $(v, 4, 1, 3)$ the necessary condition is $v \equiv 2$ or $4 \pmod 6$; Hanani [65] showed that this was sufficient.

Exercises 2.4

2.4.1 Describe 0-designs.

2.4.2 Prove that if t, v and k are any positive integers with $t \leq k \leq v$, there is a t-(v, k, λ) design with $\lambda = \binom{v-t}{k-t}$.

2.4.3* Prove that the set of all k-sets in a v-set is a t-design provided $t \leq k \leq v$. What are its parameters?

2.4.4 V consists of all n-tuples $(v_1, v_2, \ldots, v_n)$ where the v_i are integers modulo 2. Addition is defined by the law

$$(u_1, u_2, \ldots) + (v_1, v_2, \ldots) = (w_1, w_2, \ldots),$$

where $w_i \equiv u_i + v_i \pmod 2$. Prove that the blocks

$$\{x, y, z, w\} : x, y, z, w \in V, x + y + z + w = (0, 0, \ldots, 0)\}$$

form a 3-$(2^n, 4, 1)$-design.

2.4.5* Which of the following are admissible quadruples?

(i) $(11, 4, 1, 3)$
(ii) $(28, 7, 1, 5)$
(iii) $(14, 4, 1, 3)$
(iv) $(17, 6, 2, 4)$

2.4.6 Given a positive integer k, what is the value of v such that the quadruple $(v, k, 1, k-1)$ is admissible?

2.4.7 Prove that if $v \equiv 2 \pmod 6$, then$(v, 4, 1, 3)$ is an admissible quadruple.

2.4.8 Select a $(7, 7, 3, 3, 1)$-BIBD on treatment-set $S = \{1, 2, 3, 4, 5, 6, 7\}$. The blocks of this design are $B_1, B_2, \ldots, B_7$. Then append a new treatment 0 to each block; write $B_i^0 = B_i \cup \{0\}$. Then define $B_i^1 = S \backslash B_i$. Show that the blocks $B_1^0, B_1^1, B_2^0, \ldots, B_7^1$ form a 3-design. What are its parameters?

2.4.9* Suppose $\mathcal{E}$ is a t-design and x is a treatment in $\mathcal{E}$. The *leave* of x, denoted $\mathcal{E}^x$, is the design whose treatments are the treatments of $\mathcal{E}$ other than x and whose blocks are the blocks of $\mathcal{E}$ that do not contain x.

(i) Prove that $\mathcal{E}^x$ is a $(t-1)$-design.
(ii) If $\mathcal{E}$ is a t-(v, k, λ)-design, what are the parameters of $\mathcal{E}^x$?

Chapter 3

Finite Geometries

3.1 Finite Affine Planes

In this chapter we develop the theory of the finite analogs of Euclidean and projective geometry. An *incidence structure* consists of two sets, a set P of points and a set L of lines, together with a binary relation of *incidence* between elements of P and elements of L. If a point p is incident with a line l, one says "p lies on l" or "l contains p." With any line one can associate a subset of P, namely the set of all points that lie on the given line. We shall never wish to discuss geometrical situations where a line can contain no points or where two different lines can have the same set of points, so we can in fact *define* lines to be nonempty sets of points. We define a *geometry* to consist of a set P of objects called points and a set L of nonempty subsets of P called lines that satisfy the two axioms (A1) and (A2):

(A1) given any two points, there is one and only one line that contains them both;

(A2) there is a set of four points, no three of which belong to one common line.

In particular, an *affine plane* is a geometry that obeys the further axiom:

(A3) given any point p and given any line q that does not contain p, there is exactly one line that contains p and contains no point of q.

Axiom (A3) is the well-known parallel axiom, or "Euclid's fifth postulate"; if we were to add the distance properties of ordinary physical space (in a suitable form) to the axioms (A1), (A2) and (A3), we would have a set of axioms for ordinary Euclidean plane geometry. However, we shall instead impose a condition of finiteness that is inconsistent with Euclidean metrical geometry, and make the following definition:

A **finite affine plane** *is a finite set P of objects called* **points**, *together with a set L of nonempty subsets of P called* **lines**, *that satisfy the axioms* (A1), (A2) *and* (A3).

It is clear that (A1) is a "balance" axiom: Because of (A1), a geometry whose point set is finite is a pairwise balanced design with $\lambda = 1$. From (A1) it is easy to deduce the following property: Given any two lines, there is at most one point contained in both. Disjoint lines in a finite affine plane are called parallel, just as they are in ordinary (Euclidean) plane geometry. Axiom (A1) means that two points determine exactly one common line, and the remark just made means that two nonparallel lines have exactly one common point. Axiom (A3) means that if p does not lie on q, there is exactly one line through p parallel to q.

We shall use geometric terminology. In particular, lines with a common point are called *concurrent* and points that all lie on the same line are called *collinear*. We shall say "the line ab" to mean the unique line containing points a and b.

Any finite affine plane must contain at least four points, by (A3). This minimum can be realized: There exists a four-point plane, denoted $AG(2,2)$, that contains exactly six lines; their point sets are the six possible unordered pairs from the four points. This geometry has the interesting property that one can find two lines that between them contain all four points. We now prove that this characterizes $AG(2,2)$.

LEMMA 3.1

If a finite affine plane has two lines l and m that between them contain all the points of the plane, it is $AG(2,2)$.

Proof. By (A2), the plane must contain four distinct points, a, b, c, and d, say, of which no three are collinear. Clearly, two must belong to each of l and m; say

$$l = \{a, b, \ldots\}, m = \{c, d, \ldots\}.$$

Suppose l also contains another point, e. Then consider the lines ac, bd and de. Do ac and bd meet? Since all points lie on $l \cup m$, any intersection point must be either on l or on m. If it were on l, it must be the point a, which is where l and ac meet. But by the same argument it must be b. Since two lines cannot have two common points, $a = b$, which is a contradiction. So ac and bd are parallel. Similarly, we see that ac and de are parallel. But this means that there are two lines through d parallel to ac, contradicting (A3).

It follows that neither l nor m can contain a third point. So the plane has exactly four points. The six lines ab, ac, ad, bc, bd and cd must all belong to the plane. So we have a copy of $AG(2,2)$. But there can be no further lines; any such line would have exactly one point, and if (for example) $\{a\}$ were a line, then $\{a\}$ and ab would be two lines through a, parallel to cd—another contradiction. So the plane is precisely $AG(2,2)$. $\square$

LEMMA 3.2

In a finite affine plane there is a parameter n such that every line contains n points and every point lies on $n+1$ lines.

Proof. If the points of the plane are all contained in two lines, then it is $AG(2,2)$, and the lemma is satisfied with $n = 2$. So assume that no two lines contain all the points.

Now select any two lines l and m, and select a point p that lies on neither of them. Suppose that l contains points $a_1, a_2, \ldots, a_n$. Then p lies on exactly n lines that meet l, namely $pa_1, pa_2, \ldots, pa_n$. (It is easy to see that these lines are distinct.) It also lies on one line, k say, parallel to l (by (A3)).

Consider these $n+1$ lines. One will be parallel to m, and the n others will meet m in n distinct points. These n points are all the points of m (since every point of m lies on some line through p), so m contains n points. Since l and m were lines chosen at random, it follows that all lines contain n points for a fixed n.

Now p lies on $n+1$ lines. Since p could have been chosen to be any point not on the arbitrary line l, it follows that the number of lines through any point is greater by 1 than the common number of points per line. □

The constant n is called the *parameter* of the affine plane. (Some writers call it the "order," but it seems more appropriate to use that term for the number of points in the whole plane rather than in one line.) A finite affine plane with parameter n is called an $AG(2,n)$.

THEOREM 3.3

If "points" are identified with "treatments" and "lines" are identified with "blocks," a finite affine plane with parameter n is precisely a balanced incomplete block design with parameters

$$(n^2, n^2+n, n+1, n, 1). \tag{3.1}$$

Proof. Suppose that a finite affine plane is interpreted as a block design, as indicated. Lemma 3.2 says that $r = n+1$ and $k = n$, and axiom (A1) says that the design is balanced and $\lambda = 1$. So we have a balanced incomplete block design. The equations (1.1) and (1.2) must be satisfied; these yield $v = n^2, b = n^2+n$, and we have the parameters (3.1).

The converse is left as an exercise (see Exercise 3.1.2). □

Exercises 3.1

3.1.1 A trivial affine geometry is defined to be a structure that obeys (A1) and (A3) but not (A2). Describe all finite trivial affine geometries.
Hints and remarks: If there are no points, there can be no lines; if there is one point, either there is a line or not (two cases). For two points, there must be a line joining them; if there is another line, there must be exactly two more. For more than two points, there must be a line that contains all or all-but-one points.

3.1.2 Prove that a balanced incomplete block design with parameters

$$(n^2, n^2 + n, n + 1, n, 1)$$

is a finite affine plane.

3.1.3 A *finite incidence geometry* consists of a finite set P of points and a set of nonempty subsets of P called lines, that satisfy the axioms:
(F1) Two points determine exactly one line joining them;
(F2) There are at least two lines and every line contains at least two points;
(F3) There is a constant n such that every line contains n points.
(i) Prove that in any finite incidence geometry, two lines can have at most one common point.
(ii) Prove that every finite incidence geometry is a balanced incomplete block design.
(iii) Prove that in a finite incidence geometry, there is a number t such that for any point p and any line l not through p there are exactly t lines containing p that have no common point with l.

3.1.4 Prove that an $AG(2, n)$ is never a 3-design (except in the trivial case $n = 2$).

3.2 Finite Fields

In order to construct finite geometries, we shall use finite fields.

We assume the reader is familiar with fields in general, and the standard number fields, in particular the rational, real and complex numbers, usually denoted $\mathbb{Q}$, $\mathbb{R}$ and $\mathbb{C}$. Formally, a *field* $\{F, +, \times\}$ consists of a set F together with two binary operations $+$ and $\times$ that satisfy the following properties:

(F1) For every a and b in F, $a + b$ and $a \times b$ are also in F;

(F2) $\{F, +\}$ is an Abelian group, with identity 0, say;

(F3) $\{F \backslash \{0\}, \times\}$ is an Abelian group;

(F4) For any elements a, b and c in F,

$$a \times (b + c) = (a \times b) + (a \times c).$$

We shall often omit the sign $\times$, and omit brackets just as we do in ordinary arithmetic; for example, the displayed equation in (F4) would usually be written

$$a(b + c) = ab + ac.$$

We write 0 and 1 for the additive and multiplicative identity elements, respectively; $-a$ and a^{-1} denote the inverses of a under the two operations. The set $F \backslash \{0\}$ is called the *multiplicative group* of F and is often denoted F^*. Notice that 1 is a member of F^*, so $1 \neq 0$; therefore, every field has at least two elements.

LEMMA 3.4

If a and b are members of a field F, then

(i) $a0 = 0$;

(ii) $(-1)a = -a$;

(iii) $ab = 0 \Rightarrow a = 0$ *or* $b = 0$.

As we said above, we are primarily interested in *finite* fields, those with a finite number of elements. In a finite field F, write 1_n for the sum of n copies of 1: $1_0 = 0$, and for $n \geq 0, 1_{n+1} = 1_n + 1$ (n is an ordinary integer, not a member of F). Clearly,

$$1_a \times 1_b = 1_{ab},\ 1_a + 1_b = 1_{a+b},\ 1_a - 1_b = 1_{a-b}$$

(in the last case, b cannot be greater than a).

By the finiteness of F, the sequence $1_1, 1_2, 1_3, \ldots$ must contain repetitions. Suppose $1_k = 1_l$ where $k > l$. Then $1_{k-1} = 0$. So 0 occurs in the sequence. The *characteristic* of F is defined to be the smallest positive integer n such that $1_n = 0$.

THEOREM 3.5

The characteristic of a finite field is prime.

Proof. Suppose F has characteristic ab, where a and b are both integers greater than 1. Then

$$0 = 1_{ab} = 1_a \times 1_b,$$

so either $a = 0$ or $b = 0$, by Lemma 3.4(iii). But $a < ab$ and $b < ab$, which contradicts the minimality of ab. □

THEOREM 3.6

If x is a nonzero element of a field of characteristic n, the sum of k copies of x equals zero if and only if n divides k.

Proof. The sum of k copies of x satisfies

$$x + x + \ldots + x = x(1 + 1 + \ldots + 1) = x \times 1_k,$$

and $x \times 1_k = 0$ if and only if $1_k = 0$ (by Lemma 3.4(iii)). Say b is the remainder on dividing k by n: $k = an + b$ for some a, and $0 \le b < n$. Then

$$1_k = 1_{an} + 1_b = 1_a \times 1_n + 1_b.$$

Since the characteristic is minimal, 1_b equals zero if and only if $b = 0$, which is equivalent to saying n divides k. □

The most frequently encountered fields—the rational, real, and complex numbers—are infinite. The most familiar finite field is the set of integers modulo p, where p is a prime. This field has characteristic p. However, there are other finite fields. For example, the set $\{0, 1, A, B\}$ forms a field under the operations

$+$	0	1	A	B
0	0	1	A	B
1	1	0	B	A
A	A	B	0	1
B	B	A	1	0

$\times$	0	1	A	B
0	0	0	0	0
1	0	1	A	B
A	0	A	B	1
B	0	B	1	A

and that field has characteristic 2.

A *subfield* is defined to be a subset of a field which is itself a field with the same identity elements 0 and 1. If F is a field and G is a subfield of F, it is not hard to see that F is a vector space over G, using the usual definition. If both F and G are finite, F is isomorphic to the set of all k-tuples $(x_1, x_2, \ldots, x_k)$, where each x_i ranges through G. So F has gk elements, where g is the order of G.

THEOREM 3.7

A finite field of characteristic p has p^k elements for some positive integer k.

Proof. Suppose F is a finite field of characteristic p with identity 1. Write

$$P = \{1, 1_2, 1_3, \ldots, 1_p\}.$$

It is easy to verify that P is a field with identity elements $1_p(= 0)$ and 1. So P is a subfield of F. But P has p elements. So F has p^k elements, where k is the dimension of F when considered as a vector space over P. □

The field P is called the *prime field* of F.

It may be shown that any two fields with p^k elements are isomorphic. For this reason there is no confusion if we use the notation $GF(p^k)$ to mean a field with p^k elements. (GF stands for "Galois field.") The field $GF(p)$, where p is prime, is in fact the arithmetic of integers modulo p, Z_p.

We now outline a proof that there is a finite field of every prime power order. A *polynomial* over a field F is a function of the form

$$x \mapsto f(x) = a_0 + a_1x + a_2x^2 + \ldots + a_kx^k,$$

where the a_i are members of F; assuming a_k is nonzero, k is called the *degree* of the polynomial. One usually refers to the polynomial as $f(x)$. If, further,

$$g(x) = b_0 + b_1x + b_2x^2 + \ldots + b_nx^n,$$

then we define the sum and product in the obvious way:

$$\begin{aligned}(f+g)(x) &= (a_0 + b_0) + (a_1 + b_1)x + (a_2 + b_2)x^2 + \ldots;\\ (fg)(x) &= a_0b_0 + (a_0b_1 + a_1b_0)x + (a_1 + b_1)x\\ &\quad +(a_2b_0 + a_1b_1 + a_0b_2)x^2 + \ldots;\end{aligned}$$

$f + g$ has degree $\max(k, n)$ and fg has degree $k + n$. A polynomial is called *reducible* if it can be written as a product of two polynomials each of degree at least 1, and *irreducible* otherwise. One can prove the existence of at least one irreducible polynomial of every degree over the field $GF(p)$, where p is prime.

Select a prime p, and select an irreducible polynomial $f(x)$ of degree k over $GF(p)$. If $a(x)$ is any polynomial over $GF(p)$, there is a unique polynomial $\overline{a}(x)$ such that

$$a(x) = f(x)g(x) + \overline{a}(x)$$

for some polynomial $g(x)$, and $\overline{a}(x)$ has degree less than k. We shall call $\overline{a}(x)$ the *residue* of $a(x)$ (mod $f(x)$). It is not difficult to verify that these residues modulo $f(x)$ form a field with p^k elements, so we have:

THEOREM 3.8

There is a field with n elements if and only if $n = p^k$ for some prime p and some positive integer k.

The details of the proof—for example, a verification that the unique factorization exists as described—may be found in algebra textbooks. For our purpose the most important thing is to observe how a field can be constructed when it is needed. As an example we show how to construct a 9-element field.

We need a field of order 3 and an irreducible polynomial of degree 2 over it. The 3-element field is $F = \{0, 1, 2\}$, with addition and multiplication modulo 3. To calculate the irreducible quadratic polynomials over F we observe that we need only consider the *monic* polynomials (those with 1 as the coefficient of x^2) since, for example, $2x^2 + 2ax + 2b$ is irreducible if and only if $x^2 + ax + b$ is. If $x^2 + ax + b$ is reducible, then

$$x^2 + ax + b = (x + \alpha)(x + \beta)$$

for some α and β in F. Setting α and β to the various possible values we find there are six reducible monic quadratics, namely

$$x^2,\ x^2 + x,\ x^2 + 2x,\ x^2 + 2x + 1,\ x^2 + 2,\ x^2 + x + 1.$$

Hence the remaining monic quadratics over F, namely

$$x^2 + 1, x^2 + x + 2, x^2 + 2x + 2,$$

are all irreducible.

We could use any of the three irreducible polynomials to construct $GF(3^2)$; let us use $x^2 + 1$. The elements will be

$$\{0, 1, 2, x, x + 1, x + 2, 2x, 2x + 1, 2x + 2\}$$

and all calculations are reduced modulo 3 and modulo $x^2 + 1$. For example,

$$\begin{aligned}
(x + 2)(2x + 1) &= 2x^2 + 5x + 2 \\
&= 2x^2 + 2x + 2 \pmod 3 \\
&= 2(x^2 + 1) + 2x \\
&= 2x \pmod{x^2 + 1}.
\end{aligned}$$

If we abbreviate $ax + b$ to the ordered pair ab, we obtain the tables in Figures 3.1 and 3.2.

+	00	01	02	10	11	12	20	21	22
00	00	01	02	10	11	12	20	21	22
01	01	02	00	11	12	10	21	22	20
02	02	00	01	12	10	11	22	20	21
10	10	11	12	20	21	22	00	01	02
11	11	12	10	21	22	20	01	02	00
12	12	10	11	22	20	21	02	00	01
20	20	21	22	00	01	02	10	11	12
21	21	22	20	01	02	00	11	12	10
22	22	20	21	02	00	01	12	10	11

FIGURE 3.1: Addition table for $GF(3^2)$.

×	00	01	02	10	11	12	20	21	22
00	00	00	00	00	00	00	00	00	00
01	00	01	02	10	11	12	20	21	22
02	00	02	01	20	22	21	10	12	11
10	00	10	20	02	12	22	01	11	21
11	00	11	22	12	20	01	21	02	10
12	00	12	21	22	01	10	11	20	02
20	00	20	10	01	21	11	02	22	12
21	00	21	12	11	02	20	22	10	01
22	00	22	11	21	10	02	12	01	20

FIGURE 3.2: Multiplication table for $GF(3^2)$.

In the discussion preceding Theorem 3.8, it was not necessary that p be a prime; we only used the fact that a field of order p existed. In particular, if $q = p^r$, we could use an irreducible polynomial of degree s over $GF(q)$ to form a field of order $q^s = p^{rs}$. The new field would have $GF(q)$ as a subfield.

But there is only one field of order p^{rs} (up to isomorphism). So the field $GF(p^{rs})$, constructed using an irreducible polynomial of degree rs over a prime field of order p, always contains a field of p^r elements. We can deduce the following theorem.

THEOREM 3.9

The field $GF(p^k)$ contains a subfield $GF(p^r)$ whenever r divides k.

Exercises 3.2

3.2.1* Prove Lemma 3.4.

3.2.2 Prove that the field with four elements, exhibited in this section, *is* in fact a field.

3.2.3 Prove by direct construction that any field with four elements must be isomorphic to the one mentioned in Exercise 3.2.2.

3.2.4* Show that the integers $\{0, 2, 4, 6, 8\}$ form a field under addition and multiplication modulo 10.

3.2.5 F is a finite field and G is a subfield of F. Prove that F and G have the same characteristic.

3.2.6 Construct fields with 4 and 8 elements using polynomials.

3.2.7* Prove that there are exactly eight irreducible monic polynomials of degree 3 over $GF(3)$.

3.2.8 Using the irreducible polynomial $x^3 + 2x + 1$ over $GF(3)$, construct a finite field with 27 elements.

3.2.9* If G and H are two fields, then $G \times H$ is defined to be the set of all ordered pairs whose first element is in G and whose second element is in H, with two binary operations defined by

$$\begin{aligned}(x, y) + (z, t) &= (x + z, y + t)\\ (x, y) \times (z, t) &= (xz, yt),\end{aligned}$$

where the operations between x and z are the addition and multiplication of G, and the operations between y and t are those of H.

(i) Prove that $G \times H$ contains a zero element (i.e., an identity element for addition, which acts like 0 for multiplication) and a unit element (i.e., an identity element for multiplication).

(ii) Prove that $G \times H$ contains two nonzero elements whose product is zero (whence $G \times H$ is not field).

(iii) Suppose G and H both have prime order; say $|G| = p$ and $|H| = q$. Say $p \neq q$. If $0 \leq x < p$, write x for the element (x_p, x_q) where x_p means the element of G obtained when x is reduced modulo p, and similarly for x_q. Prove that $G \times H$ is essentially the arithmetic of integers modulo pq, provided x is interpreted as "$x \pmod{pq}$."

3.3 Construction of Finite Affine Geometries

Ever since the invention of Cartesian coordinates, many results of Euclidean geometry have been proven algebraically, using the properties of equations over the real field and, more recently, real linear algebra. In this section we examine analogous constructions using a finite field.

Suppose F is the field $GF(n)$ with n elements. Let V be the set of all ordered pairs of members of F. We write L for the set of all linear equations

$$ax + by + c = 0,$$

where a, b and c are elements of F and x and y are indeterminates, with the case $a = b = 0$ omitted. Two equations are equivalent if one can be obtained from the other by multiplying by a nonzero field member throughout. If l is any member of L, we write "the line l" to mean the set of all pairs (x, y) that satisfy the equation l; equivalent equations give rise to the same line. We shall verify that if we interpret the members of V as points and the "lines" derived from L as lines, we have constructed an affine plane of parameter n.

First, suppose (a, b) and (c, d) are any two points. Then there is a unique line containing both, the line

$$(x - a)(d - b) = (c - a)(y - b).$$

(Since $a = c$ and $b = d$ cannot both be true for two distinct points, this is always a line. It is easy to show that no other line contains both points.) So (A1) is satisfied.

To prove that (A2) is true, it is sufficient to find four points of which no three are collinear. But the points $(0, 0)$, $(0, 1)$, $(1, 0)$ and $(1, 1)$ always have this property (see Exercise 3.3.1).

Next consider (A3). We show that there is a unique line through (e, f) and parallel to $ax + by + c = 0$. By definition, two lines have a common point if and only if their equations have a common solution. So the lines parallel to $ax + by + c = 0$ would be the lines $ax + by + d = 0$, where $d \neq c$. If (e, f) is to lie on this line,

$$ae + bf + d = 0.$$

This will be true for exactly one value of d, namely

$$d = -ae - bf.$$

So (A3) is satisfied.

Finally, the number of points on a line is precisely n. On the line $ax+by+c = 0$, where $b \neq 0$, the n points are

$$\{(x, -b^{-1}(c+ax)) : x \in F\},$$

and on the line $ax + c = 0$ they are

$$\{(-a^{-1}c, y) : y \in F\}.$$

(In the latter case, $a = 0$ is impossible.)

Since there is a field with n elements whenever n is a prime power, we have proven the following result.

THEOREM 3.10

There is a finite affine plane $AG(2,n)$, or equivalently a balanced incomplete block design with parameters

$$(n^2, n^2+n, n+1, n, 1),$$

whenever n is a prime power.

As an example we again construct an $AG(2,2)$. We write $F = GF(2) = \{0,1\}$. The set V consists of the points

$$\{(0,0), (0,1), (1,0), (1,1)\},$$

which we abbreviate to

$$\{00, 01, 10, 11\}.$$

The possible choices of a, b and c for a line $ax+by+c = 0$ are $010, 011, 100, 101, 110, 111$. These six choices lead to the following equations:

(i) $y = 0$, with points $00, 10$;

(ii) $y + 1 = 0$, with points $01, 11$;

(iii) $x = 0$, with points $00, 01$;

(iv) $x + 1 = 0$, with points $10, 11$;

(v) $x + y = 0$, with points $00, 11$;

(vi) $x + y + 1 = 0$, with points $01, 10$.

So the $AG(2,2)$ is the $(4,6,3,2,1)$-design with blocks

$$\begin{array}{ccc} \{00,10\} & \{01,11\} & \{00,01\}, \\ \{10,11\} & \{00,11\} & \{01,10\}. \end{array}$$

The geometry developed from a finite field is called a *finite Euclidean plane geometry* and denoted $EG(2, n)$ to distinguish it from other affine planes with the same parameter. (We have not proven that all finite affine planes are Euclidean, and in fact this is not the case.)

Since every field has prime power order, an $EG(2, n)$ exists if and only if n is a prime power. The status of the existence of finite affine planes is not decided. In all known examples the parameter n is a prime power, but it has not been proven that other cases are impossible. The situation is discussed in Chapter 6.

Just as higher-dimensional Euclidean geometry can be constructed from real equations in more than two variables, so we can construct higher-dimensional finite affine geometries. We shall use the algebraic approach for the actual definition of the geometries rather than proceeding from axioms.

The **finite affine geometry** $EG(d, n)$ *of dimension* d *over* $GF(n)$ *consists of the* nd *vectors of length* d *over* $GF(n)$*, which are called points. If* W *is any* k*-dimensional subspace of (the set of points of)* $EG(d, n)$ *and* p *is any member of* $EG(d, n)$*, then* $p + V$*, defined by*

$$p + V = \{p + v : v \in V\},$$

is called a k-**flat**. *In particular,* $(k - 1)$*-flats are called* **primes** *and* 1*-flats are called* **lines**.

THEOREM 3.11

Given two points of $EG(d, n)$*, there is a unique line that contains them both.*

Proof. Suppose q and r are two points of $EG(d, n)$. Write V for the set of all points that are multiples of $r - q$. Then $q + V$ is a line that contains both q and r:

$$\begin{aligned} q &= q + 0(r - q), \\ r &= q + 1(r - q). \end{aligned}$$

Moreover, suppose $p + W$ is another line that contains both q and r; say W is the set of all multiples of a nonzero vector w. Then there are field elements a and b such that

$$\begin{aligned} q &= p + aw, \\ r &= p + bw. \end{aligned}$$

So $q - r$ is a multiple of w. It follows that V and W are the same space. It remains to show that $p + W$ and $q + W$ are the same set. But if x is any member of $p+W$, say $x = p+cw$, then $x = q+(c-a)w$ and $x \in q+W$, and conversely. So the line is unique. □

COROLLARY 3.11.1

The points of an $EG(d,n)$, interpreted as treatments, and the lines, interpreted as blocks, form a balanced incomplete block design with parameters

$$\left(n^d, \frac{n^{d-1}(n^d-1)}{n-1}, \frac{n^d-1}{n-1}, n, 1\right).$$

Proof. Balance follows from Theorem 3.11; and the points of the line $p+V$ are the n vectors

$$p + av : a \in GF(n)$$

where v is a generator of V, so the block size is a constant n. So by Theorem 2.6 we have a balanced incomplete block design, and the other parameters follow from (1.1) and (1.2). □

One can derive a great many other balanced incomplete block designs from the finite affine geometries. If we interpret k-flats as blocks, instead of lines, we obtain a design with parameters

$$\left(n^d, \frac{\lambda n^{d-k}(n^d-1)}{n^k-1}, \frac{\lambda(n^d-1)}{n^k-1}, n^k, \lambda\right), \tag{3.2}$$

where

$$\lambda = \frac{n^{d-1}-1}{n-1} \times \frac{n^{d-2}-1}{n^2-1} \times \cdots \times \frac{n^{d-k+1}-1}{n^{k-1}-1}. \tag{3.3}$$

(The proof is left as an exercise.)

Exercises 3.3

3.3.1* Verify that in any finite affine plane constructed over a field, no three of the points $(0,0)$, $(1,0)$, $(0,1)$ and $(1,1)$ are collinear.

3.3.2 Prove that if the k-flats of $EG(d,n)$ are interpreted as blocks and the points as treatments, the result is a balanced incomplete block design with parameters defined by (3.2) and (3.3).

3.3.3 Suppose A is any 2-flat of $EG(d,n)$. By a line of A we mean a line of the $EG(d,n)$ all of whose points lie in A. Prove that the points and lines of A form an $EG(2,n)$.

3.3.4 Prove that $EG(d, n)$ is never a 3-design when $n > 2$.
Hint: There are two sorts of 3-sets of points, those that are collinear and those that are not.

3.3.5 Prove that the balanced incomplete block design of Exercise 3.3.2 is a 3-design when $n = 2$ and $k \geq 2$.

3.3.6 The 3-element field $GF(3)$ consists of the three elements 1, 2, 3, with addition and multiplication modulo 3. Write down all the points and the equations of all the lines of $EG(2, 3)$.

3.4 Finite Projective Geometries

LEMMA 3.12

Suppose l is a line in a finite affine plane $AG(2, n)$. Then there are exactly $n - 1$ lines parallel to l.

Proof. Say l contains the points $p_1, p_2, \ldots, p_n$. There are n lines other than l that pass through p_i, for $i \in \{1, 2, \ldots, n\}$; call these lines $l_{i1}, l_{i2}, \ldots, l_{in}$. If $i \neq j$, there is exactly one line that contains both p_i and p_j, by (A1), and that line is l. So $l_{ix} \neq l_{jy}$ unless $i = j$ and $x = y$, and there are n^2 distinct lines l_{ix}. The lines parallel to l (those that do not meet l) are precisely those lines not equal to l or to any of the l_{ix}; their number must be

$$(n^2 + n) - (n^2 + 1),$$

that is, $n - 1$. □

COROLLARY 3.12.1

Lines that are parallel to the same line l are parallel to one another.

Proof. The number of points on each of the $n - 1$ lines parallel to l is n, so their union contains at most $n^2 - n$ points, and will contain fewer if any of the two lines have a common point. But (A3) tells us that each of the $n^2 - n$ points that do not lie on l must belong to a line parallel to l, so all the points must be in the union. Therefore, the lines parallel to l have no common points; they are parallel to each other. □

COROLLARY 3.12.2

The lines of $AG(2,n)$ may be partitioned into $n-1$ subsets of size n, called parallel classes, such that two lines meet if and only if they are in different parallel classes.

THEOREM 3.13

There exists an $AG(2,n)$, or balanced incomplete block design with parameters

$$(n^2, n^2+n, n+1, n, 1),$$

if and only if there exists a symmetric balanced incomplete block design with parameters

$$(n^2+n+1, n+1, 1). \tag{3.4}$$

Proof. Suppose there is an $AG(2,n)$ with point set P and line set L. The line set can be partitioned into $n+1$ subsets $L_1, L_2, \ldots, L_{n+1}$, where each L_i is a parallel class of lines. Since lines in a parallel class do not intersect, no point belongs to two lines in the same subset L_i; simple arithmetic shows that each point belongs to exactly one line in each L_i.

We form a new design whose treatments are the elements of P together with $n+1$ new "points" $p_1, p_2, \ldots, p_{n+1}$. The blocks are formed from the lines in L: For each line l in L, there is a unique subset L_i such that $l \in L_i$; we define a line $l^* = l \cup \{p_i\}$. The blocks of the new design are the n^2+n blocks l^*, together with l_∞, where

$$l_\infty = \{p_1, p_2, \ldots, p_{n+1}\}.$$

The new design has n^2+n+1 treatments and n^2+n+1 blocks that are $(n+1)$-sets of treatments. It remains to prove that it is balanced, with $\lambda = 1$. Now every pair of members of P belongs to precisely one member of L; they will belong to the corresponding new block, and no other. If p belongs to P, then $\{p, p_i\}$ is a subset of l^*, where l is the unique line passing through p that lies in L_i, while $\{p_i, p_j\}$ is a subset of l_∞ alone. So the new design is balanced, with $\lambda = 1$; from (1.1) we find that $r = n+1$.

Conversely, if a symmetric design with parameters $(n^2+n+1, n+1, 1)$ exists, delete from it one block and all its members. An $AG(2,n)$ is obtained; the details are left to the reader (see Exercise 3.4.1). □

The construction just given is the finite analog of the way in which real projective plane geometry may be derived from real Euclidean plane geometry by appending "points at infinity," where parallel lines meet, and a "line at infinity," consisting of all the points at infinity. For this reason, a symmetric design with the parameters (3.4) is called a *finite projective plane.* Alternatively, one may make a formal definition by axioms, such as the following.

A **finite projective plane** *consists of a finite set P of points and a set of subsets of P called lines, satisfying the axioms (P1), (P2) and (P3):*

(P1) Given two points, there is exactly one line that contains both;

(P2) Given two lines, there is exactly one point that lies in both;

(P3) There are four points, of which no three are collinear.

THEOREM 3.14

In a finite projective plane, as defined by the axioms (P1), (P2), (P3), *every line contains $n+1$ points for some parameter n. Such a plane is a symmetric balanced incomplete block design with parameters*

$$(n^2+n+1, n+1, 1).$$

The proof is left as an exercise. A finite projective plane with parameter n is denoted $PG(2,n)$.

As a first example we exhibit the geometry $PG(2,2)$. In the preceding section we derived $AG(2,2)$ from the field $GF(2)$; its points were

$$\{00, 01, 10, 11\}.$$

The six lines fall into three parallel classes that we shall denote as follows:

$$\begin{aligned}
L_1 &: \begin{cases} y &= 0, \quad \text{points } 00, 10; \\ y+1 &= 0, \quad \text{points } 01, 11; \end{cases} \\
L_2 &: \begin{cases} x &= 0, \quad \text{points } 00, 01; \\ x+1 &= 0, \quad \text{points } 10, 11; \end{cases} \\
L_3 &: \begin{cases} x+y &= 0, \quad \text{points } 00, 11; \\ x+y+1 &= 0, \quad \text{points } 01, 10. \end{cases}
\end{aligned}$$

We append new points p_1, p_2, p_3 and obtain the following lines:

$$\begin{aligned}
&\{00, 10, p_1\}, \quad \{01, 11, p_1\}, \\
&\{00, 01, p_2\}, \quad \{10, 11, p_2\}, \\
&\{00, 11, p_3\}, \quad \{01, 10, p_3\}, \\
&\{p_1, p_2, p_3\}.
\end{aligned}$$

If we perform the substitution

$$(00, 01, 10, 11, p_1, p_2, p_3) \mapsto (0, 3, 1, 6, 2, 4, 5),$$

we obtain the blocks

012	362
034	164
056	315
245	

which are the sets we first obtained for a $(7, 3, 1)$-design in Section 1.1.

As before, we can generalize to projective geometries of higher dimension; and as before the easiest technique is the algebraic one. We define a *finite projective geometry of dimension d* over $GF(n)$, or $PG(d, n)$, to be a set of points that are $(d + 1)$-vectors over $GF(n)$; however, the zero vector is not allowed, and two points are considered equal if the vector of one is a multiple of the vector of another. Alternatively, the points correspond to the vectors of the form

$$(0, 0, \ldots, 0, 1, *, \ldots, *)$$

where asterisks denote any field elements, and where the leading 1 must occur (the zero vector is not allowed). In this interpretation, a prime consists of all the points x,

$$x = (x_1, x_2, \ldots, x_{d+1}),$$

such that

$$a_1x_1 + a_2x_2 + \ldots + a_{d+1}x_{d+1} = 0$$

for some constants $a_1, a_2, \ldots, a_{d+1}$, not all zero; in other words, all the points whose vectors are in the subspace

$$ax = 0.$$

When these same rules (delete zero; all scalar multiples of a vector are equivalent) are applied to a $(k + 1)$-dimensional space, we obtain a *k-flat*; a 1-flat is a line.

It is easy to see that taking the k-flats of a $PG(d, n)$ as blocks and the points as treatments, we obtain a balanced incomplete block design with parameters

$$\left(\frac{n^{d+1}-1}{n-1}, \frac{(n^{d+1}-1)(n^{d+1}-n)}{(n^{k+1}-1)(n^{k+1}-n)}\lambda, \frac{n^{d+1}-n}{n^{k+1}-n}\lambda, \frac{n^{k+1}-1}{n-1}, \lambda\right) \tag{3.5}$$

where

$$\lambda = \frac{(n^{d+1}-n^2)(n^{d+1}-n^3)\ldots(n^{d+1}-n^k)}{(n^{k+1}-n^2)(n^{k+1}-n^3)\ldots(n^{k+1}-n^k)}. \tag{3.6}$$

Not every finite affine plane can be constructed algebraically from a field, so the field construction is not the only one for $PG(2, n)$. However, an exhaustive search shows that when $n \leq 8$, the only examples of $AG(2, n)$ are the $EG(2, n)$, so there is no $AG(2, 6)$ and the other $AG(2, n)$ with $n \leq 8$ are uniquely determined. It follows that $PG(2, 6)$ is impossible and that $PG(2, n)$ is unique for other values of n up to 8.

Those $PG(2, n)$ that are constructed from fields are called *field planes.*

Field planes have various special properties that are not necessarily enjoyed by projective planes in general. We prove one simple but surprising fact about field planes whose parameter is a power of 2.

By a *quadrangle* we mean a set of four points, no three of which are collinear. In a quadrangle $\{a, b, c, d\}$ the lines ab and cd are called opposite sides, so that a quadrangle determines three pairs of opposite sides; and the intersection of such a pair of lines is called a *diagonal point.*

THEOREM 3.15

In a field plane over $GF(2^t)$, the three diagonal points of any quadrangle are collinear.

Proof. Consider the quadrangle $\{a, b, c, d\}$. Since a, b and c are not collinear, they are independent (as vectors over $GF(2^t)$); so one can find a matrix to map them simultaneously into any desired independent set. In particular, select a matrix M such that

$$aM = (1, 0, 0),\ bM = (0, 1, 0),\ cM = (0, 0, 1).$$

Write $dM = (p, q, r)$. Since abd are not collinear, it follows that aM, bM and dM are not collinear, so $r \neq 0$. Similarly, p and q are nonzero.

It is easy to calculate the diagonal points of $\{aM, bM, cM, dM\}$:

$$\begin{aligned} aMbM \cap cMdM &= (p, q, 0) = u, \text{ say;} \\ aMcM \cap bMdM &= (p, 0, r) = v, \text{ say;} \\ aMdM \cap bMcM &= (0, q, r) = w, \text{ say.} \end{aligned}$$

These three points are collinear; they lie on

$$\frac{x_1}{p} + \frac{x_2}{q} + \frac{x_3}{r} = 0.$$

(The equation makes sense because none of p, q, r can be zero; the points satisfy it because the characteristic of the field is two.) Now u, v and w are dependent vectors, so the original diagonal points uM^{-1}, vM^{-1} and wM^{-1}, are dependent as vectors, and the points are collinear. □

The points of the quadrangle together with the diagonal points form a substructure isomorphic to $PG(2, 2)$. It is not surprising that the field planes $PG(2, 2^t)$ contain subplanes isomorphic to $PG(2, 2)$; the strength of the theorem is the fact that *every* quadrangle forms a subplane.

Exercises 3.4

3.4.1 Prove that if any one line (and all its points) is deleted from a $PG(2,n)$, the result is an $AG(2,n)$.

3.4.2 Prove Theorem 3.14.

3.4.3 We use the representation $GF(4) = \{0, 1, A, B\}$ that was given in Section 3.2. Show that the points

$$(1,1,A),\ (0,1,B),\ (1,1,B),\ (1,A,B)$$

of $PG(2,4)$ form a quadrangle. Find its diagonal points, and prove that they are collinear.

3.4.4* Consider the geometry $PG(2,5)$.

(i) What are the points of the line

$$x_1 + x_2 + 3x_3 = 0?$$

(ii) What is the equation of the line joining $(1,1,1)$ to $(2,3,2)$?

(iii) What is the point of intersection of the lines

$$x_1 + x_2 + 3x_3 = 0,\ \ 2x_1 + 3x_2 + x_3 = 0?$$

3.4.5 Prove that $PG(d,n)$ has $(n^{d+1} - 1)/(n-1)$ points.

3.4.6 Prove that by taking the k-flats of $PG(d,n)$ as blocks and the points as treatments, one obtains a balanced incomplete block design with parameters defined by (3.5) and (3.6).

3.4.7* Prove that the field plane over $GF(3)$ contains a quadrangle whose diagonal points are not collinear.

3.4.8 Prove that there is exactly one $PG(2,2)$ and exactly one $PG(2,3)$ (up to isomorphism).

Chapter 4

Some Properties of Finite Geometries

4.1 Ovals in Projective Planes

Suppose c is any set of points in a $PG(2, n)$. A line l is called an *i-secant* of c if l and c have i common points. In particular a 2-secant is simply called a secant, a 1-secant is a *tangent*, and a 0-secant is an *external line* to c. A set c that has no i-secant for $i > k$ is called a *k-arc*.

Consider a point x of a 2-arc c in a $PG(2, n)$. If c has m elements, there are $m - 1$ lines through x to the other points of c; these lines are all different (for otherwise c would have a 3-secant). So x lies on exactly $m - 1$ secants, and consequently on $n + 2 - m$ tangents. So the total number of tangents is $m(n + 2 - m)$. This cannot be negative, so $m \leq n + 2$.

We are interested in 2-arcs with either $n + 1$ points or $n + 2$ points. These are called *ovals*; the two classes are called *type I ovals* and *type II ovals*, respectively. A type I oval has one tangent through each of its points, for a total of exactly $n+1$ tangents, and a type II oval has no tangents whatsoever.

Suppose a $PG(2, n)$ contains a type II oval c. Select a point q outside the oval. Every line through q meets c in 0 or 2 points. But every point of c must lie on some line through q. It follows that c has an even number of points. So n is even.

THEOREM 4.1

If c is a type I oval in a projective plane $PG(2, n)$, where n is even, then the tangents to c are concurrent.

Proof. We first observe that every point x of the plane lies on at least one tangent to c. If x is in c, we have already noted that there is exactly one tangent at x. So assume x is not on c, and count all the intersections of c with lines through x. The answer is $n + 1$, which is odd, but it also equals the number of tangents through x plus twice the number of secants through x. So there is at least one tangent through x.

Suppose x and y lie on some secant to c. Then the tangents to c through x and y must be different (otherwise the line xy would be both a tangent and a secant). So the tangents through points of a secant are all different. Since c has exactly $n+1$ tangents, it follows that each point of a secant lies on exactly one tangent.

Now consider the point of intersection of two tangents; call it m. This point cannot lie on any secant (if it did, m would be a point of a secant and would lie on two tangents). So *all* the lines joining m to the points of c are tangents, and m lies on all $n + 1$ tangents. □

The point m is called the *nucleus* of the oval c. It is clear that $c \cup m$ is a type II oval. On the other hand, if any point m is deleted from a type II oval, the result is a type I oval with m as nucleus.

If n is odd, $PG(2, n)$ cannot contain a type II oval, so:

COROLLARY 4.1.1

If n is odd, then the tangents of a type I oval in a $PG(2, n)$ are not concurrent.

COROLLARY 4.1.2

Suppose n is even. Any type I oval in a $PG(2, n)$ can be extended to a type II oval. So a $PG(2, n)$, n even, contains a type II oval if and only if it contains a type I oval.

In particular, consider the field plane over $GF(2^t)$, and let c consist of all the points that satisfy

$$x_1x_2 + x_2x_3 + x_3x_1 = 0.$$

Clearly the points of c are $(0, 0, 1)$, $(0, 1, 0)$, and all $(1, y, y(1 + y)^{-1})$ for $y \neq 1$. So c contains $n + 1$ points, and it is easy to see that no three of them are collinear (this is left as an exercise). Therefore c is a type I oval. So:

COROLLARY 4.1.3

There is a $PG(2, 2^t)$ containing a type II oval for every positive integer t.

One reason for our interest in ovals is the following result.

THEOREM 4.2

Suppose c is a type II oval in a $PG(2,2k)$. Then there is a balanced incomplete block design with parameters

$$(2k^2 - k, 4k^2 - 1, 2k + 1, k, 1).$$

Proof. The treatments of the design will correspond to the external lines of c, and the blocks to the points of the plane that are not in c. A block contains a treatment if and only if the point corresponding to the block lies on the line corresponding to the treatment. Since c has $2k + 2$ points, it has

$$\binom{2k+2}{2} = (k+1)(2k+1)$$

secants, and therefore $4k^2 + 2k + 1 - (k + 1)(2k + 1)$ external lines. So the number of treatments is $4k^2 + 2k + 1 - (k + 1)(2k + 1) = 2k^2 - k$. Each point not on c lies on $2k + 1$ lines, of which $k + 1$ are secants, so each block of the design will contain $(2k + 1) - (k + 1) = k$ treatments. Any two external lines intersect in a unique point that is not on c, so the design is balanced with $\lambda = 1$. The other parameters follow from (1.1) and (1.2) (or they could be deduced directly). □

COROLLARY 4.2.1

There is a design with parameters

$$(2k^2 - k, 4k^2 - 1, 2k + 1, k, 1)$$

whenever k is a power of 2.

These results cannot be reversed. For example, we know a $(15, 35, 7, 3, 1)$-design, which are the parameters we would expect from a type II oval in $PG(2,6)$, but we shall show in Chapter 6 that no $PG(2,6)$ exists. But it is a useful theorem that provides an infinite class of block designs.

Ovals have other combinatorial applications. In Chapter 10 we use them, together with other tools, to prove the uniqueness of $PG(2,4)$. They have also been used in other aspects of design construction and they are interesting in their own right, in finite geometry.

Exercises 4.1

4.1.1 Verify that no three of the points

$$(0,0,1),\ (0,1,0),\ \{(1,a,a(1+a)^{-1}) : a \neq 1\}$$

are collinear over any field of characteristic 2.

4.1.2 We use the representation $GF(4) = \{0, 1, A, B\}$ that was given in Section 3.2. Show that the points

$$(1,0,0),\ (1,1,1),\ (1,1,A),\ (1,1,B),\ (0,1,A),\ (1,B,A)$$

of $PG(2,4)$ form a type II oval.

4.1.3 The points

$$(0,1,1),\ (1,1,B),\ (1,1,A),\ (1,0,B),\ (1,B,1)$$

of $PG(2,4)$ form a type I oval. Find all its tangents and find its nucleus.

4.1.4* Show that the points

$$(1,0,0),\ (1,1,0),\ (1,2,1),\ (1,2,2),\ (1,1,2),\ (1,0,1)$$

form a type I oval in $PG(2,5)$. Find all its tangents and show that they are not collinear.

4.2 The Desargues Configuration

The configuration of Desargues consists of two disjoint *triangles*, or sets of three noncollinear points, abc and def say, in which the three lines ad, be, cf meet in a seventh point p, with the further property that the three points $ab \cap de$, $ac \cap df$ and $bc \cap ef$ all lie on some line l. The configuration is illustrated in Figure 4.1. One usually says that the two triangles abc and

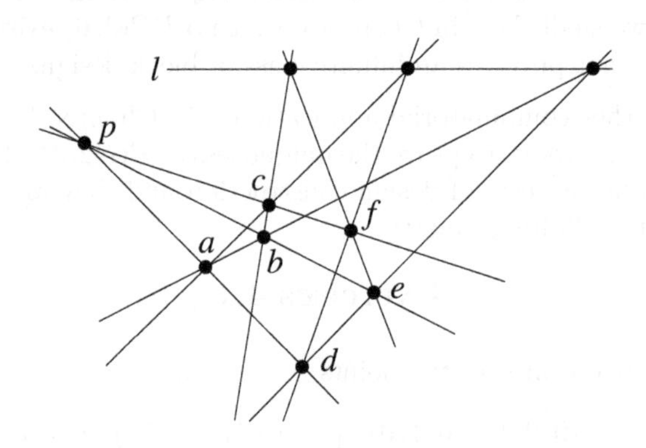

FIGURE 4.1: The Desargues Configuration.

def are *in perspective* through the point p and also in perspective through the line l.

We say a finite projective plane is *Desarguesian* if it satisfies the following law:

Whenever the six points a, b, c, d, e, f *are such that* ad, be *and* cf *are concurrent lines, and* abc *and* def *are not collinear sets, then* $ab \cap de$, $ac \cap df$ *and* $be \cap ef$ *are collinear. That is, whenever two triangles are in perspective from a point, they form a Desargues configuration.*

Suppose p were to lie on the line ab. Then ab and de are the same line, and the intersection consists of all the points of the line. One could argue that no single point $ab \cap de$ is defined, so no Desargues configuration is possible; or one could say that the line joining $ac \cap df$ to $bc \cap ef$ certainly contains a point of $ab \cap de$, and say that the triangles always yield a Desargues configuration. We shall take the latter view; if p lies on one of the sides of one of the triangles, then it lies on a common side of the two triangles and we have a trivial Desargues configuration.

THEOREM 4.3

Every field plane is Desarguesian.

Proof. Suppose abc and def are two noncollinear sets, and that ad, be and cf meet in the point p. There is no loss of generality in assuming that $a = (1, 0, 0)$, $b = (0, 1, 0)$ and $c = (0, 0, 1)$, because we can transform these into any three noncollinear points using a nonsingular matrix (compare with the proof of Theorem 3.15).

Write $p = (p_1, p_2, p_3)$. If p lies on ab, there is nothing to prove, so we can assume that $p_3 \neq 0$; similarly, $p_1 \neq 0$ and $p_2 \neq 0$. The points d, e and f lie on ap, bp and cp, respectively, and they are distinct from a, b, c, p, so we can write

$$\begin{aligned} d &= (1 + gp_1, gp_2, gp_3), \\ e &= (1 + hp_1, hp_2, hp_3), \\ f &= (1 + ip_1, ip_2, ip_3), \end{aligned}$$

for some nonzero g, h and i. The general point on de has the form

$$q_1 d + q_2 e, \tag{4.1}$$

where q_1 and q_2 are members of the field. The points of ab are the points with third coordinate zero. So the point $ab \cap de$ will have the form (4.1) with

$$q_1 g p_3 + q_2 h p_3 = 0;$$

since points are invariant under nonzero scalar multiplication, we can take $q_1 = h, q_2 = -g$, and we obtain

$$ab \cap de = (h, -g, 0)$$

and similarly

$$\begin{aligned} ac \cap df &= (-i, 0, g), \\ be \cap ef &= (0, i, -h). \end{aligned}$$

All these points lie on the line

$$gx_1 + hx_2 + ix_3 = 0,$$

so the triangles are in perspective from that line. □

THEOREM 4.4

There is a non-Desarguesian $PG(2,9)$.

Proof. We construct a plane with 91 points denoted $A_i, B_i, C_i, D_i, E_i, F_i, G_i$ for $0 \le i \le 12$, and 91 lines denoted $l_i, m_i, n_i, p_i, q_i, r_i, s_i$ for $0 \le i \le 12$.

$$\begin{aligned} l_0 &= \{A_0, A_1, A_3, A_9, B_0, C_0, D_0, E_0, F_0, G_0\}, \\ m_0 &= \{A_0, B_1, B_8, D_3, D_{11}, E_2, E_5, E_6, G_7, G_9\}, \\ n_0 &= \{A_0, C_1, C_8, E_7, E_9, F_3, F_{11}, G_2, G_5, G_6\}, \\ p_0 &= \{A_0, B_7, B_9, D_1, D_8, F_2, F_5, F_6, G_3, G_{11}\}, \\ q_0 &= \{A_0, B_2, B_5, B_6, C_3, C_{11}, E_1, E_8, F_7, F_9\}, \\ r_0 &= \{A_0, C_7, C_9, D_2, D_5, D_6, E_3, E_{11}, F_7, F_8\}, \\ s_0 &= \{A_0, B_3, B_{11}, C_2, C_5, C_6, D_7, D_9, G_1, G_8\}; \end{aligned}$$

the other lines are derived by developing the subscripts modulo 13: for example,

$$l_i = \{A_i, A_{1+i}, A_{3+i}, A_{9+i}, B_i, C_i, D_i, E_i, F_i, G_i\},$$

where subscripts are reduced mod 13 where necessary. This sort of construction, which we call a *difference method*, will be studied further in Chapter 5.)

We exhibit two triangles in perspective from a point that do not form a Desargues configuration. The triangles are $A_1B_1C_8$ and $A_3B_8E_7$, which are in perspective through A_0:

$$l_0 = A_0A_1A_3,\ m_0 = A_0B_1B_8,\ n_0 = A_0C_8E_7.$$

It is easy to check that

$$\left.\begin{aligned} A_1B_1 &= l_1 = \{A_1, A_2, A_4, A_{10}, B_1, C_1, D_1, E_1, F_1, G_1\}, \\ A_3B_8 &= q_3 = \{A_3, B_5, B_8, B_9, C_6, C_1, E_4, E_{11}, F_{10}, F_{12}\}, \end{aligned}\right\} \; l_1 \cap q_3 = C_1,$$
$$\left.\begin{aligned} A_1C_8 &= r_1 = \{A_1, C_8, C_{10}, D_3, D_6, D_7, E_4, E_{12}, F_8, F_9\}, \\ A_3E_7 &= l_7 = \{A_7, A_8, A_{10}, A_3, B_7, C_7, D_7, E_7, F_7, G_7\}, \end{aligned}\right\} \; r_1 \cap l_7 = D_7,$$
$$\left.\begin{aligned} B_1C_8 &= s_3 = \{A_3, B_6, B_1, C_5, C_8, C_9, D_{10}, D_{12}, G_4, G_{11}\}, \\ B_8E_7 &= q_6 = \{A_6, B_8, B_{11}, B_{12}, C_9, C_4, E_7, E_1, F_0, F_2\}, \end{aligned}\right\} \; s_3 \cap q_6 = C_9,$$

and C_1, D_7 and C_9 are not collinear:

$$C_1D_7 = r_5, \; C_1C_9 = q_{11}, \; D_7C_9 = r_2. \qquad \square$$

COROLLARY 4.4.1

There are nonisomorphic planes $PG(2, 9)$.

Exercises 4.2

4.2.1* Consider the following points in $PG(2, 5)$:

$$\begin{aligned} a &= (1,1,2), & b &= (1,2,1), & c &= (1,3,3), \\ d &= (1,1,3), & e &= (1,3,1), & f &= (1,4,4). \end{aligned}$$

Show that triangles abc and def are in perspective; find the coordinates of $ab \cap de$, $ac \cap df$ and $bc \cap ef$, and verify that they are collinear.

4.2.2 The *dual* of a balanced incomplete block design was defined in Section 2.2 to be the design obtained by exchanging the roles of blocks and treatments (or, equivalently, by transposing the incidence matrix). Prove that the dual of a Desarguesian $PG(2, n)$ is a Desarguesian $PG(2, n)$.

4.2.3 Verify that the lines given in the proof of Theorem 4.4 form a $PG(2, 9)$.

4.2.4 Find further pairs of triangles in the plane of Theorem 4.4 that are in perspective through a point but not through any line.

Chapter 5

Difference Sets and Difference Methods

5.1 Difference Sets

Suppose G is an Abelian group of order v, written in additive notation, and suppose B is a set of k elements of G. Then the *design generated from* B (in G) consists of all the blocks

$$\{B + g : g \in G\}.$$

It is a block design with $b = v$ and $r = k$. This process of "generation" is also called *developing* B in G; if G is Z_v, we say "developing B mod v."

LEMMA 5.1

If B contains precisely λ_i ordered pairs of elements whose difference is g_i, and if x and y are members of G such that $x - y = g_i$, then $\{x, y\}$ is a subset of precisely λ_i blocks.

Proof. Suppose the pairs of elements of B with difference g_i are

$$\{b_1, c_1\}, \{b_2, c_2\}, \ldots, \{b_{\lambda_i}, c_{\lambda_i}\};$$

for $j = 1, 2, \ldots, \lambda_i$, we have $c_j - b_j = g_i$. If $x - y = g_i$, then $\{x, y\}$ will occur as a subset of $B + (x - c_1), B + (x - c_2), \ldots$, and $B + (x - c_{\lambda_i})$. All these blocks are different; if $x - c_j = x - c_h$, then $c_j = c_h, b_j = b_h$ and the two pairs are the same. So $\{x, y\}$ is a subset of at least λ_i blocks. But if $\{x, y\}$ is a subset of $B + d$, then $(x - d) - (y - d) = g_i$ and $x - d$ and $y - d$ are in B. So $x - d = c_j$ and $y - d = b_j$ for some j, and we have already listed the pair $\{x - d, y - d\}$ above. So no further cases can arise and $\{x, y\}$ is a subset of *precisely* λ_i blocks. □

To construct a balanced design, we must make all the λ_i equal. This is effected by the following definition:

So by (i), r_k divides r_ir_j.

On the other hand, $(f_if_j)^{r_k} = 1$, so $f_i^{r_k} = (f_j^{r_k})^{-1}$. So $f_i^{r_k}$ and $f_j^{r_k}$ have the same multiplicative order, by (ii); call that order s. By (ii), s divides r_i and r_j, the orders of f_i and f_j. Since r_i and r_j are coprime, s must equal 1. Therefore $f_i^{r_k}$ has multiplicative order 1, so $f_i^{r_k} = 1$. Therefore r_i divides r_k, by (i). Similarly, r_j divides r_k. Again using the fact that r_i and r_j are coprime, r_ir_j divides r_k. So $r_k = r_ir_j$.

(iv) It is easy to see that R contains all divisors of all of its elements: If $r_i = st$, one readily verifies that f_i^s has order t. Now suppose r is the least common multiple of the elements of R and has prime factor decomposition

$$r = p_1^{a_1}p_2^{a_2}\dots p_n^{a_n}.$$

For each i, there must be some element r_j of R such that $p_i^{a_i}$ divides r_j. So $p_i^{a_i}$ is in R. By (iii), r belongs to R. Now suppose f is an element of multiplicative order r. The elements $1, f, f^2, \dots, f^{r-1}$ must all be different members of F^*, so $r \le m$. On the other hand, since each r_i divides r, each f_i is a root of the equation $x^r - 1 = 0$. Since this equation has at least m distinct roots, its degree must be at least m: $r \ge m$. So $r = m$, and the powers of f make up all of F^*. So F^* is cyclic with generator f. □

As an example, consider $GF(3^2)$. We constructed this field in Section 3.2, extending the prime field $Z_3 = \{0, 1, 2\}$ by the irreducible polynomial $x^2 + 1$. In that representation, the powers of the element $x + 1$ are

$$\begin{array}{llll} (x+1)^1 = x+1, & & (x+1)^5 = 2x+2, \\ (x+1)^2 = 2x, & & (x+1)^6 = x, \\ (x+1)^3 = 2x+1, & & (x+1)^7 = x+2, \\ (x+1)^4 = 2, & & (x+1)^8 = 1, \end{array}$$

so $x + 1$ generates the multiplicative group of $GF(3^2)$. A generator of the multiplicative group of a finite field is called a *primitive element* of the field; if the field has prime order, the term "primitive root" is often used.

Consider the equation $y^2 = q$. It has at most two roots over any field; if $y = f$ is a solution, then $y = -f$ is also a solution. It follows that $y^2 = q$ can have *precisely one* solution only if there is a root f that satisfies $f = -f$. One possibility is $f = 0$, which is a solution if and only if $q = 0$. Provided that the field has odd characteristic, there are no other cases where $f = -f$. We shall say that a nonzero element q of a field is *quadratic* if the equation $y^2 = q$ has a solution in the field, and *nonquadratic* otherwise. For example, the element 2 is nonquadratic in $GF(3)$, but is quadratic in the above representation of $GF(3^2)$. (This depends on the representation of $GF(3^2)$, which in turn depends on the irreducible polynomial used—see Exercise 5.2.3.)

Suppose $F = GF(p^n)$, where p is an odd prime. The set of all squares of the $p^n - 1$ nonzero elements must equal the set of all quadratic elements. If we list these squares, each quadratic element arises twice, once as f^2 and once as $(-f)^2$. So F has $(p^n - 1)/2$ quadratic elements. (In other words, exactly half the nonzero elements are quadratic.)

We define a map χ from a finite field F to the integers as follows:

$$\begin{aligned} \chi(x) &= 1 \text{ if } x \text{ is quadratic;} \\ \chi(0) &= 0; \\ \chi(x) &= -1 \text{ otherwise.} \end{aligned}$$

This map is called the *quadratic character* on F. The properties of χ can be deduced from the cyclic nature of F^*. Suppose a is a primitive element of F. Then the quadratic elements are $a^2, a^4, a^6, \ldots$. If the number of elements of F is even, a has odd order, and every power of a is quadratic. So we consider fields of odd order. If F has $2k + 1$ elements, a has order $2k$. There are k quadratic elements: $a^2, a^4, a^6, \ldots, a^{2k} (= 1)$. The elements $a, a^3, \ldots, a^{2k-1}$ are *nonquadratic*. From this it follows immediately that the product of two quadratic elements or two nonquadratic elements is quadratic; "mixed" products are nonquadratic. The inverse of a quadratic element is always quadratic. In terms of χ, we have

LEMMA 5.4

In any finite field,

(i) if $x \neq 0$, then $\chi(x^{-1}) = \chi(x)$;

(ii) $\chi(x)\chi(y) = \chi(xy)$.

THEOREM 5.5

If k is any nonzero element of $F = GF(p^n)$, where p is odd, then

$$\sum_{x \in F} \chi(x)\chi(x + k) = -1.$$

Proof. If $x \neq 0$, then $\chi(x^{-1}) = \pm 1$, so $\chi(x^{-1})\chi(x^{-1}) = 1$. Therefore

$$\begin{aligned} \chi(x)\chi(x + k) &= \chi(x) \times 1 \times \chi(x + k) \\ &= \chi(x)[\chi(x^{-1})\chi(x^{-1})]\chi(x + k) \\ &= [\chi(x)\chi(x^{-1})][\chi(x^{-1})\chi(x + k)] \\ &= \chi(xx^{-1})\chi(x^{-1}(x + k)) \\ &= 1 \times \chi(1 + x^{-1}k). \end{aligned}$$

Since $k \neq 0$, $x^{-1}k \neq 0$. If $x \neq y$, then $x^{-1}k \neq y^{-1}k$. Thus, as x runs through F^*, the product $x^{-1}k$ also runs through F^*. So

$$\begin{aligned}
\sum_{x \in F} \chi(x)\chi(x+k) &= 0 + \sum_{x \in F^*} \chi(x)\chi(x+k) \\
&= \sum_{x \in F^*} \chi(1 + x^{-1}k) \\
&= \sum_{x^{-1}k \in F^*} \chi(1 + x^{-1}k) \\
&= \sum_{y \in F, y \neq 1} \chi(y), \text{ where } y = 1 + x^{-1}k \\
&= \left[\sum_{y \in F} \chi(y) \right] - \chi(1).
\end{aligned}$$

Now $\chi(y) = 1$ for $(p^n - 1)/2$ values of y, $\chi(y) = -1$ for $(p^n - 1)/2$ values, and $\chi(0) = 0$. So

$$\sum_{y \in F} \chi(y) = \tfrac{1}{2}(p^n - 1) - \tfrac{1}{2}(p^n - 1) + 0 = 0;$$

since $\chi(1) = 1$, we have

$$\sum_{x \in F} \chi(1 + x^{-1}k) = -1.$$

Therefore

$$\sum_{x \in F} \chi(x)\chi(x+k) = 0 + \sum_{x \in F^*} \chi(x)\chi(x+k) = -1.$$

□

THEOREM 5.6

In $GF(p^n)$,

(i) *if $p^n \equiv 1 \pmod 4$, then -1 is quadratic;*

(ii) *if $p^n \equiv 3 \pmod 4$, then -1 is nonquadratic.*

Proof. We can assume that p^n is odd, say $p^n = 2k+1$. Consider the equation $x^2 = 1$. If a is a primitive element, then the roots of this equation are precisely a^0 and a^k. Since -1 and 1 are distinct roots of $x^2 = 1$, and since $a^0 = 1$, we have $a^k = -1$. So -1 is quadratic if k is even, and nonquadratic if k is odd. But "k even" and "k odd" are precisely $p^n \equiv 1 \pmod 4$ and $p^n \equiv 3 \pmod 4$, respectively. □

THEOREM 5.7 [111]

Suppose v is a prime power congruent to $3 \pmod 4$; for convenience, write $v = 4t - 1$. Then there is a $(4t-1, 2t-1, t-1)$ difference set.

Proof. We shall show that the set D of quadratic elements of $GF(v)$ is a difference set with the required parameters. The set has the required number of elements.

Suppose there are λ ordered pairs (x, y) of elements of D with difference 1. Then, if q is any quadratic element,

$$(y - x) = 1 \Rightarrow (qy - qx) = q,$$

and qx and qy are quadratic; different pairs (x, y) give different pairs (qx, qy). So we have λ ordered pairs of elements of D with difference q. But if $(z-t) = q$, then $(q^{-1}z - q^{-1}t) = 1$, so every pair giving difference q comes from a pair giving difference 1. We have covered the differences q where q is quadratic. But this is sufficient. Every nonzero nonquadratic element is the negative of a quadratic element, so all the nonzero differences occur in λ ways; the pairs with difference $-q$ are all the pairs (qy, qx). So D is a (v, k, λ)-difference set, with $v = 4t - 1$ and $k = 2t - 1$. From Corollary 5.2.1, $\lambda = t - 1$. □

The difference sets in Theorem 5.7 are called *Paley difference sets.*

THEOREM 5.8 [139]

If p^n and p^n+2 are both odd prime powers, then there is a difference set with parameters $(v, \frac{1}{2}(v-1), \frac{1}{4}(v-3))$, where $v = p^n(p^n+2)$.

The difference set is in the group $G = GF(p^n) \times GF(p^n+2)$ consisting of all the pairs (x, y) where x is in $GF(p^n)$ and y is in $GF(p^n+2)$; addition is defined by

$$(x, y) + (z, t) = (x + z, y + t),$$

where the addition $x + z$ is carried out in $GF(p^n)$ and $y + t$ is carried out in $GF(p^n+2)$, and multiplication by:

$$(x, y)(z, t) = (xz, yt),$$

where the products xz and yt are computed in the appropriate fields.

If a and b denote primitive elements in $GF(p^n)$ and $GF(p^n+2)$, respectively, we write $c = (a, b)$, $d = (a, 0)$, so $c^i = (a^i, b^i)$, and so on. Then

$$D = \{c^0, c, c^2, \ldots, c^{(p^{2n}-1)/2-1}, 0, d^0, d, \ldots, d^{p^n-2}\} \tag{5.1}$$

is a difference set with the required parameters. The verification, which is straightforward but messy, can be found in [139] and [168].

As an example, suppose $p = 5$ and $n = 1$, so that $p^n = 5$ and $p^n + 2 = 7$. The possible candidates for a are 2 and 3, the primitive roots modulo 5, while b can be 3 or 5. If we take $a = 2$, $b = 3$, we obtain $c = (2, 3)$, $d = (2, 0)$. Since p and $p+2$ are primes, the arithmetic here is simply modular arithmetic modulo 35. Now $(2, 3) = (17, 17)$, since $17 \equiv 2 \pmod 5$ and $17 \equiv 3 \pmod 7$; also and $(2, 0) = (7, 7)$; so we can take c and d as 17 and 7 $\pmod{35}$; $d^0 = (1, 0) = 21$. Then

$$\begin{aligned} D &= \{17^0, 17, 17^2, \ldots, 17^{11}, 0, 21, 7, 7^2, 7^3\} \\ &= \{1, 17, 9, 13, 11, 12, 29, 3, 16, 27, 4, 33, 0, 21, 7, 14, 28\} \\ &= \{0, 1, 3, 4, 7, 9, 11, 12, 13, 14, 16, 17, 21, 27, 28, 29, 33\}, \end{aligned}$$

which is the required $(35, 17, 8)$-difference set.

Observe that both the families of difference sets that we have so far constructed have the form $v = 4t - 1, k = 2t - 1, \lambda = t - 1$. They are called "Hadamard difference sets." (The symmetric balanced incomplete block designs derived from them are called "Hadamard designs"; they will be very important in Chapter 14.)

Exercises 5.2

5.2.1 Prove (in the notation of Theorem 5.3) that if $r_i = st$, then f_i^s has multiplicative order t.
Hint: Essentially use $a^q a^r = a^{q+r}$.

5.2.2 Prove Lemma 5.4.

5.2.3 Consider the representation of $GF(3^2)$ cobstructed by using the irreducible polynomial $x^2 + x + 2$ (as in Section 3.1). Show that 2 is not quadratic in that representation.

5.2.4 If p is an odd prime, show that the product of all quadratic elements in $GF(p^n)$ is 1 if $p^n \equiv 3 \pmod 4$ and -1 if $p^n \equiv 1 \pmod 4$.
Hint: Express each quadratic element, and -1*, as a power of a primitive root.*

5.2.5* Find difference sets with the parameters $(15, 7, 3)$ and $(143, 71, 35)$.

5.2.6* For what values of t up to $t = 50$ can one construct a $(4t-1, 2t-1, t-1)$-difference set using Theorems 5.7 and 5.8?

5.2.7 Suppose D is the $(4t - 1, 2t - 1, t - 1)$-difference set obtained from Theorem 5.7 for some prime power $4t - 1$. Prove that $D \cup \{0\}$ is a $(4t - 1, 2t, t)$-difference set.

5.3 Properties of Difference Sets

Consider the two sets

$$\{0,1,3\},\{2,4,5,6\}$$

modulo 7. Both are difference sets in Z_7; their parameters are $(7,3,1)$ and $(7,4,2)$ respectively. We observe that the two sets are complements in Z_7, and ask: When is the complement of a difference set again a difference set? The answer is that this is always true.

THEOREM 5.9

The complement of a (v,k,λ)-difference set is a $(v,v-k,v-2k+\lambda)$-difference set.

Proof. Suppose D is a (v,k,λ)-difference set in an Abelian group G. Consider the set of all ordered pairs of elements of G:

$$S=\{(g,h):g\in G,h\in G\}.$$

S can be broken into four disjoint parts, $S=S_1\cup S_2\cup S_3\cup S_4$, where

$$\begin{aligned}
S_1 &= \{(g,h):g\in D,h\in D\},\\
S_2 &= \{(g,h):g\in D,h\notin D\},\\
S_3 &= \{(g,h):g\notin D,h\in D\},\\
S_4 &= \{(g,h):g\notin D,h\notin D\}.
\end{aligned}$$

Write $\delta_i(x)$ for the number of elements (g,h) of S_i such that $g-h=x$. Then $\delta_1(x)=\lambda$ for all $x\neq 0$, and we wish to show that $\delta_4(x)=v-2k+\lambda$ for all $x\neq 0$.

For fixed g, the elements $g-h$ range through G as h ranges through G, so the collection of all differences arising from elements of S will contain every element of G v times:

$$\delta_1(x)+\delta_2(x)+\delta_3(x)+\delta_4(x)=v.$$

In the same way, if we check all the differences $g-h$ where $g\in D$ and $h\in G$, we get every group element k times (since D has k elements). So

$$\delta_1(x)+\delta_2(x)=k$$

for all x; and similarly,

$$\delta_1(x)+\delta_3(x)=k,$$

so $\delta_2(x) = \delta_3(x) = k - \lambda$ provided x is not zero. Thus

$$\begin{aligned}\delta_4(x) &= v - \delta_1(x) - \delta_2(x) - \delta_3(x) \\ &= v - 2k + \lambda.\end{aligned}$$

□

It may be verified that $v-k$ is never equal to k when a (v,k,λ)-difference set exists. So Theorem 5.9 tells us that every difference set we have constructed gives rise to another difference set.

Suppose D is a (v,k,λ)-difference set and consider the set $D+s$:

$$D + s = \{d + s : d \in D\}.$$

If the ordered pairs in D with difference g are $\{x_1, y_1\}, \{x_2, y_2\}, \ldots, \{x_\lambda, y_\lambda\}$, then the pairs $\{x_j+s, y_j+s\}$, for $1 \le j \le \lambda$, also have difference g. Considering all g in G, we see that $D+s$ is a (v,k,λ)-difference set. Such a set is called a *shift* of D. As an example, say one wished to construct all possible $(7,3,1)$-difference sets. Any such set must contain two elements x and $x+1$, to produce difference 1. Since $\{x, x+1, y\}$ is a shift of $\{0, 1, y-x\}$, one can start by assuming 0 and 1 to be in the difference set. Testing all the possibilities, one finds that the $(7,3,1)$-difference sets are $\{0,1,3\}$, $\{0,1,5\}$, and their shifts.

Although $\{0,1,5\}$ is not a shift of $\{0,1,3\}$, the two sets are related. If the elements $\{0,1,5\}$ are multiplied by three, the result is $\{0,3,15\}$ which reduces to $\{0,3,1\}$ modulo 7. This concept, the *multiple* of a difference set, is of course available only when the underlying additive group has a multiplicative structure as well. In theory, a difference set can arise in any Abelian group, but in practice many examples come from the modular arithmetic of the integers or from finite fields, where a suitable multiplication is available. Multiples of difference sets are discussed in Exercise 5.3.2.

Exercises 5.3

5.3.1* If D is a (v,k,λ)-difference set, denote by D' the set of additive inverses of elements of D. Prove that D' is a (v,k,λ)-difference set.

5.3.2 Suppose D is a (v,k,λ)-difference set over the integers modulo v.

(i) If m is an integer prime to v, define mD to be the set formed by multiplying the members of D by m and reducing modulo v. Prove that mD is a (v,k,λ)-difference set.

(ii) If m is prime to v, and mD equals D or some shift of D, then m is called a *multiplier* of D. Prove that the set of multipliers of D form a group under multiplication modulo v.

(iii) Prove that the condition "m is prime to v" in part (i) is necessary.

5.4 General Difference Methods

Difference sets arose from the wish that a design developed from a block should be balanced. One possible generalization is to develop two or more initial blocks and hope that the collection of all blocks formed will be a balanced design. If n initial blocks of size k are developed in a group G of order v, the resulting design will have $(v, b, r, k) = (v, nv, nk, k)$.

Suppose blocks $B_1, B_2, \ldots, B_n$ are developed in G, and that the number of ordered pairs of elements of B_j with difference g_i is λ_{ij}. It follows that if $x - y = g_i$, then $\{x, y\}$ will be a subset of λ_{ij} of the blocks generated from B_j. So we shall get a balanced design if all the sums $\sum_{j=1}^{n} \lambda_{ij}$ are equal. In this case the initial blocks are called *supplementary difference sets.*

As an example, consider the development of the initial blocks $\{0, 1, 2, 4\}$ and $\{0, 1, 4, 6\}$ modulo 9. The differences arising in the first initial block are $\pm 1, \pm 2, \pm 4, \pm 1, \pm 3, \pm 2$, while the second block yields $\pm 1, \pm 4, \pm 6, \pm 3, \pm 5$ and ± 2. It is seen (using the facts that $\pm 6 = \pm 3$ and $\pm 5 = \pm 4$) that each difference arises three times, so the resulting design is balanced with $\lambda = 3$, and we have a $(9, 18, 8, 4, 3)$-design.

Figure 5.2 shows some initial blocks that generate balanced incomplete block designs in cyclic groups. As a noncyclic example, $\{00, 01, 10, 44\}$ and $\{00, 02, 20, 33\}$ may be developed over $Z_5 \times Z_5$ to form a $(25, 50, 8, 4, 1)$-design.

v	b	r	k	λ	Initial blocks
9	18	8	4	3	$\{1, 2, 3, 5\}; \{1, 2, 5, 7\}$
41	82	10	5	1	$\{1, 10, 16, 18, 37\}; \{5, 8, 9, 21, 39\}$
13	26	12	6	5	$\{1, 2, 4, 7, 8, 12\}; \{1, 2, 3, 4, 8, 12\}$
16	80	15	3	2	$\{1, 2, 4\}; \{1, 2, 8\}; \{1, 3, 13\}; \{1, 4, 9\}; \{1, 5, 10\}$
22	44	14	7	4	$\{1, 7, 12, 16, 19, 21, 22\}; \{1, 6, 8, 9, 10, 14, 20\}$

FIGURE 5.2: Supplementary difference sets.

THEOREM 5.10

Let v be a prime power congruent to 1 modulo 4; for convenience, write $v = 4t + 1$. Then there exist two supplementary difference sets that can be developed to give a $(4t + 1, 2(4t + 1), 4t, 2t, 2t - 1)$-design.

Proof. Let a be a primitive element of the field $GF(v)$. We shall show that the sets

$$Q = \{a^{2b} : b = 1, 2, \ldots, 2t\}$$

and

$$R = \{a^{2b+1} : b = 1, 2, \ldots, 2t\}$$

of nonzero quadratic elements and nonquadratic elements respectively, are the required supplementary difference sets. Certainly, $|Q| = |R| = 2t$.

Consider the ordered pairs of elements of Q with difference 1. Say there are λ_1 such pairs, (q_{11}, q_{21}), $(q_{12}, q_{22}) \ldots$, $(q_{1\lambda_1}, q_{2\lambda_1})$, such that $q_{1i} - q_{2i} = 1$. Then the λ_1 pairs of the form $(q_{1i}q, q_{2i}q)$ have the property that $q_{1i}q - q_{2i}q = q$ for any $q \in Q$, and all these pairs belong to $Q \times Q$. Conversely, if $q_{1*}q - q_{2*}q = q$ then $(q_{1*}q^{-1}, q_{2*}q^{-1})$ must be one of the pairs (q_{1i}, q_{2i}). So there are precisely λ_1 ordered pairs of members of Q whose difference is q, for any $q \in Q$. Similarly, the number of ordered pairs of elements of R whose difference is r equals λ_1, for any r in R; the pair (q_{1i}, q_{2i}) gives rise to the pair (rq_{1i}, rq_{2i}) that satisfies $rq_{1i} - rq_{2i} = r$, and conversely.

Similarly, if there are λ_2 pairs (q_{1i}, q_{2i}) of elements of Q that satisfy

$$q_{1i} - q_{2i} = r,$$

then there will be λ_2 ordered pairs of members of Q whose difference is r, and λ_2 ordered pairs of elements of R whose difference is q, for any $r \in R$ and for any $q \in Q$.

It follows that when we go through all the differences of pairs in $Q \times Q$ and in $R \times R$, every nonzero element of $GF(v)$ will occur $\lambda_1 + \lambda_2$ times. So the two sets are supplementary difference sets. It follows immediately from (1.1) and (1.2) that $\lambda = 2t - 1$. □

It is not necessary that all blocks of a design be developed using the same arithmetical process. For example, consider the result of developing the following blocks modulo 15:

$$\{0, 1, 4\}, \{0, 2, 8\}, \{0, 5, 10\}.$$

Forty-five blocks are obtained; however, the last 15 blocks consist of the five blocks

$$\{0, 5, 10\}, \{1, 6, 11\}, \{2, 7, 12\}, \{3, 8, 13\}, \{4, 9, 14\},$$

each listed three times. If the repetitions are omitted, the resulting 35 blocks constitute a $(15, 35, 7, 3, 1)$ design.

We shall refer to this as "partial development" of the block $\{0, 5, 10\}$. In particular, most examples of this technique have arisen in cyclic cases, where the group is the set of integers under addition modulo v, and we refer to "partial circulation." We write

$$\{0, 5, 10\} \pmod{15} \text{ PC}(5)$$

to mean the set of five blocks generated by $\{0, 5, 10\}$ by partially circulating through five steps in Z_{15}. We write the full set of blocks as

$$\{0, 1, 4\}, \{0, 2, 8\} \pmod{15}, \{0, 5, 10\} \pmod{15} \text{ PC}(5).$$

Another approach to this situation follows from the fact that Z_{15} is isomorphic to the direct product $Z_3 \times Z_5$. Consider the mapping in which the integer $x \pmod{15}$ is mapped to the ordered pair $(x \pmod 3, x \pmod 5)$. (For example, $8 \mapsto 23$, as $8 \equiv 2 \pmod 3$ and $8 \equiv 3 \pmod 5$; similarly $4 \mapsto 14$ and $2 \mapsto 22$.) This is an isomorphism from Z_{15} to $Z_3 \times Z_5$. Under this mapping the given blocks map to

$$\{00, 11, 14\}, \{00, 22, 23\}, \{00, 20, 10\}.$$

We write $(\text{mod } 3, 5)$ to mean that an initial block is developed through all 15 possible values; $(\text{mod } -, 5)$ means that $\{ab\ldots\}$ goes through the five possible values obtained by developing the right-hand components modulo 5, but leaving the left-hand components fixed. So we say that the design is obtained by developing

$$\{00, 11, 14\}, \{00, 22, 23\} \pmod{3, 5}, \quad \{00, 20, 10\} \pmod{-, 5}.$$

Another generalization of difference sets and supplementary difference sets is the addition of an "infinity" element. Let G be an Abelian group with $v-1$ elements. Append to G a new element ∞; call the new set G'. The addition in G' is constructed from that of G by the additional law

$$\infty + g = \infty \text{ for all } g \in G'.$$

If B is a k-set of elements of G', the design generated from B consists of the blocks

$$B + x : x \in G$$

(there is no block "$B + \infty$" because it would contain only k copies of the element ∞). For example, if G is Z_3, the block $\{\infty, 0, 1\}$ gives rise to $\{\infty, 0, 1\}$, $\{\infty, 1, 2\}$ and $\{\infty, 2, 0\}$.

One cannot obtain a balanced incomplete block design by developing one initial block in this way: If ∞ is a member of B, it will appear in each of the $v-1$ blocks that are generated, and if it is not in B, then it never occurs; the other elements of G' appear in $k-1$ or k blocks, according as $\infty \in B$ or $\infty \notin B$. Regularity is impossible. However, one can sometimes achieve a balanced incomplete block design by developing more than one initial block. For example, the blocks

$$\{\infty, 0, 1, 3, 7\}, \{0, 1, 2, 4, 5\}$$

form a $(10, 18, 9, 5, 4)$-design when developed over $Z_9' = Z_9 \cup \{\infty\}$.

Suppose it is possible to develop n initial blocks over G', where G is a group of order $v-1$, so as to produce a (v, b, r, k, λ)-design, and suppose ∞ occurs in d initial blocks. Then ∞ lies in $d(v-1)$ of the blocks after development, so

$$r = d(v-1). \tag{5.2}$$

Since d of the initial blocks contain $k-1$ elements other than ∞, each non-infinite element will appear in $d(k-1)$ of the blocks developed from them, and they will each appear in $(n-d)k$ of the other blocks, so

$$r = d(k-1) + (n-d)k. \tag{5.3}$$

Comparing (5.2) and (5.3) we deduce that

$$dv = nk. \tag{5.4}$$

There are n sets of $(v-1)$ blocks. Since ∞ occurs in $d(v-1)$ blocks with $k-1$ other elements in each, for λ to be constant we must have

$$(v-1)\lambda = d(v-1)(k-1),$$

that is,

$$\lambda = d(k-1),$$

so the design has parameters

$$(v, n(v-1), d(v-1), k, d(k-1)). \tag{5.5}$$

Given these restrictions, the method has proven very useful in constructing individual designs. Some examples are shown in Figure 5.3.

Parameters	Initial blocks
$(10, 30, 9, 3, 2)$	$\{\infty, 0, 5\}, \{0, 1, 4\}, \{0, 2, 3\}, \{0, 2, 7\}$ (mod 9)
$(10, 18, 9, 5, 4)$	$\{\infty, 1, 3, 7\}, \{0, 1, 2, 4, 5\}$ (mod 9)
$(12, 44, 11, 3, 2)$	$\{\infty, 0, 3\}, \{0, 1, 3\}, \{0, 1, 5\}, \{0, 4, 6\}$ (mod 11)
$(15, 42, 14, 5, 4)$	$\{\infty, 0, 1, 2, 7\}, \{0, 1, 4, 9, 11\}, \{0, 1, 4, 10, 12\}$ (mod 14)

FIGURE 5.3: Initial blocks: general examples.

5.5 Singer Difference Sets

Consider the equations (3.4) and (3.5) in the case where n is a prime power and the blocks are primes in $PG(d, n)$. Since $k = d - 1$,

$$\begin{aligned}\lambda &= \frac{(n^{d+1} - n^2)(n^{d+1} - n^3)\dots(n^{d+1} - n^{d-1})}{(n^d - n^2)(n^d - n^3)\dots(n^d - n^{d-1})} \\ &= \frac{n^2 \times n^3 \times \dots \times n^{d-1}(n^{d-1} - 1)(n^{d-2} - 1)\dots(n^2 - 1)}{n^2 \times n^3 \times \dots \times n^{d-1}(n^{d-2} - 1)(n^{d-3} - 1)\dots(n - 1)} \\ &= \frac{n^{d-1} - 1}{n - 1}\end{aligned}$$

and

$$\begin{aligned}b &= \frac{(n^{d+1} - 1)(n^{d+1} - n)}{(n^d - 1)(n^d - n)}\lambda \\ &= \frac{(n^{d+1} - 1)n(n^{d-1} - 1)}{(n^d - 1)n(n - 1)} \\ &= \frac{(n^{d+1} - 1)}{(n - 1)} \\ &= v,\end{aligned}$$

so the design is symmetric. We shall prove that this design can in fact be constructed from a difference set in the cyclic group of order v. To prove this we need a fact about finite fields. Suppose x is a primitive element of $GF(n^{d+1})$. Since $GF(n^{d+1})$ can be constructed using polynomials of degree $(d+1)$ over $GF(n)$, there is an element of $GF(n)$ that satisfies an irreducible polynomial equation of degree $d+1$ over $GF(n)$. The fact we need is that x satisfies such an equation. (This is not surprising, but it needs to be proven—see, for example, algebra textbooks.)

Suppose f is an irreducible polynomial of degree $d+1$ over $GF(n)$ that has x as a root: say

$$f(y) = c_0 + c_1 y + \dots + c_d y^d + y^{d+1};$$

and $f(x) = 0$. Since f is irreducible, $c_0 \neq 0$. Then

$$x^{d+1} = -c_0 - c_1 x - \dots - c_d x^d. \tag{5.6}$$

It follows that for any exponent i, we can find elements $a_{i0}, a_{i1}, \dots, a_{id}$ of $GF(n)$ such that

$$x^i = a_{i0} + a_{i1}x + \dots + a_{id}x^d. \tag{5.7}$$

(Take $a_{ii} = 1$ and $a_{ij} = 0$ when $j \neq i$, if $i \leq d$; for larger i, use repeated substitutions of (5.6).) This establishes a correspondence between powers of

x and vectors over $GF(n)$: the $n^{d+1}-1$ different powers correspond to the $n^{d+1}-1$ different nonzero vectors. Since different powers cannot give the same vector, we have a one-to-one correspondence; and the vectors can of course be interpreted as points of $PG(d,n)$.

If we write $v = (n^{d+1}-1)/(n-1)$, then $0, 1, x^v x^{2v}, \ldots, x^{(n-1)v}$ are the n solutions of the equation $y^n = y$, so they are the elements of the subfield $GF(n)$ in $GF(n^{d+1})$. For any s, (5.7) yields

$$x^{i+sv} = (x^{sv}a_{i0}) + (x^{sv}a_{i1})x + \ldots + (x^{sv}a_{id})x^d.$$

So we see that if $a_i = (a_{i0}, a_{i1}, \ldots, a_{id})$ is the vector corresponding to x^i, then the vector corresponding to x^{i+sv} is $x^{i+sv}a_i$, which is a nonzero $GF(n)$-multiple of a_i, and the two points of $GF(d,n)$ are the same. That is, x^i and x^j correspond to the same point if and only if i and j are congruent modulo v.

The map "multiplication by x," which has the effect

$$0 \mapsto 0,\ x^i \mapsto x^{i+1},$$

can be interpreted as a mapping of the points of $PG(d,n)$, taking a_i to a_{i+1}. Now

$$\begin{aligned} x^{i+1} &= x(a_{i0} + a_{i1}x + \ldots + a_{id}x^d) \\ &= a_{i0}x + a_{i1}x^2 + \ldots + a_{id}x^{d+1} \\ &= a_{i0}x + a_{i1}x^2 + \ldots + a_{id}(-c_0 - c_1x - \ldots - c_dx^d) \\ &= -c_0a_{id} + (a_{i0} - c_1a_{id})x + \ldots + (a_{i,d-1} - c_da_{id})x^d \end{aligned}$$

using (5.6), so

$$a_{i+1} = (-c_0a_{id}, a_{i0} - c_1a_{id}, \ldots, a_{i,d-1} - c_da_{id}).$$

Therefore the map takes the point $p = (p_0, p_1, \ldots, p_d)$ to $p\psi$, where

$$p\psi = (-c_0p_0, p_01 - c_0p_d, \ldots, p_{d-1} - c_0p_d).$$

Now consider a prime h of the $PG(d,n)$. Since primes are vector spaces of dimension $d-1$, they consist of all the points whose coordinates satisfy one linear equation: There exist elements $h_0, h_1, \ldots, h_d$ of $GF(n)$ such that

$$h = \{p : \sum h_jp_j = 0\}.$$

Define $k_0, k_i, \ldots, k_d$ by

$$\begin{aligned} k_1 &= h_0, \\ k_2 &= h_1, \\ &\ldots \\ k_d &= h_{d-1}, \\ k_0 &= -c_0^{-1}(h_d + c_1h_0 + c_2h_1 + \ldots + c_dh_{d-1}). \end{aligned}$$

If p lies in h, then write q_i for i-th coordinate of $p\psi$.

$$\begin{aligned}
k_j q_j &= -k_0 q_0 p_d + \sum_{j=0}^{d-1} k_{j+1}(p_j - c_j p_d) \\
&= (h_d + c_1 h_0 + c_2 h_1 + \ldots + c_d h_{d-1}) p_d + \sum_{j=0}^{d-1} h_j (p_j - c_j p_d) \\
&= \sum_{j=0}^{d} h_j p_j \\
&= 0.
\end{aligned}$$

So $h\psi$ is the prime

$$\{d : \sum k_j q_j = 0\}.$$

Therefore, the map ψ takes primes into primes.

Suppose we replace the point a_i by the integer i modulo v. Then we have shown that the map "plus 1" maps primes into primes. It is not hard to show that all v primes can be obtained from a given prime in this way. So all v primes can be obtained by developing an initial prime modulo v. This must mean that the prime is a difference set modulo v when the identification between a_i and i is made. We have:

THEOREM 5.12 [131]

If n is a prime power, there is a cyclic difference set with parameters

$$\left(\frac{n^{d+1}-1}{n-1}, \frac{n^d-1}{n-1}, \frac{n^{d-1}-1}{n-1}\right).$$

These difference sets are called *Singer difference sets.*

In particular, when $d = 2$, a "prime" is just a line in the projective plane, so we have shown that any finite projective plane constructed over a field is cyclic when considered as a balanced incomplete block design. So there is a cyclic $(n^2 + n + 1, n + 1, 1)$-design whenever n is a prime power.

To illustrate the theorem we use the case $d = 2, n = 3$. In other words, we construct a $(13, 4, 1)$-difference set. In this case $d + 1 = 3$, and the polynomial $f(y)$ must be an irreducible polynomial of degree 3 over $GF(3)$. One example is

$$g(y) = 1 - y + y^3;$$

the reader should verify that it is irreducible. We assume that x is a root of $g(y) = 0$; in other words, x satisfies

$$x^3 = 2 + x.$$

The powers of x can now be expressed in terms of $1, x$ and x^2 and made to correspond to points in $PG(2,3)$, as follows:

$$\begin{array}{llllllcllcllllll}
x^0 & = & 1 & & & \leftrightarrow & (1,0,0) & = & 2(2,0,0) & \leftrightarrow & = & 2 & & & = & x^{13} \\
x^1 & = & & x & & \leftrightarrow & (0,1,0) & = & 2(0,2,0) & \leftrightarrow & = & & 2x & & = & x^{14} \\
x^2 & = & & & x^2 & \leftrightarrow & (0,0,1) & = & 2(0,0,2) & \leftrightarrow & = & & & 2x^2 & = & x^{15} \\
x^3 & = & 2 & +x & & \leftrightarrow & (2,1,0) & = & 2(1,2,0) & \leftrightarrow & = & 1 & +2x & & = & x^{16} \\
x^4 & = & & 2x & x^2 & \leftrightarrow & (0,2,1) & = & 2(0,1,2) & \leftrightarrow & = & & x & +2x^2 & = & x^{17} \\
x^5 & = & 2 & +x & +2x^2 & \leftrightarrow & (2,1,2) & = & 2(1,2,1) & \leftrightarrow & = & 1 & +2x & +x^2 & = & x^{18} \\
x^6 & = & 1 & +x & +x^2 & \leftrightarrow & (1,1,1) & = & 2(2,2,2) & \leftrightarrow & = & 2 & +2x & +2x^2 & = & x^{19} \\
x^7 & = & 2 & +2x & +x^2 & \leftrightarrow & (2,2,1) & = & 2(1,1,2) & \leftrightarrow & = & 1 & +x & +2x^2 & = & x^{20} \\
x^8 & = & 2 & & +2x^2 & \leftrightarrow & (2,0,2) & = & 2(1,0,1) & \leftrightarrow & = & 1 & & +x^2 & = & x^{21} \\
x^9 & = & 1 & +x & & \leftrightarrow & (1,1,0) & = & 2(2,2,0) & \leftrightarrow & = & 2 & +2x & & = & x^{22} \\
x^{10} & = & & & +x^2 & \leftrightarrow & (0,1,1) & = & 2(0,2,2) & \leftrightarrow & = & & 2x & +2x^2 & = & x^{23} \\
x^{11} & = & 2 & +x & +x^2 & \leftrightarrow & (2,1,1) & = & 2(1,2,2) & \leftrightarrow & = & 1 & +2x & +2x^2 & = & x^{24} \\
x^{12} & = & 2 & & +x^2 & \leftrightarrow & (2,0,1) & = & 2(1,0,2) & \leftrightarrow & = & 1 & & +2x^2 & = & x^{25}
\end{array}$$

For example, the first line means that $x^0 = 1$, which corresponds to $(1,0,0)$, and $x^{13} = 2$, which corresponds to $(2,0,0)$; also, $(1,0,0) = 2(2,0,0)$ over $GF(3)$ (and since one is a multiple of the other, $(1,0,0)$ and $(2,0,0)$ represent the same point of $PG(2,3)$). Now select any line of the geometry; for example, we choose the line $x_3 = 0$, which contains the points

$$(0,1,0),\ (1,0,0),\ (1,1,0),\ (1,2,0).$$

These points correspond to

$$x^1,\ x^0,\ x^9,\ x^3.$$

So we have the cyclic difference set $\{1,0,9,3\}$, or equivalently $\{0,1,3,9\}$. Any other line could have been chosen. For example, $x_1 + x_2 + x_3 = 0$ has points $(0,1,2)$, $(1,0,2)$, $(1,1,1)$, $(1,2,0)$, which correspond to x_4, x_{12}, x_6, x_3, and give difference set $\{3,4,6,12\}$; $x_1 = x_2$ has points $(0,0,1)$, $(1,1,0)$, $(1,1,1)$, $(1,1,2)$, corresponding powers x_2, x_9, x_6, x_7, and difference set $\{2,6,7,9\}$.

Exercises 5.5

5.5.1 Prove that if h is a prime of $PG(d,n)$ and ψ is the mapping defined in this section, then

$$h\psi^i \neq h\psi^j$$

unless $i \equiv j \pmod{v}$. Deduce that all primes can he constructed from h by mapping with the powers of ψ.

5.5.2* Prove that if F is any finite field and f is a reducible cubic polynomial over F, then $f(a) = 0$ for some $a \in F$. Hence prove that $1 - y + y^3$ is irreducible over $GF(3)$.

5.5.3* Find Singer difference sets to generate $PG(2,2)$ and $PG(2,5)$.

5.5.4 $GF(2^2)$ consists of the four elements $0, 1, \omega, \omega + 1$, where

$$1 + \omega + \omega^2 = 0.$$

(Note: This is equivalent to the formulation of $GF(2^2)$ in Chapter 3, with $A = \omega, B = \omega^2$).

(i) Verify that $f(y) = \omega + \omega y + \omega y^2 + y^3$ is an irreducible polynomial over $GF(2^2)$.

(ii)* Assume that x is a primitive element of $GF(2^2)$ that satisfies $f(x) = 0$. Construct a Singer difference set that generates $PG(2,4)$ (use the line $x_3 = 0$).

Chapter 6

More about Block Designs

6.1 Residual and Derived Designs

The relationship between affine and projective planes can be generalized to other block designs. If B_0 is any block of a pairwise balanced design with index λ, then any two treatments that do not belong to B_0 must occur together in λ of the remaining blocks, while any two members of B_0 must be together in $\lambda - 1$ of the remaining blocks. It follows that the blocks $B \backslash B_0$ form a pairwise balanced design of index λ when B ranges through the remaining blocks, and the blocks $B \cap B_0$ form a pairwise balanced design of index $\lambda - 1$. We shall refer to these as the *residual* and *derived* designs of the original design with respect to the block B_0. In general the replication number of a residual or derived design is not constant, and the block sizes follow no particular pattern. To solve the first of these problems it suffices to make the original design equireplicate: If a design has constant frequency r, its residual and derived designs also have constant frequencies, r and $r - 1$ respectively. The block pattern can be predicted if the size of $B \cap B_0$ is constant; the derived blocks will be of that constant size, and the blocks of the residual design will be $|B \cap B_0|$ smaller than the blocks of the original. In particular, if we start with a symmetric balanced incomplete block design, we obtain constant frequencies and block sizes.

THEOREM 6.1

The residual design of a (v, k, λ)-SBIBD is a balanced incomplete block design with parameters

$$(v - k, v - 1, k, k - \lambda, \lambda), \tag{6.1}$$

provided $\lambda \neq k - 1$. The derived design of a (v, k, λ)-SBIBD is a balanced incomplete block design with parameters

$$(k, v - 1, k - 1, \lambda, \lambda - 1), \tag{6.2}$$

provided $\lambda \neq 1$.

$\mathcal{B}$	$\mathcal{R}$	$\mathcal{D}$
1 2 3 4 5 6		
2 5 6 7 10 11	7 10 11	2 5 6
1 4 6 7 8 10	7 8 10	1 4 6
2 4 5 7 8 9	7 8 9	2 4 5
3 5 6 8 9 10	8 9 10	3 5 6
3 4 6 7 9 11	7 9 11	3 4 6
1 3 5 7 8 11	7 8 11	1 3 5
1 2 6 8 9 11	8 9 11	1 2 6
1 2 3 7 9 10	7 9 10	1 2 3
2 3 4 8 10 11	8 10 11	2 3 4
1 4 5 9 10 11	9 10 11	1 4 5

FIGURE 6.1: Residual and derived designs.

The two exceptions are made so as to exclude "trivial" designs.

Affine planes are the residual designs of projective planes; the derived designs of projective planes are trivial. As an example where both designs are nontrivial, consider the $(11, 6, 3)$-design $\mathcal{B}$ in Figure 6.1, whose residual design $\mathcal{R}$ and derived design $\mathcal{D}$ with respect to the first block are also shown. They have parameters $(5, 10, 6, 3, 3)$ and $(6, 10, 5, 3, 2)$, respectively. Suppose a design $\mathcal{D}$ has the parameters $(v - k, v - 1, k, k - \lambda, \lambda)$ for some v, k and λ. One can ask whether $\mathcal{D}$ is embeddable—that is, does there exist a (v, k, λ)-SBIBD of which $\mathcal{D}$ is a residual? In the case $\lambda = 1$, we have the parameters of an affine plane, and by Theorem 3.13 we know that a suitable symmetric design (a projective plane) necessarily exists. The corresponding result also holds for $\lambda = 2$, and slightly less can be said for greater values of λ; the proofs are too long to include here.

***THEOREM 6.2* [63, 132, 133]**

Suppose $\mathcal{D}$ is a balanced incomplete block design with parameters (6.1).

(i) If $\lambda = 1$ or $\lambda = 2$, there is a (v, k, λ)-design of which $\mathcal{D}$ is the residual.

(ii) There is a number $f(\lambda)$, depending only on λ, such that if $k \geq f(\lambda)$, there is a (v, k, λ)-design of which $\mathcal{D}$ is the residual.

A design with parameters (6.1) is called *quasi-residual*; thus Theorem 6.2(ii) says that "all sufficiently large quasi-residual designs are residual." To give an indication of the meaning of "sufficiently large," $f(3)$ is at most 90. A quasi-residual design is called *nonembeddable* if it is not the residual of some symmetric design, so our remark above might be rephrased as follows: There exist nonembeddable quasi-residual designs, but all quasi-residual designs with sufficiently large v (as a function of λ) are embeddable."

In order to prove that the theorem cannot be improved significantly, we exhibit a nonembeddable quasi-residual design with $\lambda = 3$. We start with a finite affine geometry $AG(2,3)$, whose blocks are ordered so that the first three form one parallel class, the next three form another, and so on. Say A is the 9×12 incidence matrix of this design. Define

$$C = \begin{bmatrix} 1&1&1&1&1&1&0&0&0&0&0&0 \\ 1&1&1&0&0&0&1&1&1&0&0&0 \\ 1&1&1&0&0&0&0&0&0&1&1&1 \\ 0&0&0&1&1&1&1&1&1&0&0&0 \\ 0&0&0&1&1&1&0&0&0&1&1&1 \\ 0&0&0&0&0&0&1&1&1&1&1&1 \end{bmatrix}, \quad B = \begin{bmatrix} 1&1&1&1&1&1&1&1&1 \\ 1&1&1&0&0&0&0&0&0 \\ 0&0&0&1&1&1&0&0&0 \\ 0&0&0&0&0&0&1&1&1 \end{bmatrix}.$$

Then write

$$M = \begin{bmatrix} C^T & A^T & A^T \\ O & B & J-B \end{bmatrix}.$$

THEOREM 6.3 **[156]**

M is the incidence matrix of a $(16,24,9,6,3)$-design that is quasi-residual but not residual.

Proof. It is clear that M has constant row sum 9 and constant column sum 6. To calculate MM^T, we first observe that

$$A^TA = \begin{bmatrix} 3I & J & J & J \\ J & 3I & J & J \\ J & J & 3I & J \\ J & J & J & 3I \end{bmatrix} \quad C^TC = \begin{bmatrix} 3J & J & J & J \\ J & 3J & J & J \\ J & J & 3J & J \\ J & J & J & 3J \end{bmatrix}$$

$$BB^T = \begin{bmatrix} 9&3&3&3 \\ 3&9&3&3 \\ 3&3&9&3 \\ 3&3&3&9 \end{bmatrix} \quad (J-B)(J-B)^T = \begin{bmatrix} 0&0&0&0 \\ 0&6&3&3 \\ 0&3&6&3 \\ 0&3&3&6 \end{bmatrix}.$$

Now

$$MM^T = \begin{bmatrix} C^TC + 2A^TA & A^TJ \\ JA & BB^T + (J-B)(J-B^T) \end{bmatrix}$$

and since $JA = 3J$ we have $MM^T = 6I + 3J$. So M is the incidence matrix of a design with the desired parameters.

Now consider the seventh and sixteenth blocks of the design. They each contain the four treatments corresponding to the four lines of the $AG(2,3)$ through the first point of the geometry. So these blocks have intersection size 4. If it were possible to add further treatments to this design in order to embed it in a $(25, 9, 3)$-design, we would obtain a $(25, 9, 3)$-design with pair of blocks of intersection size at least 4. (In fact, there will be nine such pairs of blocks.) This is impossible, since (by Corollary 2.10.2) two blocks of a $(25, 9, 3)$-design intersect in three treatments only. □

The first example of a quasi-residual design that is not residual was given by Bhattacharya [17] in 1947. It also had parameters $(16, 24, 9, 6, 3)$, but the construction was not as simple as the one given above.

Exercises 6.1

6.1.1* For the following parameter sets, what are the residual and derived parameters?

(i) $(16, 6, 2)$

(ii) $(22, 7, 2)$

(iii) $(45, 12, 3)$

(iv) $(13, 9, 6)$.

6.1.2 Suppose $\mathcal{D}$ is a symmetric balanced incomplete block design with parameters (v, k, λ), where $2k < v$.

(i) Prove that $2\lambda < k$.

(ii) Hence show that the residual design of $\mathcal{D}$ has no repeated blocks.

Hint: Use $\lambda(v-1) = k(k-1)$.

6.1.3* Consider the parameters of a finite projective plane: $(n^2+n+1, n+1, 1)$.

(i) What are the parameters of the complement of this design (as defined in Section 2.2)?

(ii) What are the parameters of the residual of this complementary design?

(iii) Prove that a design with these residual parameters always exists.

6.1.4 Suppose $\mathcal{D}$ is a symmetric balanced incomplete block design with support set S, and suppose $\mathcal{E}$ is its complement. Select any block B of $\mathcal{D}$. Prove that the residual design of $\mathcal{E}$ with respect to $S \backslash B$ is the complement of the derived design of $\mathcal{D}$ with respect to B.

6.1.5 Let $\mathcal{D}$ consist of all the subsets of size n of an $(n+2)$-set. Prove that $\mathcal{D}$ is a quasi-residual balanced incomplete block design.

6.2 Resolvability

We now generalize the way that the lines of an affine plane fall into *parallel classes*. We define a parallel class in a balanced incomplete block design to be a set of blocks that between them contain every treatment exactly once. A design is called *resolvable* if one can partition its blocks into parallel classes; such a partition is called a *resolution*.

Clearly, the number of blocks in a parallel class must equal vk^{-1}, and therefore k must divide v in any resolvable design. We write s for the integer vk^{-1}. Another way of putting this is to say that any resolvable design has an additional parameter s, the size of a parallel class, and an additional parameter relation

$$v = sk. \tag{6.3}$$

Not every design for which k divides v is resolvable. In fact, if we take $k = 3$, $\lambda = 2$ and $s = 2$, we obtain the parameters $(6, 10, 5, 3, 2)$. Suppose there is a resolvable design with these parameters. Without loss of generality we assume that its support is $\{1, 2, 3, 4, 5, 6\}$.

There must be precisely two blocks containing both 1 and 2; after permuting the names of symbols if necessary, these blocks can be taken either as 123, 124 or as 123, 123. If they are 123 and 124, then either 134 is a block, and the set of blocks containing 1 must be

$$123, 124, 134, 156, 156,$$

or else 134 is not a block, and the blocks containing 1 must be

$$123, 124, 135, 146, 156$$

(or an isomorphic set with 5 and 6 interchanged). If the blocks containing 1 and 2 are 123, 123, we get a set of blocks isomorphic to the first case again.

Since $s = 2$, there are only two blocks in each resolution class, so the block in the same resolution as 123 must be its complement, 456. In this way all five remaining blocks are determined. In the first case, they are

$$456, 356, 256, 234, 234,$$

and the resulting design is not balanced; in fact, the pair $\{5, 6\}$ occurs in five blocks, instead of two. In the second case, the blocks are

$$456, 356, 246, 235, 234,$$

which is again not balanced, as $\{3, 5\}$ and $\{5, 6\}$ occur three times each, while $\{3, 4\}$ occurs only once. So no resolvable balanced incomplete design exists with the parameters $(6, 10, 5, 3, 2)$.

On the other hand, a nonresolvable design with parameters $(6, 10, 5, 3, 2)$ is easy to construct. The derived design in Figure 6.1 is an example.

In 1850, the Reverend T. P. Kirkman proposed the following problem [90]: A schoolmistress has 15 girl pupils and she wishes to take them on a daily walk. The girls are to walk in five rows of three girls each. It is required that no two girls should walk in the same row more than once per week. Can this be done? The problem is generally known as "Kirkman's schoolgirl problem."

First, there are seven days in a week, and every girl will walk in the company of two others each day. As no repetitions are allowed, a girl has 14 companions over the week; since there are only 15 in the class, it follows that every pair of girls must walk together in a row at least once; in view of the requirements, every pair of girls walk together *precisely* once. Every girl walks on each day of the week. So, if we treat the girls as "treatments" and the rows in which they walk as "blocks," we are required to select 35 blocks of size 3 from a set of 15 treatments, in such a way that every treatment occurs in 7 blocks and every pair of treatments occur together in precisely one block—that is, we require a $(15, 35, 7, 3, 1)$-design. But there is a further constraint: Since all the girls must walk every day, it is necessary that the blocks be partitioned into seven groups of triples, so that every girl appears once in each group. The groups correspond to the seven days. So we require a resolvable solution. Kirkman's problem has a solution; in fact Kirkman gave one in [91], and Woolhouse [185, 186] later provided a complete list of solutions (there are seven). An example is the schedule

Monday:	$1, 2, 3$	$4, 5, 6$	$7, 8, 9$	$10, 11, 12$	$13, 14, 15$
Tuesday:	$1, 4, 11$	$2, 5, 10$	$3, 8, 13$	$6, 7, 14$	$9, 12, 15$
Wednesday:	$1, 5, 9$	$2, 6, 8$	$3, 11, 15$	$4, 12, 14$	$7, 10, 13$
Thursday:	$1, 6, 13$	$2, 4, 15$	$3, 9, 10$	$5, 7, 12$	$8, 11, 14$
Friday:	$1, 7, 15$	$2, 9, 14$	$3, 6, 12$	$4, 8, 10$	$5, 11, 13$
Saturday:	$1, 8, 12$	$2, 7, 11$	$3, 5, 14$	$4, 9, 13$	$6, 10, 15$
Sunday:	$1, 10, 14$	$2, 12, 13$	$3, 4, 7$	$5, 8, 15$	$6, 9, 11$

More generally, Kirkman also asked about resolvable designs with $k = 3, \lambda = 1$, and any value of v. We shall discuss this problem in Section 13.1.

The "schoolgirl problem" is a famous resolvable design. If v is even, say $v = 2n$, the $(2n, n(2n-1), 2n-1, 2, 1)$-design of all possible unordered pairs is also resolvable. The resolutions, called one-factorizations, were defined in Chapter 1. Affine planes constitute a third set of resolvable designs.

We define an *affine resolvable design* (or simply *affine design*) to be a resolvable design with the property that any two blocks belonging to different parallel classes have m common treatments for some constant m. So the affine planes are affine designs with $m = 1$.

THEOREM 6.4

An affine design has parameters

$$(s^2m, s^2m + \lambda s, sm + \lambda, sm, \lambda) \tag{6.4}$$

for some integers s and m, where

$$\lambda = \frac{sm - 1}{s - 1}. \tag{6.5}$$

Proof. Suppose $B_1, B_2, \ldots, B_s$ are the blocks in one parallel class in an affine design, and B is a block in another parallel class. Then $B \cap B_i$ has m elements for every i. The sets $B \cap B_i$ are disjoint (since the B_i are disjoint) and they partition B (every treatment belongs to some B_i, so obviously any element of B belongs to some B_i). So

$$k = |B| = \sum |B \cap B_i| = sm. \tag{6.6}$$

To establish the value of λ, we count the elements of a set in two ways. Given a fixed block A, we count all the ordered triples (x, y, B) where B is a block other than A and $A \cap B$ contains both x and y. There are $k(k-1)$ ordered pairs (x, y), and together they belong to λ blocks, one of which is A, so there are $\lambda - 1$ choices for B, and the count is

$$k(k-1)(\lambda - 1). \tag{6.7}$$

On the other hand, A has nonempty intersection with s blocks in each of the other $r - 1$ parallel classes; since there are m elements in each intersection there are $m(m-1)$ choices for the ordered pair (x, y), so the count is

$$s(r-1)m(m-1). \tag{6.8}$$

Equating (6.7) and (6.8) and applying (6.6), we have

$$(\lambda - 1)(sm - 1) = (r - 1)(m - 1). \tag{6.9}$$

The standard relation (1.2) yields

$$r = \lambda \frac{s^2m - 1}{sm - 1} = \lambda \frac{v - 1}{k - 1}$$

and substituting this into (6.9) we have

$$\begin{aligned}
(\lambda - 1)(sm - 1)^2 &= [\lambda(s^2m - 1) - (sm - 1)](m - 1), \\
\lambda[(sm - 1)^2 - (m - 1)(s^2m - 1)] &= (sm - 1)^2 - (m - 1)(sm - 1), \\
\lambda(s^2m^2 - 2sm - s^2m^2 + s^2m + m) &= (sm - 1)(sm - m), \\
\lambda m(s - 1)^2 &= (sm - 1)(s - 1)m, \\
\lambda &= \frac{sm - 1}{s - 1}.
\end{aligned} \tag{6.5}$$

Using (1.2) and (1.1) we get the parameters (6.4). □

COROLLARY 6.4.1

The parameters of an affine design satisfy

$$b = rs = v + r - 1.$$

The corollary is easy to verify from (6.4) and (6.5); however, observe that $b = rs$ follows directly from (6.3) and (1.1), so that part is true for any resolvable design.

As an application of affine designs we give a construction for symmetric balanced incomplete block designs that uses the existence of an affine design. Assume that we are given an affine design with parameters (6.4), and assume the blocks to have been ordered so that the first s blocks constitute one parallel class, the next s constitute another, and so on. Thus the design has incidence matrix

$$A = [A_1\, A_2\, \ldots\, A_r],$$

where each A_i is the $s^2m \times s$ incidence matrix of the set of blocks in one parallel class and $r = sm + \lambda$. It is easy to see that

$$A_iJ = J \text{ for } i \in \{1, 2, \ldots, r\} \quad (6.10)$$

$$A_i^TA_i = smI \text{ for } i \in \{1, 2, \ldots, r\} \quad (6.11)$$

$$A_i^TA_j = mJ \text{ for } i, j \in \{1, 2, \ldots, r\}, i \neq j \quad (6.12)$$

$$\sum_{i=1}^{r} A_iA_i^T = (r - \lambda)I + \lambda J \text{ for } i \in \{1, 2, \ldots, r\}. \quad (6.13)$$

(Verifications are left as an exercise.)

THEOREM 6.5 [162]

If there is an affine design with parameters (6.4), *then there is a symmetric balanced incomplete block design with parameters*

$$\left(\frac{s^2m(s^2m + s - 2)}{s - 1}, \frac{sm(s^2m - 1)}{s - 1}, \frac{sm(sm - 1)}{s - 1}\right).$$

Proof. For convenience we write r and λ for the relevant parameters of the affine design:

$$r = \frac{s^2m - 1}{s - 1}, \lambda = \frac{sm - 1}{s - 1}.$$

Defining $A_1, A_2, \ldots, A_r$ as above, we consider the square $(0,1)$-matrix

$$L = \begin{bmatrix} 0 & M_1 & M_2 & & M_r \\ M_r & 0 & M_1 & & M_{r-1} \\ M_{r-1} & M_r & 0 & \ldots & M_{r-2} \\ & & \vdots & & \\ M_1 & M_2 & M_3 & & 0 \end{bmatrix}$$

where $M_i = A_i A_i^T$. This matrix has size $s^2m(s^2m+s-2)/(s-1)$. We shall prove that

$$\begin{aligned} LL^T &= \left(\frac{sm(s^2m-1)}{s-1} - \frac{sm(sm-1)}{s-1}\right) I + \frac{sm(sm-1)}{s-1} J, \\ JL &= \frac{sm(s^2m-1)}{s-1} J. \end{aligned}$$

By Theorem 2.7 this will prove that L is the incidence matrix of the required design. Equation (6.13) tells us that

$$\sum M_i = (r-\lambda)I + \lambda J.$$

So the column sum of L equals $r-\lambda$ plus λ times the number of rows of M_i:

$$\begin{aligned} r - \lambda + \lambda s^2 m &= sm + \frac{(sm-1)s^2m}{s-1} \\ &= \frac{s^2m - sm + s^3m^2 - s^2m}{s-1} \\ &= \frac{sm(s^2m-1)}{s-1}, \end{aligned}$$

so JL has the required value.

If we assume that LL^T is partitioned in the same way as L, then the diagonal blocks of LL^T each equal $\sum M_i M_i^T$; since $M_i^T = M_i$,

$$\begin{aligned} \sum M_i M_i^T &= \sum A_i A_i^T A_i A_i^T \\ &= sm \sum A_i A_i^T \\ &= sm[(r-\lambda)I + \lambda J] \\ &= s^2m^2 I + \frac{sm(sm-1)}{s-1} J, \end{aligned}$$

using (6.11). Off the diagonal, the (i,j) block is the sum of $r-1$ terms of the form $M_pM_q^T$, where p never equals q. For any p and q,

$$\begin{aligned} M_pM_q^T &= A_pA_p^TA_qA_q^T \\ &= A_p(mJ)A_q^T \\ &= (mJ)A_q^T \\ &= mJ, \end{aligned}$$

using (6.10) and (6.12); so every off-diagonal block equals $(r-1)mJ$. It remains to verify that

$$(r-1)m = \frac{sm(sm-1)}{s-1},$$

but the left-hand side equals

$$\begin{aligned} \left(sm + \frac{sm-1}{s-1} - 1\right) &= \frac{s^2m - sm + sm - 1 - s + 1)m}{s-1} \\ &= \frac{sm(sm-1)}{s-1}. \end{aligned}$$

□

In the simplest case, if we use an $AG(2,2)$ as the affine design, we obtain a $(16,6,2)$-design.

One can construct resolvable designs by difference methods. One observation that has been useful in direct searches is the following: If a set of initial blocks contains every element of the treatment set once, the design developed from them is resolvable. If the initial blocks are supplementary difference sets, the result is a resolvable balanced incomplete block design. For example, the designs of Theorem 5.11 are resolvable. Cases with more than two initial blocks can also be constructed; as an example, the blocks

$$\begin{array}{ccc} \{010\ 020\ 101\ 201\} & \{011\ 021\ 102\ 202\} & \{012\ 022\ 100\ 200\} \\ \{210\ 120\ 221\ 111\} & \{211\ 121\ 222\ 122\} & \{212\ 122\ 220\ 120\} \\ & \{\infty\ \ 000\ 001\ 002\} & \end{array}$$

form a resolvable $(28,63,9,4,1)$-design when developed mod $(3,3,-)$.

Although we have discussed resolvability and affine resolvability only in the case of balanced incomplete block designs, or 2-designs, there is no reason why one should not consider general t-designs in this way; the properties only involve block intersections, and not the degree of balance of the design.

Exercises 6.2

6.2.1 Prove that developing the following blocks modulo 17 produces a resolvable $(18,102,17,3,2)$-design:

$$\begin{array}{l} \{\infty\ 0\ 11\},\{1\ 9\ 14\},\{\ 2\ \ 5\ \ 7\}, \\ \{\ 3\ 4\ 10\},\{6\ 8\ 15\},\{12\ 13\ 17\}. \end{array}$$

6.2.2* Find a partition of $Z_{11} \cup \{\infty\}$ into three 4-sets whose development forms a $12, 33, 11, 4, 3)$-design.

6.2.3 Suppose A is the incidence matrix of a resolvable (v, b, r, k, λ) design. Prove that A has rank at least $b - r + 1$. Hence prove that $b \geq v + r - 1$.

6.2.4* A *PB2-design* with parameters $(v; b_1, b_2; r_1, r_2; k_1, k_2; \lambda)$ is defined to be a design on v objects, whose "blocks" are b_1 subsets of size k_1 and b_2 subsets of size k_2, such that every object belongs to r_1 of the blocks of size k_1 and to r_2 of the blocks of size k_2.

(i) Prove that the design formed by deleting all members of one block from a balanced incomplete block design with parameters $(v, b, r, k, 1)$ is a $PB2$-design $(v - k; b_1, b_2; r_1, r_2; k, k - 1; 1)$ for some b_1, b_2, r_1, and r_2; and derive expressions for these four parameters in terms of v, b, r, and k. Does a similar result hold if the balanced incomplete block design has $\lambda = 2$ instead of $\lambda = 1$?

(ii) Prove that the existence of a resolvable balanced incomplete block design with parameters (v, b, r, k, λ) and a balanced incomplete block design with parameters $(k, h, s, t, 1)$ implies the existence of a $PB2$-design $(v; b - vk^{-1}, hvk^{-1}; r - 1, s; k, t; \lambda)$. More generally, show that they imply the existence of $PB2$-designs with parameters

$$(v; b - mvk^{-1}, mhvk_{-1}; r - m, ms; k, t; \lambda)$$

for $m = 1, 2, \ldots, r - 1$.

6.2.5* Suppose one were to insert all the 3-sets on $3n$ objects in a square array, so that every object occurs exactly once per row and exactly once per column. How big must the array be? Does such an array exist in the case $n = 2$? For $n = 3$?

6.2.6 Prove equations (6.10), (6.11), (6.12) and (6.13).

6.2.7 Prove the existence of a symmetric balanced incomplete block design with parameters

$$(\lambda^2(\lambda + 2), \lambda(\lambda + 1), \lambda)$$

whenever λ is a prime power.

Chapter 7

The Main Existence Theorem

7.1 Sums of Squares

We shall now study the decomposition of positive integers into sums of integer squares. Theorems 7.8 and 7.12 will be needed later; the proofs may be omitted at a first reading.

We first ask: Which integers n have the form

$$n = x^2 + y^2$$

where x and y are integers? If

$$\begin{aligned} q &= x^2 + y^2, \\ r &= z^2 + t^2, \end{aligned}$$

then

$$qr = (x^2 + y^2)(z^2 + t^2) = (xz + yt)^2 + (xt - yz)^2. \tag{7.1}$$

So:

LEMMA 7.1

If each prime factor of the positive integer n can be written as the sum of two integer squares, then n can be written in that way.

So we consider the primes.

LEMMA 7.2

Let p be a prime congruent to $1 \pmod 4$*. Then there exist integers x, y and m, with $0 < m < p$, such that*

$$x^2 + y^2 = mp.$$

Proof. By Theorem 5.6, -1 is a quadratic element modulo p. Choose y such that $y^2 \equiv -1 \pmod{p}$ and $0 < y < p/2$ (if $y > p/2$, replace y by $p - y$). Then

$$1 + y^2 \equiv 0 (\bmod\, p),$$

so $1 + y^2 = mp$ for some positive m. Then

$$mp = 1 + y^2 < 1 + \tfrac{1}{4}p^2 < p^2,$$

so $m < p$. Therefore the numbers $x = 1, y, m$ satisfy the conditions. □

LEMMA 7.3

Let p be a prime congruent to $1 \pmod 4$. *If there exist integers* x, y, m, *with* $1 < m < p$, *such that*

$$x^2 + y^2 = mp,$$

then there exist integers X, Y, M, *with* $1 \le M < m$, *such that*

$$X^2 + Y^2 = Mp.$$

Proof. (i) If m is even, then $x \equiv y \pmod 2$. Since $x^2 + y^2 = mp$, we have

$$\left(\frac{x+y}{2}\right)^2 + \left(\frac{x-y}{2}\right)^2 = \frac{m}{2}p$$

and we choose

$$X = \frac{x+y}{2}, \quad Y = \frac{x-y}{2}, \quad M = \frac{m}{2}.$$

(ii) If m is odd, there are unique integers a and c such that

$$x = am + c, \ |c| < m/2.$$

Similarly, there are unique b and d satisfying

$$y = bm + d, \ |d| < m/2.$$

Since $x^2 + y^2 = mp$,

$$mp = (am + c)^2 + (bm + d)^2 = c^2 + d^2 + 2(ac + bd)m + (a^2 + b^2)m^2 \quad (7.2)$$

which implies that m divides $(c^2 + d^2)$. So $c^2 + d^2 = Mm$, for some $M \ge 0$. Hence, by (7.2),

$$mp = Mm + 2(ac + bd)m + (a^2 + b^2)m,$$

so

$$p = M + 2(ac + bd) + (a^2 + b^2)m$$

and

$$\begin{aligned} Mp &= M^2 + 2(ac+bd)M + (a^2+b^2)Mm \\ &= M^2 + 2(ac+bd)M + (a^2+b^2)(c^2+d^2) \\ &= M^2 + 2(ac+bd)M + (ac+bd)^2 + (ad-bc)^2 \\ &= (M+ac+bd)^2 + (ad-bc)^2. \end{aligned}$$

So Mp has the required form. If $M = 0$, then $c = d = 0$, which means that m divides x, m divides y, and hence m^2 divides $x^2 + y^2$, or in other words m divides mp. But now m divides p; since p is prime and $1 < m < p$, we have a contradiction. So $M \geq 1$. Also,

$$Mm = c^2 + d^2 < \frac{m^2}{4} + \frac{m^2}{4} < m,$$

so $M < m$.

Hence $X = M + ac + bd$, $Y = ad - bc$, and M satisfy the conditions. □

THEOREM 7.4

Let p be a prime congruent to $1 \pmod 4$. Then p can be represented as the sum of two integer squares.

Proof. By Lemma 7.2, there exist integers x, y and m, with $1 \leq m < p$, such that $x^2 + y^2 = mp$. Let $m = k$ be the smallest possible value of m such that this is true. If $k > 1$, then by Lemma 7.3 there is a smaller possible value of m, which is a contradiction. So $k = 1$ and $p = x^2 + y^2$. □

The integers x and y are essentially unique (see Exercise 7.1.4).

Primes congruent to $3 \pmod 4$ do not have any such decomposition:

LEMMA 7.5

No number congruent to $3 \pmod 4$ is a sum of two integer squares.

Proof. If x is even, $x^2 \equiv 0 \pmod 4$. If x is odd, $x^2 \equiv 1 \pmod 4$. So the sum of two squares can only possibly be congruent to 0 $(= 0 + 0)$, 1 $(= 1 + 0)$ or 2 $(= 1 + 1)$ modulo 4. □

We define a *proper representation* of an integer n as a sum of two squares to be a representation

$$n = x^2 + y^2$$

where x and y are *relatively prime.*

LEMMA 7.6

Let p be a prime congruent to $3 \pmod 4$ and let n be a positive multiple of p. Then n has no proper representation as the sum of two squares.

Proof. Suppose the contrary: $n = x^2 + y^2$, where x and y are integers such that $(x, y) = 1$. Now p divides n; if p divides x, then p divides y also, so p divides (x, y), a contradiction. Hence p does not divide x. Therefore x has an inverse modulo p; suppose u is such that $ux \equiv 1 \pmod p$. Then $uxy \equiv y \pmod p$, so

$$n = x^2 + y^2 \equiv x^2 + u^2x^2y^2 \equiv x^2(1 + u^2y^2) \pmod p.$$

But p divides n, so $x^2(1 + u^2y^2) \equiv 0 \pmod p$. Since p does not divide x, $1 + u^2y^2 \equiv 0 \pmod p$. Therefore $-1 \equiv (uy)^2$, that is, -1 is quadratic modulo p. But this is impossible, since $p \equiv 3 \pmod 4$. □

LEMMA 7.7

If $n = p^c m$, where p is a prime congruent to $3 \pmod 4$, c is odd and p does not divide m, then n has no representation as the sum of two squares.

Proof. Suppose otherwise: $n = x^2 + y^2$. Let $(x, y) = d$, so that $x = dX$ and $y = dY$ for some X and Y satisfying $(X, Y) = 1$. Then $n = Nd^2$, where $N = X^2 + Y^2$. Now p^c divides n and c is odd, so p must divide N. But N cannot have the proper representation $X^2 + Y^2$ by Lemma 7.6, so we have a contradiction. □

THEOREM 7.8

A positive integer n can be represented as the sum of two integer squares if and only if each of its prime factors congruent to $3 \pmod 4$ appears to an even power in its prime power decomposition.

Proof. By Lemma 7.7 the condition is necessary. Now suppose the prime power decomposition of n is

$$n = 2^a p_1^{b_1} p_2^{b_2} \dots p_r^{b_r} q_1^{2c_1} q_2^{2c_2} \dots q_s^{2c_s}$$

where $p_1, p_2, \dots, p_r$ are the primes congruent to 1 modulo 4 that divide n and $q_1, q_2, \dots, q_s$ are those congruent to 3 modulo 4 that divide n. By Theorem

7.4, each p_i is a sum of two squares. Clearly, each q_i^2 can be considered as a sum $0 + q_i^2$. Also, $2 = 1^2 + 1^2$. So the result follows from Lemma 7.1, except for the trivial case $n = 1$; but $1 = 0^2 + 1^2$. □

The following corollary is now easy to prove, but it is surprisingly useful.

COROLLARY 7.8.1

If the positive integer n can be expressed as the sum of two rational squares, it equals the sum of two integer squares.

Proof. Suppose

$$n = \left(\frac{x}{a}\right)^2 + \left(\frac{y}{b}\right)^2$$

where x, y, a and b are integers and a and b are nonzero. Then na^2b^2 is a sum of two integer squares:

$$na^2b^2 = (bx)^2 + (ay)^2.$$

So no prime congruent to 3 modulo 4 occurs to an odd power in the prime power decomposition of na^2b^2. Since a^2b^2 is a perfect square, n must also have this property. The result now follows from Theorem 7.8. □

It is easy to show, by an argument similar to Lemma 7.5, that no integer congruent to 7 (mod 8) is a sum of three squares. But we shall show that every positive integer is a sum of four squares.

If $q = a^2+b^2+c^2+d^2$ and $r = e^2+f^2+g^2+h^2$, then $qr = A^2+B^2+C^2+D^2$, where

$$\begin{array}{rclrcl} A &=& ae + bf + cg + dh, & B &=& af - be + ch - dg, \\ C &=& ag - bh - ce + df, & D &=& ah + bg - cf - de. \end{array} \tag{7.3}$$

So:

LEMMA 7.9

If every prime can be represented as the sum of four squares, then every positive integer can be expressed as the sum of four squares.

LEMMA 7.10

If p is an odd prime, then there exist integers x, y, z, m with $0 < m < p$ such that

$$x^2 + y^2 + z^2 = mp.$$

Proof. Consider the congruence

$$x^2 + y^2 + z^2 \equiv 0 \pmod{p}. \tag{7.4}$$

It has the trivial solution $x = y = z = 0$. If $p \equiv 1 \pmod 4$ then, by Theorem 7.4, we can find integers X and Y such that $p = X^2 + Y^2$, so (7.4) also has the nontrivial solution $x = X, y = Y, z = 0$.

Suppose (7.4) has only the trivial solution. This means that $p \equiv 3 \pmod 4$ and hence that -1 is not quadratic modulo p. Let a be a quadratic element modulo p. If $-(a+1)$ is also a quadratic element, we can choose x, y and z such that $x^2 \equiv 1$, $y^2 \equiv a$ and $z^2 \equiv -(a+1) \pmod{p}$, giving a nontrivial solution to (7.4). So if a is quadratic, $-(a+1)$ is not quadratic. So $a+1 = -1 \times (-(a+1))$ is quadratic. So the hypothesis that (7.4) has only the trivial solution implies that whenever a is quadratic, $a+1$ is also quadratic. But 1 is always quadratic, implying that 2 is quadratic, and in turn that 3 is quadratic, and so on. Hence all the integers are quadratic modulo p, which is impossible. So (7.4) must have a nontrivial solution.

Let such a solution be given by x, y, z. We can assume that $x < p/2$, for if $x > p/2$ we can replace it by $p - x$. Similarly, $y < p/2$ and $z < p/2$, so $0 < x^2 + y^2 + z^2 < 3p^2/4 < p^2$. Since $x^2 + y^2 + z^2 = mp$, we have $0 < m < p$, as required. □

We really only need the above lemma in the case of *four* squares, but it is just as easy to prove for three squares. Of course, if $mp = x^2 + y^2 + z^2$, then $mp = x^2 + y^2 + z^2 + 0^2$.

LEMMA 7.11

If p is an odd prime and if there exist integers w, x, y, z and m, with $1 < m < p$, such that

$$w^2 + x^2 + y^2 + z^2 = mp,$$

then there exist integers W, X, Y, Z, and M with $1 \leq M < m$ such that

$$W^2 + X^2 + Y^2 + Z^2 = Mp.$$

Proof. (i) If m is even, we use the easily checked fact that

$$\left(\frac{w+x}{2}\right)^2 + \left(\frac{w-x}{2}\right)^2 + \left(\frac{y+z}{2}\right)^2 + \left(\frac{y-z}{2}\right)^2 = \frac{w^2 + x^2 + y^2 + z^2}{2}$$

for any w, x, y, z. This implies that the given w, x, y, z satisfy

$$\left(\frac{w+x}{2}\right)^2 + \left(\frac{w-x}{2}\right)^2 + \left(\frac{y+z}{2}\right)^2 + \left(\frac{y-z}{2}\right)^2 = \frac{m}{2}p; \tag{7.5}$$

as m is even, then either w, x, y, z are all odd, or all even, or two of them are odd and two are even, and in any case we can assume that $w \equiv x \pmod 2$ and $y = z \pmod 2$, without loss of generality. Then

$$W = \frac{w+x}{2},\ X = \frac{w-x}{2},\ Y = \frac{y+z}{2},\ Z = \frac{y-z}{2},\ M = \frac{m}{2}$$

are integers that satisfy the lemma.

(ii) If m is odd, we can find a, b, c, d, e, f, g, h such that

$$w = am + e,\ x = bm + f,\ y = cm + g,\ z = dm + h,$$

where $|e| < m/2$, $|f| < m/2$, $|g| < m/2$, $|h| < m/2$. Substituting into (7.5), we find that (in the notation of (7.3))

$$e^2 + f^2 + g^2 + h^2 + 2Am + (a^2 + b^2 + c^2 + d^2)m^2 = mp, \tag{7.6}$$

which implies that m divides $e^2 + f^2 + g^2 + h^2$. Define M by $Mm = e^2 + f^2 + g^2 + h^2$; $M \geq 0$.

If $M = 0$, then $e = f = g = h = 0$, so m^2 divides $w^2 + x^2 + y^2 + z^2$, which implies that m^2 divides mp and hence that m divides p. But $1 < m < p$ and p is prime, which is a contradiction. So $M \geq 1$. Also, $e^2 + f^2 + g^2 + h^2 < 4(m^2/4) = m^2$, so $M < m$. So $1 \leq M < m$.

Substituting into (7.6) we have

$$Mm + 2Am + (a^2 + b^2 + c^2 + d^2)m^2 = mp;$$

multiplying by M/m gives

$$M^2 + 2AM + A^2 + B^2 + C^2 + D^2 = Mp$$

in the notation of (5.2). In other words

$$(M + A)^2 + B^2 + C^2 + D^2 = Mp$$

so the integers $W = M + A$, $X = B$, $Y = C$, $Z = D$ and M satisfy the conditions. □

Finally, we prove

THEOREM 7.12

Every positive integer can be written as the sum of four squares.

Proof. One can apply the same proof as that of Theorem 7.4 to the results of Lemmas 7.10 and 7.11 to show that every odd prime is a sum of four squares.

So the theorem follows from Lemma 7.9, provided that we note:

$$\begin{aligned} 2 &= 1^2 + 1^2 + 0^2 + 0^2 \\ 1 &= 1^2 + 0^2 + 0^2 + 0^2. \end{aligned}$$

□

Exercises 7.1

7.1.1 Verify Equation (7.1).

7.1.2* In each case, either express the integer as a sum of two integer squares, or show it is impossible.

(i) 31	(ii) 49	(iii) 261
(iv) 637	(v) 525	(vi) 62
(vii) 185	(viii) 125	(ix) 99

7.1.3 Suppose p is a prime congruent to 1 (mod 4).
(i) Prove that there exist integers x and y such that $p = x^2 + y^2$ where x and y are *relatively prime* and $x < y$.
(ii) Suppose q is another prime congruent to 1 (mod 4). Use part (i) and (7.1) to show that pq can be written as a sum of two squares in two different ways.

7.1.4* Suppose $p = x^2 + y^2$ and also $p = a^2 + b^2$ where p is an odd prime, x, y, a, b are positive integers, $x < y$ and $a < b$.
(i) Show that each of x, y, a, b is smaller than $\sqrt{p}$.
(ii) Prove $x^2b^2 \equiv y^2a^2 \pmod{p}$.
(iii) Show that either $xb - ya = 0$ or $xb + ya = p$.
(iv) Show that $xb + ya = p$ is impossible, so $xb = ya$.
(v) Prove $x = a$ and $y = b$.

7.1.5 Verify Equation (7.3).

7.1.6* In each case, find all ways to express the integer as a sum of four integer squares $a^2 + b^2 + c^2 + d^2$, with $0 \leq a \leq b \leq c \leq d$.

(i) 7	(ii) 10	(iii) 11
(iv) 18	(v) 20	(vi) 32

7.1.7 Use (7.3) and the results of Exercise 7.1.6 to express each of the following as the sum of four integer squares:

(i) 70	(ii) 77	(iii) 770
(iv) 100	(v) 308	(vi) 110

7.2 The Bruck-Ryser-Chowla Theorem

So far we have seen various families of designs, but know of no parameters (v, b, r, k, λ) that satisfy (1.1) and (1.2) but for which no design exists. In this section we prove that those two conditions are not sufficient. The result is usually called the *Bruck-Ryser-Chowla Theorem.*

THEOREM 7.13 **[26, 35, 129]**

If there exists a symmetric balanced incomplete block design with parameters (v, k, λ), *then:*

(i) if v *is even,* $k - \lambda$ *must be a perfect square;*

(ii) if v *is odd, there must exist integers* x, y *and* z, *not all zero, such that*

$$x^2 = (k - \lambda)y^2 + (-1)^{(v-1)/2}\lambda z^2. \tag{7.7}$$

(Note: For those with a knowledge of linear algebra that includes matrix congruence, a shorter proof of this theorem is given in Section 7.3.)

Proof. (i) Suppose v is even, and suppose there exists a (v, k, λ)-design with incidence matrix A. Since A is square it has a determinant, $\det(A)$ say, and $\det(A^T) = \det(A)$, so

$$\det(A) = \sqrt{\det(AA^T)}$$

But from Lemma 2.8,

$$\det(AA^T) = k^2(k - \lambda)^{v-1},$$

so

$$\det(A) = \pm k(k - \lambda)^{\frac{1}{2}(v-1)}.$$

Now A is an integer matrix, so $\det(A)$ is an integer. Since k and λ are integers, and v is even, it follows that $\sqrt{k - \lambda}$ is an integer also.

(ii) Now suppose v is odd. It is important to notice that all the calculations in this proof take place over the rational numbers.

From Theorem 7.12, $k-\lambda$ can be written as the sum of four integer squares; say

$$k-\lambda = a^2+b^2+c^2+d^2$$

where a, b, c, d are nonnegative integers.

We consider the matrix

$$B = \begin{bmatrix} a & -b & -c & -d \\ b & a & -d & c \\ c & d & a & -b \\ d & -c & b & a \end{bmatrix}. \tag{7.8}$$

Since $BB^T = (k-\lambda)I$, $\det(BB^T) = (k-\lambda)^4$ and $\det B = (k-\lambda)^2$, which is nonzero, so B is nonsingular.

Now let W be any rational column vector of length 4, and write

$$U = BW,$$

so that

$$\begin{aligned} u_1 &= aw_1 - bw_2 - cw_3 - dw_4, \\ u_2 &= bw_1 + aw_2 - dw_3 + cw_4, \\ u_3 &= cw_1 + dw_2 + aw_3 - bw_4, \\ u_4 &= dw_1 - cw_2 + bw_3 + aw_4. \end{aligned}$$

Then

$$U^TU = (BW)^T \cdot BW = W^TB^TBW;$$

in other words,

$$\begin{aligned} u_1^2+u_2^2+u_3^2+u_4^2 &= (a^2+b^2+c^2+d^2)(w_1^2+w_2^2+w_3^2+w_4^2) \\ &= (k-\lambda)(w_1^2+w_2^2+w_3^2+w_4^2). \end{aligned} \tag{7.9}$$

Suppose there is a symmetric balanced incomplete block design with parameters (v,k,λ), whose incidence matrix is $A=(a_{ij})$. Use A to define v homogeneous linear functions $A_1, A_2, \ldots, A_v$ as follows:

$$A_j = A_j(X) = \sum_{i=1}^{v} x_i a_{ij}$$

where X is the row vector of v variables $(x_1, x_2, \ldots, x_v)$. Then

$$XAA^TX^T = \sum_{j=1}^{v} A_j^2. \tag{7.10}$$

Now (2.11) tells us $AA^T = (k-\lambda)I + \lambda J$, so (7.10) yields

$$\begin{aligned}\sum_{j=1}^{v} A_j^2 &= (k-\lambda)XX^T + \lambda XJX^T \\ &= (k-\lambda)\sum_{j=1}^{v} x_j^2 + \lambda(\sum_{j=1}^{v} x_j)^2 \qquad (7.11)\end{aligned}$$

Now assume that $v \equiv 1 \pmod 4$, so that $(-1)^{\frac{1}{2}(v-1)} = 1$. Define a new set of variables $y_1, y_2, \ldots y_v$ by

$$\begin{bmatrix} y_1 \\ y_2 \\ y_3 \\ y_4 \end{bmatrix} = B \begin{bmatrix} x_1 \\ x_2 \\ x_3 \\ x_4 \end{bmatrix}, \begin{bmatrix} y_5 \\ y_6 \\ y_7 \\ y_8 \end{bmatrix} = B \begin{bmatrix} x_5 \\ x_6 \\ x_7 \\ x_8 \end{bmatrix}, \ldots, \begin{bmatrix} y_{v-4} \\ y_{v-3} \\ y_{v-2} \\ y_{v-1} \end{bmatrix} = B \begin{bmatrix} x_{v-4} \\ x_{v-3} \\ x_{v-2} \\ x_{v-1} \end{bmatrix}, \; y_v = x_v,$$

and write σ for $\sum x_j$. Then, from (7.9),

$$(k-\lambda)\sum_{j=1}^{v-1} x_j^2 = \sum_{j=1}^{v-1} y_j^2$$

so (7.11) becomes

$$\sum_{j=1}^{v} A_j^2 = \left(\sum_{j=1}^{v-1} y_j^2\right) + (k-\lambda)y_v^2 + \lambda\sigma^2. \qquad (7.12)$$

As B is nonsingular, each x_i is a rational linear function of the y_i, so A_1 is also a rational linear function of the y_i. Say

$$A_1 = \sum_{i=1}^{v} e_i y_i$$

where the e_i are some rational numbers. Further, the nonsingularity of B implies that by proper choice of the variables x_j, we can give any values we wish to the variables y_j, so that (7.12) remains true whatever values the y_j may take. In particular, it will be true if we restrict ourselves to values such that

$$y_1 = (1-e_1)^{-1}\sum_{i=2}^{v} e_i y_i,$$

or, in the particular case where $e_1 = 1$ and $(1-e_1)^{-1}$ does not exist,

$$y_1 = (-1-e_1)^{-1}\sum_{i=2}^{v} e_i y_i.$$

It is easy to see that this value of y_1 can be specified by the choice of x_1, whatever values of $x_2, x_3, \ldots x_v$ are chosen.

In both cases we have $A_1^2 = y_1^2$, so equation (7.12) becomes

$$\sum_{j=2}^{v} A_j^2 = \left(\sum_{j=2}^{v-1} y_j^2\right) + (k-\lambda)y_v^2 + \lambda\sigma^2. \tag{7.13}$$

Each of the A_j, and also σ, are linear homogeneous functions of the variables $y_1, y_2, \ldots, y_v$. But y_1 has been made into a linear homogeneous function of the remaining y_j. So the A_j, and σ, can be considered to be linear functions of $y_2, y_3, \ldots, y_v$, and they will all have zero constant terms.

One can carry out a similar reduction on (7.13), obtaining an equation for $\sum_{j=3}^{v} A_j^2$ in terms of $y_3, y_4, \ldots, y_v$ only, and so on. Eventually,

$$A_v^2 = (k-\lambda)y_v^2 + \lambda\sigma^2. \tag{7.14}$$

No value for y_v has yet been specified; that choice will be made shortly. The process of eliminating variables has been such that A_v and σ are homogeneous linear rational functions of y_v; for some rational numbers $\frac{p}{q}$ and $\frac{r}{s}$ we have

$$A_v = \frac{p}{q}y_v, \quad \sigma = \frac{r}{s}y_v.$$

Now we may give y_v any value. We choose to put $y_v = qs$. Then (7.14) becomes

$$(\frac{p}{q}qs)^2 = (k-\lambda)(qs)^2 + \lambda(\frac{r}{s}qs)^2$$

whence

$$(ps)^2 = (k-\lambda)(qs)^2 + \lambda(qr)^2.$$

So there are integers x, y and z such that

$$x^2 = (k-\lambda)y^2 + \lambda z^2,$$

namely, $x = ps$, $y = qs$, $z = qr$. Since q and s are the denominators of two rational numbers, y is nonzero. So the theorem is proved for the case $v \equiv 1 \pmod 4$.

In the case $v \equiv 3 \pmod 4$ we introduce a new variable x_{v+1} and continue the change of variables as far as

$$\begin{bmatrix} y_{v-2} \\ y_{v-1} \\ y_v \\ y_{v+1} \end{bmatrix} = B \begin{bmatrix} x_{v-4} \\ x_{v-1} \\ x_v \\ x_{v+1} \end{bmatrix}.$$

Then (7.12) becomes

$$\left(\sum_{j=1}^{v} A_j^2\right) + (k-\lambda)x_{v+1}^2 = \left(\sum_{j=1}^{v} y_j^2\right) + y_{v+1}^2 + \lambda\sigma^2$$

and rather than (7.14) we obtain

$$y_{v+1}^2 = (k-\lambda)x_{v+1}^2 - \lambda\sigma^2.$$

The proof then proceeds as in the case $v \equiv 1 \pmod 4$; the details are left as an exercise. □

The application of part (i) is quite easy, and we see immediately that many sets of parameters are impossible: For example, $(22,7,2)$, $(46,10,2)$, $(52,18,6)$, and so on. Part (ii) requires some number-theoretic results in many cases. However, a few designs can be excluded quite simply. As an example we consider the parameters $(141,21,3)$.

Suppose a $(141,21,3)$-design exists. Then there are integers x, y and z, not all zero, such that

$$x^2 = 18y^2 + 3z^2.$$

Without loss of generality we can assume that x, y, z have no mutual common factor. Clearly, 3 must divide x^2, so 3 divides x: say $x = 3e$. Then

$$9e^2 = 18y^2 + 3z^2$$

and z must also be divisible by 3; say $z = 3f$. We have

$$\begin{aligned} 9e^2 &= 18y^2 + 27f^2, \\ e^2 &= 2y^2 + 3f^2. \end{aligned}$$

Reducing modulo 3, we find

$$e^2 \equiv 2y^2 \pmod 3.$$

If y is not divisible by 3, $2y^2$ is congruent to 2, which is not quadratic: a contradiction. So 3 divides y, and x, y and z have common factor 3: another contradiction. So no $(141,21,3)$-design can exist.

Part (ii) is especially interesting in the case $\lambda = 1$; the designs are finite projective planes. In the case of a $PG(2,n)$ we have $v = n^2+n+1$, which is always odd, and $(v-1)/2 = n(n-1)/2$, which is even when n or $n+1$ is divisible by 4 and odd otherwise. So when $n \equiv 0$ or $3 \pmod 4$, (7.7) becomes

$$x^2 = ny^2 + z^2$$

which always has the solution $x = z = 1, y = 0$, and the theorem gives us no information. But when $n \equiv 1$ or $2 \pmod 4$, (7.9) is

$$x^2 = ny^2 - z^2,$$

or equivalently

$$ny^2 = x^2 + z^2. \tag{7.15}$$

So ny^2 is the sum of two integer squares. By Theorem 7.8, any prime congruent to $3 \pmod 4$ that divides ny^2 must appear to an even power in the factorization of ny^2. Since y^2 is obviously a square, such a prime must divide n itself to an even power. Applying Corollary 7.8.1, we see that n itself can be written as the sum of two integer squares, leading to the following result.

COROLLARY 7.13.1

If there is a $PG(2, n)$, and $n \equiv 1$ or $2 \pmod 4$, then

$$n = a^2 + b^2$$

for some integers a and b.

The first few integers congruent to 1 or 2 modulo 4 are 1, 2, 5, 6, 9, 10, 13, 14. Of these each can be expressed as a sum of two integer squares except 6 and 14. So no $PG(2, 6)$ or $PG(2, 14)$ can exist. The theorem gives no information about the other six cases. Case $n = 1$ is trivial and 2, 5, 9 and 13 are prime powers, but the existence or otherwise of a $PG(2, 10)$ was undecided until Lam and his colleagues completed a large computer-aided search, and showed that no such design exists (see [94]).

We know of no finite projective plane $PG(2, n)$ where n is not a prime or prime power. Some researchers think no such design exists.

Everything we have said so far applies only to symmetric designs. Very little is known about the nonexistence of nonsymmetric balanced incomplete block designs, except for the following obvious consequence of Theorems 6.2 and 7.13.

COROLLARY 7.13.2

If there is a $(v - k, v - 1, k, k - \lambda, \lambda)$-design where $\lambda = 1$ or 2, then v, k, λ must satisfy the conditions of Theorem 7.13.

It follows that no $(15, 21, 7, 5, 2)$-design can exist; nor can $AG(2, 6)$ or $AG(2, 14)$. Theorem 6.2 enables us to say something about quasi-residual designs with $\lambda \geq 3$, but only when k is sufficiently large. For example, it gives us no information about the existence or otherwise of $(120, 140, 21, 18, 3)$-designs.

v	k	λ
22	7	2
43	7	1
46	10	2

v	k	λ
67	12	2
34	12	4
53	13	3
92	14	2

v	k	λ
211	15	1
106	15	2
43	15	5

FIGURE 7.1: Small parameters ruled out by Theorem 7.13.

v	b	r	k	λ
51	86	10	6	1
61	122	12	6	1
40	52	13	10	3
144	156	13	12	1
157	157	13	13	1
85	170	14	7	1
136	204	15	10	1
46	69	15	10	3
86	216	16	6	1

v	b	r	k	λ
113	225	16	8	1
177	236	16	12	1
65	60	16	13	3
105	120	16	14	2
225	240	16	15	1
241	241	16	16	1
121	121	16	16	2
81	81	16	16	3

FIGURE 7.2: Small undecided parameters for a BIBD.

Apart from these results, we know of only two sets of parameters where no BIBD exists: $(46, 69, 9, 6, 1)$ [75] and $(22, 33, 12, 8, 4)$ [18]. However, the number of "undecided" parameter sets is high. Perhaps there are further theorems that would show that certain sets of parameters are unrealizable.

For $r \leq 16$, Figure 7.1 shows the possible parameters for symmetric designs that are ruled out by Theorem 7.13, and Figure 7.2 shows the parameters for which the existence of a BIBD is in doubt. The data comes from [37].

Exercises 7.2

7.2.1 Complete the proof of Theorem 7.13 in the case $v \equiv 3 \pmod 4$.

7.2.2 Suppose there were a symmetric balanced incomplete block design with $\lambda = 3$ and $k = 21 + 27t$ for some positive integer t.

(i)* According to (1.1) and (1.2), what is the corresponding value of v?

(ii) Prove that no such design exists.

Hint: Generalize the argument that shows $(141, 21, 3)$ *cannot be realized.*

7.2.3 Nonembeddable designs were defined in Section 6.1. Use Exercise 6.1.3 to prove the existence of a family of nonembeddable quasi-residual designs.

7.2.4 Consider the design $\mathcal{D}$ of Exercise 6.1.5, in the case where $n \equiv 0$ or $1 \pmod 4$ and n is not a perfect square. Prove that $\mathcal{D}$ is nonembeddable.

7.3 Another Proof

In this section we present a shorter proof of Theorem 7.13, using the algebra of matrix congruence. It was originally discovered independently by Lenz [96] and Ryser [125].

Suppose A and B are symmetric matrices over a field F. Then we say that A and B are *congruent* over F if there is a nonsingular matrix S over F such that

$$A = S^T BS.$$

In the particular case where F is the field of rational numbers we simply say "congruent," without specifying the field, and write

$$A \sim B.$$

We require some standard results on congruence.

LEMMA 7.14

Every symmetric matrix is congruent to a diagonal matrix.

LEMMA 7.15

If $a_1, a_2, \ldots, a_m$, $b_1, b_2, \ldots, b_m$, $c_1, c_2, \ldots, c_m$ *are any rational numbers, then*

$$\mathrm{diag}\,(a_1, a_2, \ldots, a_m, b_1, b_2, \ldots, b_m) \sim \mathrm{diag}\,(a_1, a_2, \ldots, a_m, c_1, c_2, \ldots, c_m)$$

is true if and only if

$$\mathrm{diag}\,(b_1, b_2, \ldots, b_m) \sim \mathrm{diag}\,(c_1, c_2, \ldots, c_m).$$

(Lemma 7.15 is called the *Witt cancellation law.*)

LEMMA 7.16

If n *is any positive integer, then*

$$\mathrm{diag}\,(n, n, n, n) \sim \mathrm{diag}\,(1, 1, 1, 1).$$

Proof. Since n is a positive integer there exist integers a, b, c, d, such that

$$n = a^2 + b^2 + c^2 + d^2.$$

If B is the matrix defined in (7.8), then

$$B^T I B = B^T B = nI,$$

so $nI \sim I$. Since the matrices are 4×4, we have the result. □

LEMMA 7.17

Suppose A is a nonsingular symmetric matrix, and B is a symmetric matrix derived from A by appending a row and column. Then

$$\begin{bmatrix} A & 0 \\ 0 & c \end{bmatrix} \sim B, \text{ where } c = \left(\frac{\det(B)}{\det(A)}\right) t^2,$$

for any $t \neq 0$.

We can now prove Theorem 7.13. The proof of part (i) proceeds as before. Suppose v is odd and suppose there is a (v, k, λ)-design with incidence matrix A. Then

$$AA^T = (k - A)I + AJ.$$

If S is the $(v+1) \times (v+1)$ matrix with all diagonal entries 1, last row all 1's and all other entries zero,

$$S^T[\operatorname{diag}(k-\lambda, k-\lambda, \ldots, k-\lambda, \lambda)]S = \left[\begin{array}{cccc|c} & & & & \lambda \\ & & & & \lambda \\ & AA^T & & & \cdots \\ & & & & \cdots \\ \hline \lambda & \lambda & \cdots & \cdots & \lambda \end{array}\right]$$

so, by Lemma 7.17,

$$\operatorname{diag}(k-\lambda, k-\lambda, \ldots, k-\lambda, \lambda) \sim \begin{bmatrix} AA^T & 0 \\ 0 & c \end{bmatrix}, \tag{7.16}$$

where

$$\begin{aligned} c &= \frac{\det(\operatorname{diag}(k-\lambda, k-\lambda, \ldots, k-\lambda, \lambda))}{k^2(k-\lambda)^{v-1}} t^2 \\ &= \frac{(k-\lambda)^v \lambda}{k^2(k-\lambda)^{v-1}} t^2 \\ &= \frac{\lambda(k-\lambda)t^2}{k^2} \end{aligned}$$

and putting $t = k$, we see that we can choose $c = \lambda(k - \lambda)$.

Now $AA^T = (A^T)^T I(A^T) \sim I$, so (7.16) yields

$$\operatorname{diag}(k-\lambda, k-\lambda, \ldots, k-\lambda, \lambda) \sim \operatorname{diag}(1, 1, \ldots, 1, \lambda(k-\lambda)). \tag{7.17}$$

Suppose $v \equiv 1 \pmod 4$. By repeated use of Lemma 7.16 (with $n = k - \lambda$) and Lemma 7.15 we get

$$\operatorname{diag}(k-\lambda, \lambda) \sim \operatorname{diag}(1, \lambda(k-\lambda)),$$

so for some p, q, r, s we have

$$\begin{bmatrix} 1 & 0 \\ 0 & \lambda(k-\lambda) \end{bmatrix} = \begin{bmatrix} p & q \\ r & s \end{bmatrix} \begin{bmatrix} k-\lambda & 0 \\ 0 & \lambda \end{bmatrix} \begin{bmatrix} p & r \\ q & s \end{bmatrix}$$

and comparing $(1, 1)$ entries we get

$$1 = (k - \lambda)p^2 + \lambda q^2.$$

Now p and q are rationals, so $p = y/x$ and $q = z/x$ for some integers x, y and z, whence

$$x^2 = (k - \lambda)y^2 + \lambda z^2$$

and the solution is nontrivial since x cannot be zero.

If $v \equiv 3 \pmod 4$, we obtain

$$\operatorname{diag}(k-\lambda, k-\lambda, k-\lambda, \lambda) \sim \operatorname{diag}(1, 1, 1, (k-\lambda)\lambda),$$

so by Lemma 7.16,

$$\operatorname{diag}(k-\lambda, k-\lambda, k-\lambda, k-\lambda, \lambda) \sim \operatorname{diag}(k-\lambda, 1, 1, 1, \lambda(k-\lambda)).$$

On the other hand, applying Lemma 7.16 to Lemma 7.15 yields

$$\operatorname{diag}(k-\lambda, k-\lambda, k-\lambda, k-\lambda, \lambda) \sim \operatorname{diag}(1, 1, 1, 1, \lambda),$$

and comparing these results

$$\operatorname{diag}(1, 1, 1, 1, \lambda) \sim \operatorname{diag}(k-\lambda, 1, 1, 1, \lambda(k-\lambda))$$

whence

$$\operatorname{diag}(1, \lambda) \sim \operatorname{diag}(k-\lambda, \lambda(k-\lambda));$$

for some rational p, q, r, s we have

$$\begin{bmatrix} k-\lambda & 0 \\ 0 & \lambda(k-\lambda) \end{bmatrix} = \begin{bmatrix} p & q \\ r & s \end{bmatrix} \begin{bmatrix} 1 & 0 \\ 0 & \lambda \end{bmatrix} \begin{bmatrix} p & r \\ q & s \end{bmatrix}$$

and comparing $(1, 1)$ entries

$$k - \lambda = p^2 + \lambda q^2;$$

putting $p = x/y$ and $q = z/y$ we have

$$x^2 = (k - \lambda)y^2 - \lambda z^2.$$

This completes the proof of Theorem 6.4. $\square$

Chapter 8

Latin Squares

8.1 Latin Squares and Subsquares

A *Latin square* of side (or order) n is an $n \times n$ array based on some set S of n symbols (treatments), with the property that every row and every column contains every symbol exactly once. In other words, every row and every column is a permutation of S. Since the arithmetical properties of the symbols are not used, the nature of the elements of S is immaterial; unless otherwise specified, we take them to be $\{1, 2, \ldots, n\}$. As an example, let G be a finite group with n elements $g_1, g_2, \ldots, g_n$. Define an array $A = (a_{ij})$ by $a_{ij} = k$ where k is the integer such that $g_k = g_i g_j$ (the *multiplication table* of G). We look at column j of A. Consider all the elements $g_i g_j$, for different elements g_i. Recall that g_j has an inverse element in G, written g_j^{-1}. If i and k are any two integers from 1 to n, then $g_i g_j = g_k g_j$ would imply $g_i g_j g_j^{-1} = g_k g_j g_j^{-1}$, whence $g_i = g_k$, so $i = k$. So the elements $g_1 g_j, g_2 g_j, \ldots, g_n g_j$ must all be different. This means that the $(1, j), (2, j), \ldots, (n, j)$ entries of the array are all different, so column j contains a permutation of $\{1, 2, \ldots, n\}$. This is true of every column. A similar proof applies to rows. So A is a Latin square. This example gives rise to infinitely many Latin squares. For example, at sides 1, 2 and 3 we have

$$\begin{array}{|c|}\hline 1 \\ \hline\end{array} \quad \begin{array}{|cc|}\hline 1 & 2 \\ 2 & 1 \\ \hline\end{array} \quad \begin{array}{|ccc|}\hline 1 & 2 & 3 \\ 2 & 3 & 1 \\ 3 & 1 & 2 \\ \hline\end{array}$$

There are Latin squares of side 3 that look quite different from the one just given, for example

$$\begin{array}{|ccc|}\hline 1 & 2 & 3 \\ 3 & 1 & 2 \\ 2 & 3 & 1 \\ \hline\end{array}$$

but this can be converted into the given 3×3 square by exchanging rows 2 and 3.

Suppose A and B are Latin squares of side n. We shall say that A and B are *equivalent* if it is possible to reorder the rows of A, reorder the columns of

A and/or relabel the treatments of A in such a way as to produce the array B. An obvious question is: "Are there Latin squares of the same side that are inequivalent?" The answer is "Yes." Another obvious question is, "Is there a Latin square of side n that is not equivalent to the square derived from some group of order n?" (If not, the study of Latin squares would be a part of group theory.) Again the answer is "Yes."

The smallest inequivalent squares have side 4 and the smallest square that does not arise from a group has side 5. It is not hard to see that the following Latin squares of side 4 are inequivalent:

$$\begin{array}{|cccc|}\hline 1&2&3&4\\ 4&1&2&3\\ 3&4&1&2\\ 2&3&4&1\\ \hline\end{array}\quad \begin{array}{|cccc|}\hline 1&2&3&4\\ 2&1&4&3\\ 3&4&1&2\\ 4&3&2&1\\ \hline\end{array}\,.$$

Both of these can be derived from groups (from the two nonisomorphic groups of order 4). The following Latin square of side 5 cannot be derived from any group; the proof appears in the exercises:

$$\begin{array}{|ccccc|}\hline 1&2&3&4&5\\ 2&1&5&3&4\\ 3&4&1&5&2\\ 4&5&2&1&3\\ 5&3&4&2&1\\ \hline\end{array}\,.$$

A *partial Latin square* of side n is an $n \times n$ array in which some cells are filled with members of some n-set while others are empty, such that no row or column contains a repeated element. In 1960, Evans [52] conjectured that any partial Latin square of side n that has $n-1$ or fewer cells occupied can be *completed*, that is, the empty cells can be filled in so that the result is a Latin square. This turned out to be a difficult problem, but in 1981 Smetaniuk [135] showed that the Evans conjecture was true. This is the best possible result; it is easy to find an n-element example that cannot be completed.

Among partial Latin squares, the *Latin rectangle* is of special interest. A Latin rectangle of size $k \times n$ is a $k \times n$ array with entries from $\{1, 2, \ldots, n\}$ such that every row is a permutation and the columns contain no repetitions. For example, the first k rows of an $n \times n$ Latin square form a Latin rectangle. Clearly, k can be no larger than n.

THEOREM 8.1

If A is a $k \times n$ Latin rectangle, then one can append $(n-k)$ further rows to A so that the resulting array is a Latin square.

Proof. If $k = n$, the result is trivial. So assume that $k < n$. Write S for $\{1, 2, \ldots, n\}$ and define S_j to be the set of members of S not occurring in column j of A. Each S_j has k elements. Now consider any element i of S. The k occurrences of i in A are in distinct columns, so i belongs to exactly $(n-k)$ of the sets S_j.

Suppose there were r of the sets S_j whose union contained fewer than r members, for some r. If we wrote a list of all members of each of these r sets, we would write down $r(n-k)$ symbols, since each S_j is an $(n-k)$-set. On the other hand, we would write at most $r-1$ different symbols, and each symbol occurs in exactly $n-k$ sets, so there are at most $(r-1)(n-k)$ symbols written. Since $k < n, (r-1)(n-k) < r(n-k)$, a contradiction. So there is no set of r of the S_j that contain between them fewer than r symbols. By Theorem 1.3, there is a system of distinct representatives for the S_i. Say the representative of S_j is i_j. Then $(i_1, i_2, \ldots, i_n)$ is a permutation of S. If we append this to A as row $k+1$, we have a $(k+1) \times n$ Latin rectangle. The process may be repeated. After $(n-k)$ iterations we have a Latin square with A as its first k rows. □

A *subsquare* of side s in a Latin square of side n is an $s \times s$ subarray that is itself a Latin square of side s. A necessary and sufficient condition for an $s \times s$ subarray to be a subsquare is that it should contain only s different symbols.

COROLLARY 8.1.1

If L is a Latin square of side s and $n \geq 2s$, then there is a Latin square of side n with L as a subsquare.

Proof. Say $n = s + t$. Define M to be the $s \times t$ array

$$M = \begin{bmatrix} s+1 & s+2 & \ldots & s+t \\ s+2 & s+3 & \ldots & s+1 \\ \vdots & \vdots & & \vdots \\ s+s & s+s+1 & \ldots & s+s-1 \end{bmatrix}$$

with entries reduced modulo t, where necessary, to ensure that they lie in the range $(s+1, s+2, \ldots, s+t)$. Write A for the array constructed by laying row i of M on the right of row i of L, for $1 \leq i \leq s$:

$$A = \boxed{L \ M}.$$

Then A is an $s \times n$ Latin rectangle; if we embed it in an $n \times n$ Latin square, that square has L as a subsquare. □

This is the best possible result (see Exercise 8.1.7), provided one ignores the trivial case $s = n$.

Exercises 8.1

8.1.1* Prove that there are exactly two different Latin squares of side 4, up to equivalence.

8.1.2 Prove that there exist at least

$$n!(n-1)!\dots(n-k+1)!$$

$k \times n$ Latin rectangles and therefore at least

$$n!(n-1)!\dots 2!1!$$

Latin squares of side n.
Hint: Use induction on k.

8.1.3 Use SDRs to complete the following Latin square.

1	2	3	4	5
5	1	2	3	4
4	5	1	2	3

8.1.4* A *reduced Latin square* of side n is one in which the first row and first column are $(1, 2, \dots, n)$ in that order.

(i) Prove that every Latin square is equivalent to a reduced Latin square.

(ii) Say there are r_n different reduced Latin squares of side n. Prove that the number of Latin squares of side n is

$$n!(n-1)!r_n.$$

(iii) Find r_n for $n = 3$, 4 and 5.

8.1.5 Show that the array

1			
			2
		3	

can be completed to a Latin square.

8.1.6 Find a 4×4 array containing four entries, one each of 1, 2, 3, 4, that cannot be completed to a Latin square. Generalize your result to show that the Evans conjecture is the best possible result.

8.1.7* Prove that if a Latin square of side n has a subsquare of side s, where $s < n$, then $n \geq 2s$.

8.1.8 If a Latin square A contains a 2×2 subsquare based on the treatments $\{x, y\}$, we say that x and y form an *intercalate* in A.

(i) Suppose A is derived from the multiplication table of a finite group $G = \{g_1, g_2, \ldots, g_n\}$, where g_1 is the identity, and suppose 1 and 2 form an intercalate in A. Prove that $g_2^2 = g_1$, and consequently that n is even.

(ii) Prove that if every symbol is a member of an intercalate in a Latin square A, and if A is derived from the multiplication table of a finite group, then A has an even side.

(iii) Prove that the following Latin square A is not derived from the multiplication table of any finite group:

$$\begin{array}{|ccccc|} \hline 1 & 2 & 3 & 4 & 5 \\ 2 & 1 & 5 & 3 & 4 \\ 3 & 4 & 1 & 5 & 2 \\ 4 & 5 & 2 & 1 & 3 \\ 5 & 3 & 4 & 2 & 1 \\ \hline \end{array}\ .$$

8.2 Orthogonality

We now discuss the important concept of *orthogonality.* Two Latin squares A and B of the same side n are called orthogonal if the n^2 ordered pairs (a_{ij}, b_{ij}), the pairs formed by superimposing one square on the other, are all different. We say "A is orthogonal to B" or "B is orthogonal to A"; clearly, the relation of orthogonality is symmetric. More generally, one can speak of a set of k *mutually orthogonal* Latin squares, Latin squares $A_1, A_2, \ldots, A_k$ such that A_i is orthogonal to A_j whenever $i \neq j$.

A set of n cells in a Latin square is called a *transversal* if it contains one cell from each row and one cell from each column and if the n entries include every treatment precisely once. Using this idea we get an alternative definition of orthogonality: A and B are orthogonal if and only if for every treatment x in A, the n cells where A has entry x form a transversal in B. The symmetry of the concept is not so obvious when this definition is used.

Latin squares were first defined by Euler [51] in 1782. He discussed orthogonality, and in particular he considered the following old puzzle. Thirty-six military officers wish to parade in a square formation. They represent six regiments, and the six officers from a regiment hold six different ranks (the same six ranks for each regiment). Is it possible for them to parade so that no two officers of the same rank, or the same regiment, are in the same row or column? If the ranks and regiments are numbered 1, 2, 3, 4, 5, 6, define two 6×6 arrays, A and B, by setting a_{ij} and b_{ij} equal to the regiment number and rank number, respectively, of the officer in the (i, j) position. Clearly, the parade formation has the desired properties if and only if A and B are orthogonal Latin squares of side 6.

Euler could not find a solution and concluded that one was impossible. He further conjectured ("Euler's conjecture") that no pair of orthogonal Latin squares of side n can exist when n is congruent to 2 modulo 4. His intuition about the case $n = 6$ was proven correct by Tarry [150], who carried out a complete census of Latin squares of side 6, in 1900. However, the rest of Euler's conjecture is spectacularly wrong; Bose, Parker, and Shrikhande ([24]; see also [54]) proved in 1959–1960 that there is a pair of orthogonal Latin squares of every side greater than 6. Our main purpose in this chapter is to prove this theorem (in Section 8.4).

First we prove an upper bound on the number of mutually orthogonal Latin squares.

THEOREM 8.2

If there are k mutually orthogonal Latin squares of side n, $n > 1$, then $k < n$.

Proof. Suppose A is a Latin square on the symbols $\{1, 2, \ldots, n\}$. There is no loss of generality in assuming that the first row is

$$\boxed{1\ 2\ 3\ \ldots\ n} \tag{8.1}$$

in ascending order; if not, we simply permute the names of symbols. Similarly, if B is orthogonal to A, we can permute the symbols in B so as to achieve first row (8.1), and this does not affect the orthogonality. We say a Latin square with first row (8.1) is *standardized* or in *standard form.* Now suppose $A_1, A_2, \ldots, A_k$ are Latin squares of side n, each of which is orthogonal to each other one. Without loss of generality, assume that each has been standardized by symbol permutation, so that each has first row (8.1). Assuming that $n > 1$, write a_i for the $(2, 1)$ entry in A_i. No a_i can equal 1 (since the first columns can contain no repetition), and the a_i must be different (if $a_i = a_j$, then the n cells that contain a_i in A_i must contain a repetition in A_j; both the $(1, a_i)$ and $(2, 1)$ cells of A_j contain a_i). So $\{a_1, a_2, \ldots, a_k\}$ contains k distinct elements of $\{2, 3, \ldots, n\}$, and $k < n$. □

Let us write $N(n)$ for the number of squares in the largest possible set of mutually orthogonal Latin squares of side n. In this notation, we have just shown

$$N(n) \leq n - 1 \text{ if } n > 1.$$

(If $n = 1$, we can take $A_1 = A_2 = \ldots$ and it makes sense to write "$N(1) = \infty$" in some situations. Whenever a theorem requires the existence of at least k mutually orthogonal Latin squares of side n, the conditions are satisfied by $n = 1$ for any k. Some authors only define orthogonality for side greater than 1.)

For example, $N(4) \leq 3$. In fact, $N(4) = 3$; one set of three mutually orthogonal Latin squares of side 4 is

1	2	3	4
2	1	4	3
3	4	1	2
4	3	2	1

1	2	3	4
3	4	1	2
4	3	2	1
2	1	4	3

1	2	3	4
4	3	2	1
2	1	4	3
3	4	1	2

It is easy to see that the upper bound $n - 1$ is attained whenever n is a prime power. If we write

$$GF(n) = \{f_1, f_2, \ldots, f_n\},$$

where f_n is the zero element, and define

$$a_{ij}^h = f_i + f_h f_j,$$

then $A_h = (a_{ij}^h)$ is a Latin square when $1 \leq h \leq n - 1$, and the $n - 1$ Latin squares are orthogonal. Another proof is a corollary of the following theorem.

THEOREM 8.3 [23, 97]

$N(n) = n - 1$ *if and only if there is a balanced incomplete block design with parameters* $(n^2, n^2 + n, n + 1, n, 1)$.

Proof. Assume a block design with the stated parameters exists. From Section 3.1 the set of blocks of the design can be partitioned into $n+1$ parallel classes of n blocks each; blocks in the same parallel class have no common elements, while two blocks in different classes have one element in common. Suppose the parallel classes have been numbered $0, 1, \ldots, n$ in some order, and the blocks in class i have been labeled $B_{i1}, B_{i2}, \ldots, B_{in}$. Select two parallel classes for reference purposes, say class 0 and class n. Then construct a square array from class i, where $1 \leq i \leq n - 1$, as follows. Find the point of intersection of B_{0x} with B_{ny}. The point will lie on precisely one block of class i. If it lies in B_{ir}, put r in position (x, y) of the array. Column y of the array will contain all the numbers of the blocks in class i that contain a point of B_{ny}. Since the points of B_{ny} lie one on each of the blocks in class i, the column will contain the elements of $\{1, 2, \ldots, n\}$ in some order. A similar argument applies to rows. So the array is a Latin square.

Write L_i and L_j for the Latin squares obtained from parallel classes i and j, respectively, where $i \neq j$. Those cells where L_i has entry r all come from points on block B_{ir}. The elements of B_{ir} consist of one element of B_{j1}, one from $B_{j2}, \ldots$, and one from B_{jn}. So the cells contain $1, 2, \ldots, n$ once each. So L_i is orthogonal to L_j.

Conversely, assume a set of $n-1$ mutually orthogonal Latin squares exists. The construction above is readily reversed to obtain the design. □

COROLLARY 8.3.1

If n is a prime power, then $N(n) = n - 1$.

The theorem in fact tells us more: In view of the Bruck-Chowla-Ryser Theorem (Theorem 7.13), we know that $N(n) < n - 1$ for infinitely many values of n: 6, 14, and so on.

It is easy to see that the taking of direct products preserves orthogonality; if A_1 is orthogonal to A_2 and B_1 is orthogonal to B_2, then $A_1 \times B_1$ is orthogonal to $A_2 \times B_2$.

$$N(nr) \geq \min\{N(n), N(r)\}. \tag{8.2}$$

So we have the following result (MacNeish's theorem):

THEOREM 8.4 [104]

Suppose

$$n = p_1^{a_1} p_2^{a_2} \ldots p_r^{a_r},$$

where the p_i are distinct primes and each $a_i \geq 1$. Then

$$n - 1 \geq N(n) \geq \min(p_i^{a_i}).$$

One can discuss orthogonality of subsquares; if we say that "A and B are orthogonal Latin squares with orthogonal subsquares S and T," we not only imply that S is a subsquare of A and T is a subsquare of B, and that S and T are orthogonal, but also that S and T occupy the same set of cells in A and B.

Exercises 8.2

8.2.1 Verify that the two definitions of orthogonality of Latin squares given in the text are equivalent.

8.2.2 Suppose $f(x)$ is a polynomial function of two variables x, y over $GF(n)$, n prime. We say $f(x)$ *generates* the Latin square L if $l_{ij} = f(i,j)$ for all i and j.

(i) Find polynomials that generate the Latin squares

1	2	0
2	0	1
0	1	2

1	2	0
0	1	2
2	0	1

over $GF(3)$. (Assume the rows are numbered 0, 1, 2.)

(ii) Does this method work when n is not a prime?

8.2.3 Prove that the Latin square A of Exercise 8.1.8 has no orthogonal mate.

8.2.4 The t mutually orthogonal Latin squares $A_1, A_2, \ldots, A_t$ of side n have mutually orthogonal subsquares $S_1, S_2, \ldots, S_t$ occupying their upper left $s \times s$ corners. Prove that $n \geq (t+1)s$.

8.2.5 If A and B are orthogonal Latin squares whose top left-hand corners are $r \times r$ subsquares S and T, prove that S is orthogonal to T.

8.2.6* A Latin square is called *self-orthogonal* if it is orthogonal to its own transpose.

(i) Prove that the diagonal elements of a self-orthogonal Latin square of side n must be $1, 2, \ldots, n$ in some order.

(ii) Prove that there is no self-orthogonal Latin square of side 3.

(iii) Find self-orthogonal Latin squares of sides 4 and 5.

8.2.7 For any positive integer n, define two matrices

$$R = \begin{bmatrix} 1 & 1 & \ldots & 1 \\ 2 & 2 & \ldots & 2 \\ \ldots & \ldots & & \ldots \\ n & n & \ldots & n \end{bmatrix}, \quad C = \begin{bmatrix} 1 & 2 & \ldots & n \\ 1 & 2 & \ldots & n \\ \ldots & \ldots & & \ldots \\ 1 & 2 & \ldots & n \end{bmatrix}.$$

Show that an $n \times n$ array, based on $\{1, 2, \ldots, n\}$, is a Latin square if and only if it is simultaneously orthogonal to R and to C.

8.2.8 Suppose $L_1, L_2, \ldots, L_{n-2}$ are mutually orthogonal Latin squares of side n. Prove that there exists a Latin square of side n orthogonal to each of the L_i.

8.2.9 Suppose there exists a balanced incomplete block design with parameters (v, b, r, k, λ), and suppose there is a set of $k-2$ mutually orthogonal Latin squares of side v. Show that there exists a balanced incomplete block design with parameters $(vk, b^*, r^*, k, \lambda)$ and evaluate b^* and r^*.

8.2.10 Say t mutually orthogonal Latin squares of side v exist; suppose each has first row $1\ 2\ \ldots\ v$. For $2 \le i \le v$ and $1 \le j \le v$, let B_{ij} denote the set of t numbers obtained by taking the symbols in the (i, j) cell in the t squares. Show that the sets B_{ij} are the blocks of a BIBD on v elements with block size t. Write down the parameters of this design.

8.3 Idempotent Latin Squares

In this section we use pairwise balanced designs to construct sets of orthogonal Latin squares. An $n \times n$ Latin square is called a *transversal square* if its diagonal entries are all different. In particular a transversal square whose main diagonal is

$$(1, 2, \ldots, n) \tag{8.3}$$

is called an *idempotent Latin square.*

LEMMA 8.5

There exists a set of $N(n) - 1$ mutually orthogonal idempotent Latin squares of side n.

Proof. Suppose $A_1, A_2, \ldots, A_{N(n)}$ are mutually orthogonal Latin squares of side n based on $1, 2, \ldots, n$. By permuting the columns, transform $A_{N(n)}$ so that the element 1 appears in all the diagonal cells. Carry out the same column permutation on the other squares. The diagonals of the first $N(n) - 1$ new squares will be transversals. Now, in each square, permute the names of the treatments so as to produce the diagonal (8.3). We have the required orthogonal idempotent Latin squares. □

THEOREM 8.6

Suppose there exists a $PB(v; K; 1)$, and suppose there exists a set of t mutually orthogonal idempotent Latin squares of side k for every element k of K. Then there exists a set of t mutually orthogonal idempotent Latin squares of side v.

Proof. For every k in K, let L_k be an idempotent Latin square of side k. If S is any ordered set of k integers, write $L_k(S)$ to mean the array derived from L_k by replacing the number 1 throughout with the first element of S, replacing 2 with the second element of S, and so on. We construct a Latin square A of side v as follows. Select one block B of the pairwise balanced design; say its order is k, and its elements are $b_1, b_2, \ldots, b_k$, where the b_i have been ordered so that $b_1 < b_2 \ldots < b_k$. Then the entry in position (b_i, b_j) of A is the (i, j) entry of $L_k(B)$. This process is carried out on every block of the design in turn.

If x and y are any two integers in the range from 1 to v, then x and y occur together in precisely one block of the $PB(v; K; 1)$, and the (x, y) entry of A is allotted when and only when that block is processed. This means that one and only one entry is placed in position (x, y). When we investigate the diagonal entries, we see that position (x, x) is assigned an entry every time a block containing x is processed, but the entry is always x. So the (x, x) entry is x. Thus the process defines an array A, whose diagonal has the proper form.

To prove that A is a Latin square, we observe that as x and y occur together in one block, they will occur together in one of the squares $L_k(B)$. So y will appear in the row of $L_k(B)$ that has diagonal element x, and consequently it will appear in row x of A. Therefore, each row of A contains each of the integers $1, 2, \ldots, v$ at least once; as it has only v elements, it must contain each of $1, 2, \ldots, v$ exactly once, and be a permutation of $\{1, 2, \ldots, v\}$. A similar argument applies to columns. So A is a Latin square.

Let $L_k^1, L_k^2, \ldots, L_k^t$ be the t mutually orthogonal idempotent Latin squares of side k, for each k in K. Write A^i for the Latin square of side v constructed as above, with L replaced by L_k^i for every k. It is easy to see that the A^i are orthogonal. □

As an example we shall construct a pair of orthogonal idempotent Latin squares of side 17 using the $PB(17; \{4, 5\}; 1)$ with blocks

$$\begin{array}{lllll}
01234 & 159d & 16bg & 17ce & 18af \\
05678 & 26ae & 25cf & 28bd & 279g \\
09abc & 37bf & 389e & 35ag & 36cd \\
0defg & 48cg & 47ad & 469f & 45be
\end{array}$$

and the pairs of orthogonal Latin squares

$$L_4^1 = \begin{array}{|cccc|}\hline 1 & 3 & 2 & 4 \\ 4 & 2 & 1 & 3 \\ 2 & 4 & 3 & 1 \\ 3 & 1 & 4 & 2 \\ \hline\end{array}\,, \quad L_4^2 = \begin{array}{|cccc|}\hline 1 & 4 & 2 & 3 \\ 3 & 2 & 4 & 1 \\ 4 & 1 & 3 & 2 \\ 2 & 3 & 1 & 4 \\ \hline\end{array}$$

of side 4 and

$$L_5^1 = \boxed{\begin{array}{ccccc} 1&4&2&5&3\\ 4&2&5&3&1\\ 2&5&3&1&4\\ 5&3&1&4&2\\ 3&1&4&2&5 \end{array}}, \; L_5^2 = \boxed{\begin{array}{ccccc} 1&3&5&2&4\\ 5&2&4&1&3\\ 4&1&3&5&2\\ 3&5&2&4&1\\ 2&4&1&3&5 \end{array}}$$

of side 5. For each block of the design we construct a pair of Latin squares. For example, from the block 159d we construct two squares by the substitution $(1, 2, 3, 4) \mapsto (1, 5, 9, d)$, obtaining

$$L_4^1(159d) = \boxed{\begin{array}{cccc} 1&9&5&d\\ d&5&1&9\\ 5&d&9&1\\ 9&1&d&5 \end{array}}, \; L_4^2(159d) = \boxed{\begin{array}{cccc} 1&d&5&9\\ 9&5&d&1\\ d&1&9&5\\ 5&9&1&d \end{array}}$$

Then these elements are placed in 17×17 squares A^1 and A^2, in all the positions (x, y) with x and y in $\{1, 5, 9, d\}$. Similarly, from the block $\{05678\}$ we construct the two squares

$$L_5^1(05678) = \boxed{\begin{array}{ccccc} 0&7&5&8&6\\ 7&5&8&6&0\\ 5&8&6&0&7\\ 8&6&0&7&5\\ 6&0&7&5&8 \end{array}}, \; L_5^2(05678) = \boxed{\begin{array}{ccccc} 0&6&8&5&7\\ 8&5&7&0&6\\ 7&0&6&8&5\\ 6&8&5&7&0\\ 5&7&0&6&8 \end{array}}$$

and from $\{47\mathrm{ad}\}$ we get

$$L_4^1(47ad) = \boxed{\begin{array}{cccc} 4&a&7&d\\ d&7&4&a\\ 7&d&a&4\\ a&4&d&7 \end{array}}, \; L_4^2(47ad) = \boxed{\begin{array}{cccc} 4&d&7&a\\ a&7&d&4\\ d&4&a&7\\ 7&a&4&d \end{array}}.$$

The result of using these squares is shown in Figure 8.1. The other 17 blocks give rise to 34 other squares, and when they are all used we end with the Latin squares in Figure 8.2.

Let us write $I(n)$ for the maximal cardinality of a set of mutually orthogonal idempotent Latin squares of side n. Using Lemma 8.5, we see that for every n,

$$N(n) - 1 \le I(n) \le N(n). \tag{8.4}$$

If n is a prime or prime power, equality holds on the left-hand side of (8.4). The only known case where $I(n) = N(n)$ is

$$I(6) = N(6) = 1.$$

0					7	5	8	6								
	1				9				d				5			
				4			a			d			7			
7	d				5	8	6	0	1				9			
5					8	6	0	7								
8				d	6	0	7	5		4		1	a			9
6					0	7	5	8								
	5				d				9				1			
				7			d			a			4			
	9			a	1		4		5	7			d			

0					6	8	5	7								
	1					d			5				9			
				4			d			7			a			
8	9				5	7	0	6	d				1			
7					0	6	8	5								
6				a	8	5	7	0		d			4			
5					7	0	6	8								
	d				1				9				5			
				d			4			a			7			
	5			7	9		a		1	4			d			

FIGURE 8.1: Result of substituting for three blocks.

0	3	1	4	2	7	5	8	6	b	9	c	a	f	d	g	e
3	1	4	2	0	9	b	c	a	d	f	g	e	5	7	8	6
1	4	2	0	3	c	a	9	b	g	e	d	f	8	6	5	7
4	2	0	3	1	a	c	b	9	e	g	f	d	6	8	7	5
2	0	3	1	4	b	9	a	c	f	d	e	g	7	5	6	8
7	d	f	g	e	5	8	6	0	1	3	4	2	9	b	c	a
5	g	e	d	f	8	6	0	7	4	2	1	3	c	a	9	b
8	e	g	f	d	6	0	7	5	2	4	3	1	a	c	b	9
6	f	d	e	g	0	7	5	8	3	1	2	4	b	9	a	c
b	5	7	8	6	d	f	g	e	9	c	a	0	1	3	4	2
9	8	6	5	7	g	e	d	f	c	a	0	b	4	2	1	3
c	6	8	7	5	e	g	f	d	a	0	b	9	2	4	3	1
a	7	5	6	8	f	d	e	g	0	b	9	c	3	1	2	4
f	9	b	c	a	1	3	4	2	5	7	8	6	d	g	e	0
d	c	a	9	b	4	2	1	3	8	6	5	7	g	e	0	f
g	a	c	b	9	2	4	3	1	6	8	7	5	e	0	f	d
e	b	9	a	c	3	1	2	4	7	5	6	8	0	f	d	g

0	2	4	1	3	6	8	5	7	a	c	9	b	e	g	d	f
4	1	3	0	2	g	d	e	f	5	8	6	7	9	c	a	b
3	0	2	4	1	f	e	g	d	7	6	8	5	b	a	c	9
2	4	1	3	0	g	d	f	e	8	5	7	6	3	9	b	a
1	3	0	2	4	e	f	d	g	6	7	5	8	a	b	9	c
8	9	c	a	b	5	7	0	6	d	g	e	f	1	4	2	3
7	b	a	c	0	9	6	8	5	f	e	g	d	c	2	4	1
6	c	9	b	a	8	5	7	0	g	d	f	e	4	1	3	2
5	a	b	9	c	7	0	6	8	e	f	d	g	2	3	1	4
c	d	g	e	f	1	4	2	3	9	b	0	a	5	8	6	7
b	f	e	g	d	3	2	4	1	0	a	c	9	7	8	6	5
a	d	g	f	e	4	1	3	2	c	9	b	0	8	5	7	6
9	e	f	d	g	2	3	1	4	b	0	a	c	6	7	5	8
c	5	8	6	7	9	c	a	b	1	4	2	3	d	f	0	e
f	7	6	8	5	b	a	c	9	3	2	4	1	0	e	g	d
e	8	7	5	6	c	9	b	a	4	1	3	2	g	d	f	0
d	6	7	5	8	a	b	9	c	2	3	1	4	f	0	e	g

FIGURE 8.2: Final arrays.

(We know that $N(6) = 1$, so $I(6) = 1$ or 0. Proving the existence of an idempotent Latin square of side 6 is left as an exercise.) As easy generalizations of the results of Section 8.2, we have

THEOREM 8.7

$I(n) \leq n - 2$, with equality when n is a prime power.

THEOREM 8.8

For every m and n, $I(mn) \geq \min(I(m), I(n))$. If n has prime power decomposition

$$n = p_1^{a_1} p_2^{a_2} \dots p_k^{a_k},$$

then $I(n) \geq \min(p_i^{a_i} - 1) - 1$.

From Theorem 8.6 we have

COROLLARY 8.8.1

If there exists a $PB(v; K; 1)$, then:

(i) if $I(k) \geq t$ for all $k \in K$, then $I(v) \geq t$;

(ii) if $N(k) \geq t$ for all $k \in K$, then $I(v) \geq t - 1$.

Exercises 8.3

8.3.1* Construct an idempotent Latin square of side 6.

8.3.2* Use the $PB(10; \{4, 3\}; 1)$ with blocks

$$\begin{array}{llll} B_1 = 1348, & B_4 = 127, & B_7 = 159, & B_{10} = 160, \\ B_2 = 2568, & B_5 = 369, & B_8 = 230, & B_{11} = 249, \\ B_3 = 7890, & B_6 = 450, & B_9 = 467, & B_{12} = 357, \end{array}$$

to construct a Latin square of side 10 with diagonal

$$(1, 2, 3, 4, 5, 6, 7, 8, 9, 0).$$

8.3.3* Self-orthogonal Latin squares were defined in Exercise 8.2.6.

(i) Modify the proof of Theorem 8.6 to show that if there is a $PB(v; K; 1)$ and if there is a self-orthogonal Latin square of side k for every $k \in K$, there is a self-orthogonal Latin square of side v.

(ii) Construct a self-orthogonal Latin square of side 17.

8.4 Transversal Designs

A *simple group divisible design* or SGDD is constructed from a pairwise balanced design with $\lambda = 1$ by selecting a *parallel class*, a set of blocks that between them contain every treatment exactly once, and deleting those blocks. (Of course, not every pairwise balanced design with $\lambda = 1$ contains a parallel class, so they cannot all be used to construct SGDDs.) The deleted blocks are called "groups"; thus two treatments either determine a common block or a common group, but not both. An SGDD whose sets of treatments, groups, and blocks are X, $\mathcal{G}$, and $\mathcal{A}$, respectively, will sometimes simply be denoted by the triple $(X, \mathcal{G}, \mathcal{A})$.

By $TD(k,n)$ we denote a *transversal design* or *transversal system*, which is an SGDD with uniform block size k, with uniform group size n, and with k groups. It follows that each block is a *transversal* of the groups: Each block contains precisely one element from each group. There are kn treatments. It will be convenient to accept the existence of a (degenerate) $TD(k,0)$ with no points, no groups, and k empty blocks.

Consider groups G_1 and G_2 selected from a $TD(k,n)$. Given x belonging to G_1 and y belonging to G_2, there will be exactly one block containing both. Letting y range through G_2, we see that x lies on exactly n blocks (since every block must contain one member of G_2); and as x ranges through G_j, we see that the $TD(k,n)$ has exactly n^2 blocks.

THEOREM 8.9

If $n \geq 2$ and there exists a $TD(k,n)$, then $k \leq n+1$. If $k = n+1$, any two blocks have a point of intersection. If $k \leq n$, every block has another block disjoint from it.

Proof. Consider any block A of a $TD(k,n)$, $n \geq 2$. Through each of its k points, there pass $n-1$ other blocks, so the number of blocks that meet A is $k(n-1)$, and there are $n^2 - k(n-1) - 1$ blocks disjoint from it. If $k > n+1$, this number is negative, a contradiction; if $k = n+1$, it is zero; if $k \leq n$, it is positive. □

The relationship between transversal designs and Latin squares is very important:

THEOREM 8.10

The existence of a set of $k-2$ mutually orthogonal Latin squares of order n is equivalent to the existence of a $TD(k,n)$.

Proof. Suppose a $TD(k,n)$ with groups $G_1, G_2, \ldots, G_k$ is given. Relabel the elements so that

$$G_h = \{x_{h1}, x_{h2}, \ldots, x_{hn}\}.$$

Then for every h, i, j with $1 \le h \le k-2$ and $1 \le i, j \le n$, define

$$a_{ij}^h = m,$$

where m is the integer such that x_{hm} is the (unique) member of G_h in the (unique) block of the $TD(k,n)$ that contains both $x_{k-1,i}$ and $x_{k,j}$. It is easy to verify that the array A_h with (i,j) element a_{ij}^h is a Latin square of side n, and that $A_1, A_2, \ldots, A_{k-2}$ are orthogonal. The construction may be reversed. □

Assume a $TD(k,n)$ exists. Then it is easy to see that a $PB(kn; \{k,n\}; 1)$ exists: One simply takes a pairwise balanced design whose blocks are the blocks and the groups of the transversal design. By adding a common further element to all the blocks of size n, one obtains a $PB(kn+1; \{k, n+1\}; 1)$. So we have:

COROLLARY 8.10.1

If there exists a set of $k-2$ mutually orthogonal Latin squares of side n, then there exist a $PB(kn, \{k,n\}, 1)$ and a $PB(kn+1, \{k, n+1\}, 1)$.

In particular, the existence of a Latin square of side n for every positive integer n proves that there is a $PB(3n, \{3,n\}, 1)$ for every n, a fact that was foreshadowed in Section 2.1. Theorem 8.10 shows that in order to construct orthogonal Latin squares, we may proceed by constructing transversal designs. Our main tool is the following theorem.

THEOREM 8.11

Suppose $(X, \mathcal{G}, \mathcal{A})$ is a $TD(k+l, t)$ with groups $G_1, G_2, \ldots, G_k, H_1, H_2, \ldots, H_l$. Let S be any given subset of $H_1 \cup H_2 \ldots \cup H_l$ and m any given nonnegative integer. Say there exist transversal designs of the following kinds:

(i) *for each $i = 1, 2, \ldots, l$, a $TD(k, h_i)$, where $h_i = |S \cap H_i|$;*

(ii) *for each block $A \in \mathcal{A}$, a $TD(k, m+u_A)$ that contains a set of u_A pairwise disjoint blocks, where $u_A = |S \cap A|$.*

Then there exists a $TD(k, mt+u)$, where $u = |S|$.

Proof. We first set up some notation. We set $X_0 = G_1 \cup G_2 \cup \ldots \cup G_k$, so that S is a subset of $X \backslash X_0 = H_1 \cup H_2 \cup \ldots \cup H_l$; and we write A' for $A \cap S$ when $A \in \mathcal{A}$, write M for some m-set, and write K for $\{1, 2, \ldots, k\}$. We shall construct a transversal design whose treatments are ordered pairs, the elements of $X_0 \cup M$ and $S \times K$.

For each $A \in \mathcal{A}$, there exists a $TD(k, m+u_A)$ with u_A disjoint blocks. For each of those u_A blocks allocate an element of A'; in the block that receives the element p, relabel the elements as

$$\{(p,1),(p,2),\ldots,(p,k)\},$$

where (p,i) is the element in group i. Then relabel the remaining elements of the design in such a way that the other m elements of the i-th group are $\{(d,s) : s \in M\}$, where d is the member of A in G_i. We denote by $\mathcal{B}(A)$ the set of blocks in the design *other than* the u_A disjoint blocks.

For each set $S \cap H_i, i \in \{1,2,\ldots,l\}$, there is a corresponding $TD(k,h_i)$, where $h_i = |S \cap H_i|$. Take the elements of the first group in this design to be the elements (p,j), where p ranges through $S \cap H_i$. The set of blocks of this design is denoted $\mathcal{C}_i$.

We now put together these ingredients. We obtain a design with elements, groups, and blocks $(X', \mathcal{G}', \mathcal{A}')$, where

$$\begin{aligned} X' &= (X_0 \times M) \cup (S \times K); \\ \mathcal{G}' &= (G'_1, G'_2, \ldots, G'_k), \text{ where } G'_i = (G_i \times M) \cup (S \times \{i\}); \\ \mathcal{A}' &= \{\cup_{A \in \mathcal{A}} \mathcal{B}(A)\} \cup \mathcal{C}_1 \cup \mathcal{C}_2 \cup \ldots \cup \mathcal{C}_l. \end{aligned}$$

We now prove that this design is a $TD(k, mt+u)$. Clearly, G' is a partition of X' into $(mt+u)$-sets, and in each block the treatment that belonged to the i-th group in the relevant component design is a member of G'_i. This makes it clear that the blocks are transversals of the G'_i. We show that any two members x and y of different blocks lie in a unique common block, which completes the proof of the theorem. We distinguish four cases.

Case 1. $x = (a,s)$ and $y = (b,t)$, where $a \in G_i, b \in G_j, s \in M$, and $t \in M$. Clearly, neither x nor y is in any block of any $\mathcal{C}_r$. There is a unique block of the $TD(k+l,t)$ that contains both a and b. Call this block A. Then there is a unique block of $\mathcal{B}(A)$ that contains both x and y; if $D \neq A$, no block of $\mathcal{B}(D)$ contains both.

Case 2. $x = (a,s)$ and $y = (\theta, j)$, where $a \in G_i, s \in M, \theta \in S$, and $1 \leq j \leq k$ but $j \neq i$. Again x is not in any $\mathcal{C}_r$. Let A be the unique member of $\mathcal{A}$ that contains both a and θ. Then $\mathcal{B}(A)$ contains a unique block with both x and y; again, if $D \neq A$, no block of $\mathcal{B}(D)$ contains both.

Case 3. $x = (\theta, i)$ and $y(\phi, j)$ where $\theta \in S \cap H_p$, $\phi \in S \cap H_q$ $(q \neq p)$, and $1 \leq i < j \leq k$. Since $p \neq q$, no block of any $\mathcal{C}_r$ contains both x and y. As in case 1, there is exactly one block in one $\mathcal{B}(A)$ that contains both.

Case 4. $x = (\theta, i)$ and $y(\phi, j)$ where $\theta, \phi \in S \cap H_p$, and $1 \leq i < j \leq k$. No block in any $\mathcal{B}(A)$ contains both x and y, since no A contains both θ and ϕ. Since $\theta \notin H_r$ when $r \neq p$, no $\mathcal{C}_r$ except $\mathcal{C}_p$ can contain them; and $\mathcal{C}_p$ has x and y together in precisely one block. □

COROLLARY 8.11.1

If $0 \le u \le t$, *then*

$$N(mt+u) \ge \min\{N(m), N(m+1), N(t)-1, N(u)\}.$$

Proof. Write $k-2$ for the minimum on the right-hand side. Then

$$\begin{aligned} N(m) \ge k-2 &\Rightarrow TD(k,m) \text{ exists;} \\ N(m+1) \ge k-2 &\Rightarrow TD(k,m+1) \text{ exists;} \\ N(t)-1 \ge k-2 &\Rightarrow TD(k+1,t) \text{ exists;} \\ N(u) \ge k-2 &\Rightarrow TD(k,u) \text{ exists.} \end{aligned}$$

Consider Theorem 8.11 with $l = 1$ and with S any u-subset of the special group H_1. For any block A, $u_A = 0$ or $u_A = 1$. So the required $TD(k+l,t)$ is a $TD(k+1,t)$, which exists; the $TD(k,h_i)$ boil down to $TD(k,u)$, which exists; the $TD(k,m+u_A)$ are either $TD(k,m)$ or $TD(k,m+1)$, which exist (the requirement "a set of one disjoint block" is clearly vacuous). Therefore, $TD(k,mt+u)$ exists, and $N(mt+u) \ge k-2$, as required. □

COROLLARY 8.11.2

If $0 \le u, v \le t$, *then*

$$N(mt+u+v) \ge min\{N(m), N(m+1), N(m+2), N(t)-2, N(u), N(v)\}.$$

Proof. Again, let $k-2$ be the minimum of the right-hand side. We know the following designs exist:

$$TD(k,m), TD(k,m+1), TD(k,m+2), TD(k+2,t), TD(k,u), TD(k,v).$$

As in Corollary 8.11.1, we have all the ingredients—this time, for the case $l = 2$. However, we require that the $TD(k,m+2)$ should have a pair of disjoint blocks. But $k-2 \le N(m)$, so certainly $k-2 \le m-1, k \le m+1$, and Theorem 8.10 assures us of disjoint blocks. □

We are now in a position to prove that the Euler conjecture is wrong for all n greater than 6.

THEOREM 8.12

There exists a pair of orthogonal Latin squares of side n *for all* $n > 6$.

Proof. From Theorem 8.4 there will exist a pair of orthogonal Latin squares of side n unless $n \equiv 2 \pmod 4$. We can treat nearly all the remaining cases

v_1 :	2	6	10	14	18	22	26	30	34
t_1 :	−1	1	1	1	5	7	7	7	11
u :	5	3	7	11	3	1	5	9	1

FIGURE 8.3: For Theorem 8.12: $v = 36 + v_1, t = 12 + t_1$.

using Corollary 8.11.2 with $m = 3$. Essentially, we work modulo 36. Figure 8.3 shows, for every $v \equiv 2 \pmod 4$, a representation in terms of the residue of v modulo 36, a value of t and a value of u. For conciseness, the table shows v_1, t_1 and u, where $v = 36 + v_1$ and $t = 12 + t_1$. In each case, t is prime to 6, so $N(t) \geq 3$. Also $N(n) \geq 2$. So $N(v) \geq 2$ provided $t \geq u$. The only cases where $t < u$ are $v = 2, 6, 10, 14$, and 30. We know that $N(2) = N(6) = 1$. Latin squares of sides 10 and 14 are presented in Figure 8.4; each of these squares is self-orthogonal (orthogonal to its own transpose), so $N(10) \geq 2$ and $N(14) \geq 2$. Finally, $30 = 10 \cdot 3$. So $N(30) \geq 2$, by (8.2). □

0	3	9	7	8	1	2	5	6	4
1	6	0	9	2	3	5	4	7	8
2	5	1	6	9	4	0	7	8	3
3	4	7	5	1	9	8	6	2	0
4	0	8	2	7	5	9	3	1	6
5	8	6	3	4	2	7	9	0	1
6	7	3	1	0	8	4	2	9	5
9	1	2	0	5	6	3	8	4	7
8	9	5	4	6	7	1	0	3	2
7	2	4	8	3	0	6	1	5	9

1	9	4	13	10	3	6	11	7	12	2	5	14	8
14	2	10	5	1	11	4	7	12	8	13	3	6	9
7	14	3	11	6	2	12	5	8	13	9	1	4	10
5	8	14	4	12	7	3	13	6	9	1	10	2	11
3	6	9	14	5	13	8	4	1	7	10	2	11	12
12	4	7	10	14	6	1	9	5	2	8	11	3	13
4	13	5	8	11	14	7	2	10	6	3	9	12	1
13	5	1	6	9	12	14	8	3	11	7	4	10	2
11	1	6	2	7	10	13	14	9	4	12	8	5	3
6	12	2	7	3	8	11	1	14	10	5	13	9	4
10	7	13	3	8	4	9	12	2	14	11	6	1	5
2	11	8	1	4	9	5	10	13	3	14	12	7	6
8	3	12	9	2	5	10	6	11	1	4	14	13	7
9	10	11	12	13	1	2	3	4	5	6	7	8	14

FIGURE 8.4: Self-orthogonal Latin squares of sides 10 and 14.

Exercises 8.4

8.4.1 We construct a $TD(6, 46)$. Write X for $\{x_c : c = 7, 12, 10, 33, 9, 26, 34, 16, 1\}$. (The nine values of c are the perfect fourth powers (quartic residues) modulo 37.) Z denotes Z_{37}. If Y is any set, Y^i denotes Y with a superscript i appended to each element. (Multiplicative powers are not used in the construction, so no confusion arises.) The $TD(6, 46)$ has $Z^i \cup X^i$ for its i-th group. The blocks consist of all the blocks of a

$TD(6,9)$ with i-th group X^i, together with the blocks

$$\begin{array}{l}
\{x_c^1, (c+g)^2, (13c+g)^3, (21c+g)^4, (14c+g)^5, (34c+g)^6\} \\
\{(34c+g)^1, x_c^2, (c+g)^3, (13c+g)^4, (21c+g)^5, (14c+g)^6\} \\
\{(14c+g)^1, (34c+g)^2, x_c^3, (c+g)^4, (13c+g)^5, (21c+g)^6\} \\
\{(21c+g)^1, (14c+g)^2, (34c+g)^3, x_c^4, (c+g)^5, (13c+g)^6\} \\
\{(13c-g)^1, (21c+g)^2, (14c+g)^3, (34c+g)^4, x_c^5, (c+g)^6\} \\
\{(c+g)^1, (13c+g)^2, (21c+g)^3, (14c+g)^4, (34c+g)^5, x_c^6\}
\end{array}$$

for all $g \in Z$ and all $x_c \in X$. Prove that the design is a $TD(6,46)$.

8.4.2 Two tennis clubs, each with k members, wish to play a competition in which doubles matches are played, with two players from one club opposing two players from the other club in each match, subject to the following conditions:

(a) Every player must play against every member of the other club exactly once.

(b) Two members of the same club can be partners in at most one match.

Prove that k must be even. If $k = 2n$, prove that there is a solution if $N(n) \geq 2$.

8.4.3* For the tournament in Exercise 8.4.2, prove that there is no solution when $n = 2$. Is there a solution when $n = 6$?

8.4.4* Suppose a $(v, b, r, k, 1)$-design exists. One treatment is deleted from the design and the blocks that previously contained it are interpreted as groups.

(i) Prove that the result is a simple group divisible design with constant group size and block size.

(ii) Can the design ever be a transversal design? If so, when? If not, why not?

Chapter 9

More about Orthogonality

9.1 Spouse-Avoiding Mixed Doubles Tournaments

In our discussion of sets of three mutually orthogonal Latin squares we are led to consider another type of design, one of several types of apparently unrelated designs that are related to orthogonal Latin squares; it also has some interest in its own right as an experimental design. Many of the results in this section are from [174].

A *spouse-avoiding mixed doubles tournament* (or *SAMD*) is an arrangement for couples to play mixed doubles tennis so that no player is partnered by, or opposes, his or her spouse; otherwise, every player has each other player as an opponent exactly once and has each other player of the opposite sex as a partner exactly once. The tournament is *sharply resolvable* if its matches are arranged into rounds such that either every player takes part in every round or else precisely one couple sits out of each round (according as the number of couples is even or odd). A sharply resolvable spouse-avoiding mixed-doubles tournament for n couples is denoted $SR(n)$.

We write $[A, B \mid C, D]$ to denote a match in which Mr. A and Mrs. B play against Mr. C and Mrs. D. (Observe that $[A, B \mid C, D]$ and $[C, D \mid A, B]$ are two ways of denoting the same match.) Given an $SR(n)$ whose couples are labeled $1, 2, \ldots, n$, we construct two $n \times n$ arrays, L and S, as follows. If n is odd, reorder the rounds so that couple i sits out of round i. Set $s_{ii} = i$ if n is odd and $s_{ii} = n$ if n is even. Set $l_{ii} = i$ for all n. If $[i, j \mid k, r]$ occurs in round m, then set $l_{ik} = j$, $l_{ki} = r$, and $s_{ik} = s_{ki} = m$. Then L and S are orthogonal Latin squares. Moreover, L is *self-orthogonal* (orthogonal to its transpose) and S is symmetric. So S, L and L^T are a set of three pairwise orthogonal Latin squares of order n.

As first examples of these designs we exhibit an $SR(4)$ and an $SR(8)$. The $SR(4)$ has rounds

$$\begin{array}{cc} [4,1 \mid 2,3] & [1,4 \mid 3,2] \\ [4,2 \mid 3,1] & [2,4 \mid 1,3] \\ [4,3 \mid 1,2] & [3,4 \mid 2,1] \end{array}$$

corresponding to the Latin squares

$$L = \begin{bmatrix} 1 & 3 & 4 & 2 \\ 4 & 2 & 1 & 3 \\ 2 & 4 & 3 & 1 \\ 3 & 1 & 2 & 4 \end{bmatrix}, \quad S = \begin{bmatrix} 4 & 2 & 1 & 3 \\ 2 & 4 & 3 & 1 \\ 1 & 3 & 4 & 2 \\ 3 & 1 & 2 & 4 \end{bmatrix}.$$

The $SR(8)$ has rounds:

$$\begin{array}{cccc}
[1,4 \mid 2,7] & [3,1 \mid 4,5] & [5,3 \mid 6,8] & [7,6 \mid 8,2] \\
[1,7 \mid 3,6] & [2,5 \mid 7,3] & [4,2 \mid 5,8] & [6,1 \mid 8,4] \\
[1,6 \mid 4,8] & [2,4 \mid 8,5] & [3,7 \mid 5,2] & [6,3 \mid 7,1] \\
[1,3 \mid 5,4] & [2,8 \mid 6,7] & [3,5 \mid 8,6] & [4,1 \mid 7,2] \\
[1,5 \mid 6,2] & [2,1 \mid 3,8] & [4,3 \mid 8,7] & [5,6 \mid 7,4] \\
[1,8 \mid 7,5] & [2,3 \mid 4,6] & [3,2 \mid 6,4] & [5,7 \mid 8,1] \\
[1,2 \mid 8,3] & [2,6 \mid 5,1] & [3,4 \mid 7,8] & [4,7 \mid 6,5]
\end{array}.$$

THEOREM 9.1

There exists an $SR(n)$ whenever n is prime to 6.

Proof. Interpret the labels $1, 2, \ldots, n$ as integers modulo n. Then the matches

$$\{[3i, i + 2j \mid 3j, 2i + j] : 0 \leq i < j \leq n\}$$

form a spouse-avoiding mixed-doubles tournament, and for every $a \pmod n$ the matches with $i = a + k$ and $j = a - k$, $0 \leq k \leq n - 1$, that is, the matches

$$\left\{[3a + 3k, 3a - k \mid 3a - 3k, 3a + k] : 0 < k \leq \tfrac{1}{2}(n - 1)\right\},$$

form a round in which couple $3a$ is omitted. □

THEOREM 9.2

If there is an $SR(n)$, where n is even, then there is an $SR(4n)$.

Proof. Suppose the $SR(n)$ has couples labeled $1, 2, \ldots, n$ and rounds labeled $R(1), R(2), \ldots, R(n - 1)$. We construct an $SR(4n)$ with couples labeled $1_1, 2_1, \ldots, n_1, 1_2, 2_2, \ldots, n_1, n_2, \ldots, n_4$. Define $R(i)_j$ to be the set of all matches $[a_j, b_j \mid c_j, d_j]$ such that $[a, b \mid c, d]$ is a match in $R(i)$. If we define $R^*(i) = R(i)_1 \cup R(i)_2 \cup R(i)_3 \cup R(i)_4$ then $R^*(i)$ contains every player exactly once. We shall take the $R^*(i)$, $1 \leq i \leq n - 1$, as rounds in the tournament of size $4n$; these rounds precisely cover the requirements for meetings between players whose labels have the same subscript.

Now define

$$\begin{aligned} S^1(i) &= \{[a_1, b_2 \mid c_3, d_4]\,,\ [a_2, b_1 \mid c_4, d_3]\ :\ [a, b \mid c, d] \in R(i)\}\,; \\ S^2(i) &= \{[a_1, b_3 \mid c_4, d_2]\,,\ [a_3, b_1 \mid c_2, d_4]\ :\ [a, b \mid c, d] \in R(i)\}\,; \\ S^3(i) &= \{[a_1, b_4 \mid c_2, d_3]\,,\ [a_4, b_1 \mid c_3, d_2]\ :\ [a, b \mid c, d] \in R(i)\}\,. \end{aligned}$$

The $3n - 3$ rounds $S^j(i)$, for $j = 1, 2, 3$ and $1 \le i \le n - 1$ cover all the remaining requirements except those where both players have symbols k_i and k_j, where $i \neq j$. (In verifying this, bear in mind that each match in $R(i)$ gives rise to *four* matches in $S^j(i)$; for example, in $S^1(i)$, $[a, b \mid c, d]$ gives rise not only to $[a_1, b_2 \mid c_3, d_4]$ and $[a_2, b_1 \mid c_4, d_3]$, but also to $[c_1, d_2 \mid a_3, b_4]$ and $[c_2, d_1 \mid a_4, b_3]$, because $[c, d \mid a, b] = [a, b \mid c, d]$.)

Finally, write

$$\begin{aligned} T^1 &= \{[i_1, i_2 \mid i_3, i_4], [i_2, i_1 \mid i_4, i_3] : 1 \le i \le n\} \\ T^2 &= \{[i_1, i_3 \mid i_4, i_2], [i_3, i_1 \mid i_2, i_4] : 1 \le i \le n\}\,. \\ T^3 &= \{[i_1, i_4 \mid i_2, i_3], [i_4, i_1 \mid i_3, i_2] : 1 \le i \le n\} \end{aligned}$$

The $4n - 1$ rounds

$$\begin{aligned} R^*(i) &: 1 \le i \le n - 1, \\ S^j(i) &: 1 \le i \le n - 1, 1 \le j \le 3, \\ T^j &: 1 \le j \le 3 \end{aligned}$$

form the required $SR(4n)$. □

Exercises 9.1

9.1.1 Write down the Latin squares L and S corresponding to the $SR[8]$ given in this section.

9.1.2* An $SR(n)$ is called *symmetric* if, whenever $[a, b \mid c, d]$ is a match in a round, $[b, a \mid d, c]$ is a match in the same round.

(i) Verify that there is a symmetric $SR(4)$.

(ii) Is there a symmetric $SR(8)$?

(iii) Prove that if there is a symmetric $SR(n)$, where n is even, then there is a symmetric $SR(4n)$.

9.1.3 Prove that there exists an $SR(2^k)$ when $k \ge 2$.

9.1.4 If G is an Abelian group with $2k + 1$ elements, a *starter sequence* in G is a way of ordering the nonzero elements of G as a sequence, say $(a_1, a_2, \ldots, a_{2k})$, so that:

(a) the $2k$ differences $(a_2 - a_1), (a_3 - a_2), \ldots, (a_{2k} - a_{2k-1}), (a_1 - a_{2k})$ are all distinct and nonzero;

(b) the $2k$ differences $\pm(a_{k+1} - a_1), \pm(a_{k+2} - a_2), \ldots, \pm(a_{2k} - a_k)$ are all distinct and nonzero; and

(c) the $2k$ differences $(a_{k+2} - a_1), (a_{k+3} - a_2), \ldots, (a_{2k} - a_{k-1}), (a_{k+1} - a_k)$ and $(a_2 - a_{k-1}), (a_3 - a_k), \ldots, (a_k - a_{2k-1}), (a_1 - a_{2k})$ are all distinct and nonzero.

Prove that if there is a starter sequence in some group of order $2k+1$ then there is an $SR(2k+1)$.
(Hint: The first round consists of the k matches $[a_i, a_{i+1} \mid a_{i+k}, a_{i+k+1}]$*.)*

9.1.5 Using the definition in Exercise 9.1.4, prove that there is a starter sequence in the additive group of the field $GF(2k+1)$ whenever $2k+1$ is a prime power greater than 3.

9.2 Three Orthogonal Latin Squares

None of the results in the preceding sections enable us to construct three orthogonal Latin squares of sides 12, 14, or 15. We discuss them as special cases.

THEOREM 9.3 **[83]**

$N(12) \geq 5$.

Proof. One of the Latin squares is

12	1	2	3	4	5	6	7	8	9	10	11
1	2	3	4	5	12	7	8	9	10	11	6
2	3	4	5	12	1	8	9	10	11	6	7
3	4	5	12	1	2	9	10	11	6	7	8
4	5	12	1	2	3	10	11	6	7	8	9
5	12	1	2	3	4	11	6	7	8	9	10
6	7	8	9	10	11	12	1	2	3	4	5
7	8	9	10	11	6	1	2	3	4	5	12
8	9	10	11	6	7	2	3	4	5	12	1
9	10	11	6	7	8	3	4	5	12	1	2
10	11	6	7	8	9	4	5	12	1	2	3
11	6	7	8	9	10	5	12	1	2	3	4

.

Each of the other squares is obtained from this by column permutation; the four permutations give the following four first rows:

$$\begin{matrix}[\,12 & 6 & 8 & 2 & 7 & 1 & 9 & 11 & 4 & 10 & 5 & 3\,], \\ [\,12 & 3 & 6 & 1 & 9 & 11 & 2 & 8 & 5 & 4 & 7 & 10\,], \\ [\,12 & 8 & 1 & 11 & 5 & 9 & 3 & 10 & 2 & 7 & 6 & 4\,], \\ [\,12 & 4 & 11 & 10 & 2 & 7 & 8 & 6 & 9 & 1 & 3 & 5\,].\end{matrix}$$

□

THEOREM 9.4 [154]

$$N(14) \geq 3.$$

Proof. We construct three squares L_1, L_2 and L_3 based on the symbols $\infty, 0, 1, \ldots, 12$. These are interpreted as the integers modulo 13, with an "infinity element" appended: $\infty + i = \infty$ for all i.

Each square has $\infty, 0, 1, 2, 3, 4, 5, 6, 7, 8, 9, 10, 11, 12$ as its first column. The first rows are specified separately from the rest of the squares; they are:

$$\begin{matrix} L_1: & [\,\infty & 0 & 1 & 2 & 3 & 4 & 5 & 6 & 7 & 8 & 9 & 10 & 11 & 12\,], \\ L_2: & [\,\infty & 2 & 3 & 4 & 5 & 6 & 7 & 8 & 9 & 10 & 11 & 12 & 0 & 1\,], \\ L_3: & [\,\infty & 12 & 0 & 1 & 2 & 3 & 4 & 5 & 6 & 7 & 8 & 9 & 10 & 11\,]. \end{matrix}$$

The second rows are

$$\begin{matrix} L_1: & [\,0 & \infty & 2 & 11 & 10 & 9 & 8 & 5 & 4 & 1 & 7 & 12 & 6 & 3\,], \\ L_2: & [\,0 & 8 & 12 & 9 & 6 & 3 & 1 & 10 & 7 & 5 & 2 & 11 & \infty & 4\,], \\ L_3: & [\,0 & 1 & \infty & 7 & 5 & 11 & 2 & 6 & 10 & 12 & 4 & 3 & 9 & 8\,] \end{matrix}$$

respectively. The later rows of the squares are constructed recursively from the second rows using the following rule: The $(i+1, j+1)$ entry is obtained by adding 1 to the (i, j) entry $(\bmod 13)$; if necessary, row and column numbers are reduced modulo 13 to the range $\{0, 1, \ldots, 12\}$. □

THEOREM 9.5 [128]

$$N(15) \geq 4.$$

Proof. The first rows of the squares are

$$\begin{matrix}[\,1 & 15 & 2 & 14 & 3 & 13 & 4 & 12 & 5 & 11 & 6 & 10 & 7 & 9 & 8\,], \\ [\,1 & 14 & 3 & 11 & 6 & 9 & 8 & 7 & 10 & 4 & 13 & 12 & 5 & 15 & 2\,], \\ [\,1 & 10 & 7 & 13 & 4 & 2 & 15 & 6 & 11 & 9 & 8 & 3 & 14 & 12 & 5\,], \\ [\,1 & 6 & 11 & 10 & 7 & 15 & 2 & 5 & 12 & 14 & 3 & 9 & 8 & 4 & 13\,].\end{matrix}$$

The later rows of each square are formed by developing the first row modulo 15. □

Another set of small values is handled by the following construction.

THEOREM 9.6 [174]

There exist three orthogonal Latin squares for each of the orders $n = 18, 22, 26, 30, 34, 38$ *and* 42.

Proof. In each case we construct a Latin square L and a symmetric Latin square S such that L, L^T and S are mutually orthogonal. That is, we show there is an $SR(n)$. In each case there are subarrays such that

$$L = \begin{bmatrix} L_1 & L_2 \\ L_3 & L_4 \end{bmatrix}, \quad S = \begin{bmatrix} S_1 & S_2 \\ S_3 & S_4 \end{bmatrix}$$

and L_1, L_1^T and S_1 are orthogonal Latin squares, so the corresponding tournament has a subtournament.

For $n = 18, 22, 26, 30$, the subtournaments have four couples, so L_1 and S_1 are 4×4 arrays. We use the arrays corresponding to those given in Section 9.1 for an $SR(4)$; for convenience, replace entries 1, 2, 3, 4 by a_1, a_2, a_3, a_4 respectively. Assume the rows and columns of L and S are labeled a_1, a_2, a_3, a_4, 0, 1, ..., $n-5$ in order, so that rows and columns a_1 to a_4 correspond to the subtournament and 0 through $n-5$ correspond to L_4 and S_4. The array L_4 has (i, j) element l_{ij}, and is defined in terms of its first row by

$$l_{ij} = l_{0,j-i} + i,$$

where the addition of subscripts is mod $(n-4)$, and the addition of elements is mod $(n-4)$ if $l_{0,j-i} \in Z_{n-4}$, and $a_j + i = a_j$ for $j = 1, 2, 3, 4$ in L_4, but in S_4

$$\begin{aligned} a_1 + i &= a_1, \quad a_4 + i = a_4, \\ a_2 + i &= a_2 \quad \text{if } i \text{ is even}, \\ a_2 + i &= a_3 \quad \text{if } i \text{ is odd}, \\ a_3 + i &= a_3 \quad \text{if } i \text{ is even}, \\ a_3 + i &= a_2 \quad \text{if } i \text{ is odd}. \end{aligned}$$

In the other four arrays only members of Z_{n-4} appear. Row i of L_3 is obtained from row 0 by adding i (mod $(n-4)$) to each element; and similarly for the columns of L_2 and S_2. (Obviously $S_3 = S_2^T$.)

The cases $n = 34, 38, 42$ are handled similarly, but the subtournament has order 8. Suitable arrays for L_1 and S_1 are derived from

$$L_1 : \begin{bmatrix} 1&4&7&6&3&5&8&2 \\ 7&2&1&3&6&8&5&4 \\ 6&8&3&1&7&2&4&5 \\ 8&6&5&4&2&7&1&3 \\ 4&1&2&8&5&3&6&7 \\ 2&7&4&5&8&6&3&1 \\ 5&3&8&2&4&1&7&6 \\ 3&5&6&7&1&4&2&8 \end{bmatrix}, \quad S_1 : \begin{bmatrix} 8&1&2&3&4&5&6&7 \\ 1&8&5&6&7&4&2&3 \\ 2&5&8&1&3&6&7&4 \\ 3&6&1&8&2&7&4&5 \\ 4&7&3&2&8&1&5&6 \\ 5&4&6&7&1&8&3&2 \\ 6&2&7&4&5&3&8&1 \\ 7&3&4&5&6&2&1&8 \end{bmatrix}.$$

The generation of the squares follows the earlier pattern; in S_4, the rule is

$$\begin{array}{lll} a_1 + i = a_1, & a_8 + i = a_8, & \\ a_j + i = a_j, & 2 \leq j \leq 7, & \text{if } i \text{ is even,} \\ a_2 + i = a_5, & a_5 + i = a_2 & \text{if } i \text{ is odd,} \\ a_3 + i = a_6, & a_6 + i = a_3 & \text{if } i \text{ is odd,} \\ a_4 + i = a_7, & a_7 + i = a_4 & \text{if } i \text{ is odd.} \end{array}$$

So all the squares are fully determined by the first rows and columns of their components. A suitable collection is shown in Figure 9.1. □

THEOREM 9.7

There exist three mutually orthogonal Latin squares of every side except 2, 3, 6, *and possibly* 10.

Proof. From Theorem 8.4, $N(n) \geq 3$ unless either 2 divides n but 4 does not or 3 divides n but 9 does not. So we need only discuss those sides n congruent to one of the following numbers (mod 24): $0, 2, 3, 6, 9, 10, 12, 14, 15, 18, 21, 22$. If t is any integer prime to 6, $N(t) \geq 4$ by Theorem 8.4. So we can interpret Corollary 8.11.1 in the case $m = 4$ in the following way:

> If $0 \leq u \leq t$, where t and 6 are coprime, then $N(4t + u) \geq \min\{3, N(u)\}$.

In Figure 9.2 we tabulate the 12 residue classes (modulo 24) that are to be considered. In each case we give a value of t and u such that $(t, 6) = 1$ and $N(u) \geq 3$. The column "N.C." shows the necessary condition forced by the demand that $u \leq t$, and the final column lists all sides for which this necessary condition is not satisfied. So $N(n) \geq 3$ except possibly when n is in the last column of Figure 9.2.

Cases $n = 2$ and $n = 3$ are clearly impossible by Theorem 8.2 and we know that $N(6) = 1$. Theorems 9.3 through 9.6 and Exercise 8.4.1 show that

$n = 18$ $L_2 : (13, 3, 12, 2)^T$ $L_3 : (8, 5, 4, 6)$ $S_3 : (0, 10, 8, 1)$
$L_4 : (0, 7, 13, 12, 11, 10, 2, 1, 9, a_4, a_3, a_2, a_1, 3)$
$S_4 : (a_4, a_2, 13, 6, 9, 7, 4, a_1, 12, 2, 5, 3, 11, a_3)$
$n = 22$ $L_2 : (7, 5, 4, 16)^T$ $L_3 : (2, 3, 10, 12)$ $S_3 : (11, 0, 6, 1)$
$L_4 : (0, 9, 17, 16, 15, 14, 8, 6, 11, 1, 4, 7, 13, a_4, a_3, a_2, a_1, 5)$
$S_4 : (a_4, a_2, 17, 10, 2, 13, 9, 12, 4, a_1, 14, 5, 3, 8, 16, 7, 15, a_3)$
$n = 26$ $L_2 : (10, 20, 7, 3)^T$ $L_3 : (3, 6, 4, 2)$ $S_3 : (18, 5, 20, 4)$
$L_4 : (0, 9, 21, 20, 19, 18, 15, 12, 10, 13, 16, 1, 11, 14,$
$8, 7, 5, a_4, a_3, a_2, a_1, 17)$
$S_4 : (a_4, a_2, 21, 9, 16, 13, 7, 17, 0, 11, 3, a_1, 15, 2, 14, 10, 1, 8, 12, 6, 19, a_3)$
$n = 30$ $L_2 : (13, 8, 15, 18)^T$ $L_3 : (2, 3, 6, 16)$ $S_3 : (22, 7, 3, 8)$
$L_4 : (0, 7, 25, 24, 23, 22, 4, 11, 15, 18, 21, 12, 14, 1, 8, 20, 19, 13, 17, 5,$
$10, a_4, a_3, a_2, a_1, 17)$
$S_4 : (a_4, a_2, 25, 9, 2, 20, 17, 19, 18, 13, 5, 1, 0, a_1, 14, 16, 21, 4, 10, 12,$
$11, 15, 24, 6, 23, a_3)$
$n = 34$ $L_2 : (3, 15, 18, 23, 7, 25, 4, 10)^T$ $L_3 : (13, 7, 4, 2, 3, 6, 10, 8)$
$S_3 : (0, 19, 13, 12, 10, 7, 2, 1)$
$L_4 : (0, 25, 23, 22, 21, 18, 17, 15, 20, 5, 16, 12, 14, 1, 19, 24, a_8, 11, a_5,$
$9, a_7, a_6, a_4, a_3, a_2, a_1)$
$S_4 : (a_8, 25, 22, 18, 9, 8, 23, a_2, 14, a_3, 21, a_4, 16, a_1, 4, a_7, 11, a_6, 6,$
$a_5, 17, 3, 5, 15, 20, 24)$
$n = 38$ $L_2 : (3, 15, 8, 11, 29, 22, 12, 5)^T$ $L_3 : (2, 3, 4, 6, 7, 8, 10, 11)$
$S_3 : (4, 13, 19, 3, 12, 15, 16, 25)$
$L_4 : (0, 29, 27, 26, 25, 22, 24, 16, 21, 5, 20, 13, 18, 17, 28, 1, 23, 14, 19,$
$9, a_8, 15, a_7, 12, a_5, a_6, a_4, a_3, a_2, a_1)$
$S_4 : (a_8, 29, 26, 21, 11, 10, 20, a_7, 8, a_2, 2, a_3, 9, 6, 1, a_1, 17, 23, 27,$
$a_6, 22, a_5, 0, a_4, 14, 5, 7, 18, 24, 28)$
$n = 42$ $L_2 : (12, 31, 22, 3, 6, 14, 15, 9)^T$ $L_3 : (5, 8, 7, 2, 10, 11, 13, 3)$
$S_3 : (18, 0, 16, 10, 2, 15, 3, 26)$
$L_4 : (0, 33, 30, 32, 27, 31, 25, 28, 24, 20, 6, 12, 17, 26, 21, 23, 18, 1, 22,$
$29, 19, 4, 15, 9, 14, 16, a_7, a_5, a_6, a_4, a_3, a_2, a_1)$
$S_4 : (a_8, 33, 25, 31, 8, 29, 11, 13, 20, a_7, 17, 30, 14, a_6, 1, a_2, 9, a_1, 27,$
$a_5, 21, a_3, 2, 19, 7, a_4, 12, 6, 5, 24, 4, 28, 23, 32)$

FIGURE 9.1: Initial rows and columns for Theorem 9.6.

$N(n) \geq 3$ for $n = 12, 14, 15, 18, 22, 26, 30, 34, 38, 42$ and 46. There remain 18 exceptions to be considered: sides

$$9, 10, 27, 36, 39, 50, 54, 58, 62, 66, 70, 74, 78, 86, 94, 98, 102, 126.$$

Six of these exceptions are handled by Theorem 8.4, or by 8.2 and the facts

n	t	u	N.C.	Missing
$24r+2$	$6r-5$	22	$r \geq 5$	$2, 26, 50, 74, 98$
$24r+3$	$6r-1$	7	$r \geq 2$	$3, 27$
$24r+6$	$6r-5$	26	$r \geq 6$	$6, 30, 54, 78, 102, 126$
$24r+9$	$6r+1$	5	$r \geq 1$	9
$24r+10$	$6r-1$	14	$r \geq 3$	$10, 34, 58$
$24r+12$	$6r+1$	8	$r \geq 2$	$12, 36$
$24r+14$	$6r-1$	18	$r \geq 4$	$14, 38, 62, 86$
$24r+15$	$6r+1$	11	$r \geq 2$	$15, 39$
$24r+18$	$6r+1$	14	$r \geq 3$	$18, 42, 66$
$24r+21$	$6r+5$	1	$r \geq 0$	
$24r+22$	$6r+1$	18	$r \geq 4$	$22, 46, 70, 94$
$24r$	$6r-1$	4	$r \geq 1$	

FIGURE 9.2: Exceptions to be checked.

that $N(14) \geq 3$ and $N(18) \geq 3$:

$$\begin{array}{rclrclrcl} 9 &=& 3^2 & 36 &=& 2^2 \times 3^2 & 98 &=& 7 \times 14 \\ 27 &=& 3^3 & 70 &=& 5 \times 14 & 126 &=& 7 \times 18. \end{array}$$

Seven more follow from Corollary 8.11.1:

m	t	u	$mt+u$		m	t	u	$mt+u$
4	8	7	39		7	11	1	78
7	7	1	50		7	11	9	86
7	7	5	54		18	5	4	94
					7	13	11	102.

Finally, we use Corollary 8.11.2 for four values:

m	t	u	v	$mt-u+v$
7	8	1	1	58
7	8	5	1	62
7	8	5	5	66
7	9	7	4	74

and only the case n = 10 is undecided.

COROLLARY 9.7.1

There is a pair of orthogonal idempotent Latin squares of every side except 2, 3 and 6.

Proof. For $n \neq 10$, select three orthogonal Latin squares A, B, C of side n. (This is possible by Theorem 9.7.) Reorder the columns of A so that it has all its entries 1 on the main diagonal, and use the same ordering to reorder the columns of B and C. The two squares obtained will have transversals on their main diagonals, so they yield orthogonal idempotent Latin squares after symbol permutation. So we need only consider side 10. But the square of side 10 given in Figure 8.4 and its transpose are orthogonal idempotent Latin squares of that side. □

9.3 Bachelor Squares

Even if there exists a set of k mutually orthogonal Latin squares of side n, it does not follow that every Latin square of side n is a member of such a set. A Latin square without an orthogonal mate is called a *bachelor square.* Alternatively, a Latin square of side n is a bachelor if and only if its cells *cannot* be partitioned into n mutually disjoint transversals.

We already know of bachelor squares of sides 2 and 6. (In fact, any Latin square of these orders must be a bachelor.) At side 3, there is no bachelor.

Norton [110] (see also [126]) listed the isomorphism classes of Latin squares of side 7; there are 147 in all. There is one set of six mutually orthogonal squares (all six squares are isomorphic); no other set of three or more mutually orthogonal square exists, and only five of the remaining classes contain squares with mates. So there are 141 isomorphism classes of bachelor squares. This led to speculation that bachelor squares might be very common, but empirical evidence suggests that for orders larger than 7, almost all Latin squares have mates.

This said, we shall show that there is a bachelor square of every order greater than 3.

For orders congruent to 1 (mod 4) we use a technique involving subsquares due to Mann [108].

LEMMA 9.8

A Latin square of side $4t+1$ containing a subsquare of side $2t$ has no orthogonal mate.

Proof. Suppose L is a Latin square of side $4t+1$, on symbols $1, 2, \ldots, 4t+1$, whose top left $2t \times 2t$ subarray is a Latin subsquare, A say. Assume A uses symbols $1, 2, \ldots, 2t$. Write

$$L = \begin{array}{|c|c|} \hline A & B \\ \hline C & D \\ \hline \end{array}.$$

Each row of B and each column of C contains precisely one copy of each member of $S = \{2t+1, 2t+2, \ldots, 4t+1\}$, which makes $4t$ occurrences of each of those symbols, so D contains each member of S once.

If a transversal of L contains a cells from A, it contains $2t - a$ entries from each of B and C, which makes an *even* number of elements of S. As $|S|$ is odd, each transversal of L intersects D in an *odd* number of elements; this must be at least 1. So we cannot have more than $2t + 1$ disjoint transversals. Therefore L has no orthogonal mate. □

THEOREM 9.9

There is a bachelor Latin square of side $4t + 1$ *whenever* $t \geq 1$.

Proof. We exhibit a Latin square of side $4t + 1$ whose top $2t \times 2t$ subarray is a Latin square. Take A to be a Latin square of side $2t$ on symbols $1, 2, \ldots, 2t$. Take B to be the $2t \times 2t + 1$ array

$$\begin{array}{|cccc|} \hline 2t+2 & 2t+3 & \ldots & 4t+1 \;\; 2t+1 \\ 2t+3 & 2t+4 & \ldots & 2t+1 \;\; 2t+2 \\ & & \ldots & \\ 4t+1 & 2t+1 & \ldots & 4t-1 \;\;\;\; 4t \\ \hline \end{array}$$

and C to be B^T. Define

$$D = \begin{array}{|ccccccc|} \hline 2t+1 & 1 & 2 & 3 & \ldots & 2t-1 & 2t \\ 2t & 2t+2 & 1 & 2 & \ldots & 2t-2 & 2t-1 \\ 2t-1 & 2t & 2t+3 & 1 & \ldots & 2t-3 & 2t-2 \\ & & & & \ldots & & \\ 1 & 2 & 3 & 4 & \ldots & 2t & 4t+1 \\ \hline \end{array}.$$

From Lemma 9.8,

$$L = \begin{array}{|c|c|} \hline A & B \\ \hline C & D \\ \hline \end{array}$$

is a bachelor. □

If L is of side $4t + 3$, the same argument could be used with A of size $2t + 1$ to show that D must contain an *even* number of elements from $\{2t + 2, 2t +$

$3, \ldots, 4t+3\}$. So any particular member of this set can occur in at most $2t+1$ transversals. If there are k mutually orthogonal Latin squares, all orthogonal to L, then every cell of D must occur in a transversal used in constructing each of the k squares. So $k \leq 2t+1$. So if a Latin square of side $4t+3$ has a subsquare of side $2t+1$, the largest set of mutually orthogonal Latin squares containing L has at most $2t+2$ members.

This method can be extended to show that there exists a Latin square of any side $4n+3$ that does not belong to any set of *three* mutually orthogonal squares.

We prove the existence of bachelors of sides congruent to 0, 2 or 3 (mod 4) by using squares related to back-circulant Latin squares. We define a function that measures how much an entry differs from an entry of the circulant square: If $A = (a_{ij})$ is an array with entries in Z_n, we define $\Delta_A(i,j) \equiv i + j - a_{ij}$ (mod n).

LEMMA 9.10

If T is a transversal in A, the sum of $\Delta_A(i,j)$ over the entries a_{ij} of T is $\frac{1}{2}n$ if n is even and 0 if n is odd.

Proof. Suppose cells $(1,j_1), (2,j_2), \ldots, (n,j_n)$ form a transversal. As they form a transversal, their entries are $\{1,2,\ldots,n\}$ in some order. So

$$\begin{aligned}\sum_{i=1}^{n} \Delta_A(i,j_i) &= (1+2+\ldots+n) + (1+2+\ldots+n) - (1+2+\ldots+n) \\ &= (1+2+\ldots+n) \\ &= \tfrac{1}{2}n(n+1).\end{aligned}$$

□

THEOREM 9.11

Let A be the Latin square of side $2t$ defined by

$$a_{ij} \equiv i + j \pmod{2t}, 1 \leq a_{ij} \leq 2t.$$

Then A contains no transversal.

Proof. Suppose cells $(1,j_1), (2,j_2), \ldots, (2t, j_{2t})$ form a transversal. From Lemma 9.10

$$\sum \Delta_A(i,j_i) = t.$$

However, $a_{ij_i} \equiv i + j_i$, so

$$\sum \Delta_A(i,j_i) \equiv 2(1+2+\ldots+2t) \equiv 0 \pmod{2t},$$

a contradiction. □

Obviously these squares have no orthogonal mates.

Ryser [124] conjectured that the total number of transversals in a Latin square of side n is congruent to $n \pmod 2$. Balasubramanian [9] proved this for even n in 1990; it is false for odd side (over one-third of the classes listed by Norton have an even number of transversals), but it seems possible that every Latin square of odd side has at least one transversal. So there is little hope of extending Theorem 9.11 to odd orders. But Wanless has found a way to use the function Δ when n is odd.

THEOREM 9.12 [175]

There is a bachelor Latin square of side $4t + 3$ whenever $t \geq 2$.

Proof. For this proof it will be convenient to number rows $0, 1, 2, \ldots$, rather than $1, 2, 3, \ldots$.

We shall define a Latin square $A = (a_{ij})$ of side $n = 4t+3$ and show that one of its entries never lies on a transversal. In most cases $a_{i,j} \equiv i+j \pmod{4t+3}$, so $\Delta_A(i, j_i) = 0$. However, there are the following exceptions:

$$\begin{array}{llll}
a_{0,0} & = 1 & \text{so} \ \Delta_A(0,0) & = -1; \\
a_{0,1} & = 0 & \text{so} \ \Delta_A(0,1) & = 1; \\
a_{2t+1,0} & = 0 & \text{so} \ \Delta_A(2t+1,0) & = 2t+1; \\
a_{2t+1,2t+2} & = 2t+1 & \text{so} \ \Delta_A(2t+1,2t+2) & = 2t+2; \\
a_{4t+2,1} & = 1 & \text{so} \ \Delta_A(4t+2,1) & = -1; \\
a_{4t+2,2} & = 2t+1 & \text{so} \ \Delta_A(4t+2,2) & = 2t+3; \\
a_{4t+2,2t+2} & = 0 & \text{so} \ \Delta_A(4t+2,2t+2) & = 2t+1; \\
 & & \text{so} & \\
\end{array}$$

and for $i = 1, 3, 5, \ldots, 2t-1$,

$$\begin{array}{llll}
a_{i,0} & = i+2 & \text{so} \ \Delta_A(i,0) & = -2; \\
a_{i,2} & = i & \text{so} \ \Delta_A(i,2) & = 2.
\end{array}$$

If a set of n entries are to form a transversal, there must be one entry per row and one per column, the n symbols must be all different, and the Δ values (n of them) must sum to $0 \pmod n$. Entry $a_{4t+2,2t+2}$ is 0, and $\Delta_A(4t+2, 2t+2) = 2t+1$. To find a transversal through this entry we need to find entries from the other rows and columns whose Δ values sum to $2t \pmod n$; moreover neither the $(0,1)$ entry nor the $(2t+1, 0)$ entry can be included, as $a_{0,1} = a_{2t+1,0} = a_{4t+2,2t+2} = 0$. The only nonzero possibilities with Δ nonzero are at most one element from column 0, with $\Delta = -2$, and

$$A = \begin{array}{|ccccccccccc|}\hline 1 & 0 & 2 & 3 & 4 & 5 & 6 & 7 & 8 & 9 & \times \\ 3 & 2 & 1 & 4 & 5 & 6 & 7 & 8 & 9 & \times & 0 \\ 2 & 3 & 4 & 5 & 6 & 7 & 8 & 9 & \times & 0 & 1 \\ 5 & 4 & 3 & 6 & 7 & 8 & 9 & \times & 0 & 1 & 2 \\ 4 & 5 & 6 & 7 & 8 & 9 & \times & 0 & 1 & 2 & 3 \\ 0 & 6 & 7 & 8 & 9 & \times & 5 & 1 & 2 & 3 & 4 \\ 6 & 7 & 8 & 9 & \times & 0 & 1 & 2 & 3 & 4 & 5 \\ 7 & 8 & 9 & \times & 0 & 1 & 2 & 3 & 4 & 5 & 6 \\ 8 & 9 & \times & 0 & 1 & 2 & 3 & 4 & 5 & 6 & 7 \\ 9 & \times & 0 & 1 & 2 & 3 & 4 & 5 & 6 & 7 & 8 \\ \times & 1 & 5 & 2 & 3 & 4 & 0 & 6 & 7 & 8 & 9 \\ \hline \end{array} \qquad \Delta_A = \begin{array}{|ccccccccccc|}\hline \times & 1 & 0 & 0 & 0 & 0 & 0 & 0 & 0 & 0 & 0 \\ 9 & 0 & 2 & 0 & 0 & 0 & 0 & 0 & 0 & 0 & 0 \\ 0 & 0 & 0 & 0 & 0 & 0 & 0 & 0 & 0 & 0 & 0 \\ 2 & 0 & 2 & 0 & 0 & 0 & 0 & 0 & 0 & 0 & 0 \\ 0 & 0 & 0 & 0 & 0 & 0 & 0 & 0 & 0 & 0 & 0 \\ 5 & 0 & 0 & 0 & 0 & 0 & 6 & 0 & 0 & 0 & 0 \\ 0 & 0 & 0 & 0 & 0 & 0 & 0 & 0 & 0 & 0 & 0 \\ 0 & 0 & 0 & 0 & 0 & 0 & 0 & 0 & 0 & 0 & 0 \\ 0 & 0 & 0 & 0 & 0 & 0 & 0 & 0 & 0 & 0 & 0 \\ 0 & 0 & 0 & 0 & 0 & 0 & 0 & 0 & 0 & 0 & 0 \\ 0 & \times & 7 & 0 & 0 & 0 & 5 & 0 & 0 & 0 & 0 \\ \hline \end{array}$$

FIGURE 9.3: Bachelor square A of order 11 ($\times$ stands for 10).

at most one from column 2, with $\Delta = 2$. There is no solution with sum $2t \pmod{n}$. □

For convenience, Figure 9.3 shows the array A for the 11×11 case, as well as the array of Δ values.

A *partial transversal* of size t in a Latin square is a set of t cells, no two of which are in the same row or column, and no two of which contain the same symbol. Although some Latin squares contain no transversals, they must contain partial transversals of a certain size. Koksma [92] proved that every Latin square of side 7 or greater has a transversal of size at least $(2n+1)/3$ (a complete search shows that this result also holds for all squares of order between 3 and 6). Koksma's proof is outlined in Exercise 9.3.3. Brualdi has conjectured that every Latin square of side n has a transversal of length at least $n-1$.

Exercises 9.3

9.3.1* Show that there exists a pair of Latin squares of side 6 based on $S = \{1, 2, 3, 4, 5, 6\}$, L and M say, such that the 36 ordered pairs (l_{ij}, m_{ij}) include 32 of the ordered pairs of elements of S once each and two pairs twice each.

9.3.2 Show that there is a Latin square of side 6 that contains 4 pairwise disjoint transversals.

9.3.3 Suppose the Latin square S of side n contains a partial transversal $T = (1\ 2\ \ldots\ t)$ of size t, and all other partial transversals in S are of size less than or equal to t. (We call this a *maximal* partial transversal.)

Without loss of generality, write

$$S = \begin{array}{|c|c|} \hline A & B \\ \hline C & D \\ \hline \end{array}$$

where T is the main diagonal of A.

(i) Prove that if the symbol x occurs in the subarray D, then x is a member of $\{1, 2, \ldots, t\}$. Consequently show that C contains at most $(n-t)^2$ symbols from the set $\{t+1, \ldots, n\}$.

(ii) Suppose $t + \sqrt{t} < n$. Show that there exists at least one column j of A, $1 \leq j \leq t$, that contains two symbols selected from the set $\{t+1, \ldots, n\}$. Use this and the fact that T is a maximal partial transversal to prove:

(a) j occurs $n - t$ times in B,

(b) if j occurs in row k of B, then the symbol in cell (j, k) of A is a member of $\{1, 2, \ldots, t\}$;

(c) all symbols in row j of B are chosen from the set $\{1, \ldots, t\}$.

Deduce that $2(n-t) + 1 \leq t$.

(i) Now suppose $t + \sqrt{t} \geq n \geq 7$. Prove that $t \geq (2n+1)/3$.

9.3.4 A *k-transversal* in an $n \times n$ Latin square is a set of kn cells that between them contain k members from each row, k members from each column and k copies from each symbol. Show that there is a Latin square containing a 2-transversal for every even order.

Chapter 10

One-Factorizations

10.1 Basic Ideas

Suppose a balanced incomplete block design on v treatments has block size 2. When we say the elements x and y occur together in λ blocks, this is precisely the same as saying that $\{x, y\}$ occurs as a block λ times. So the blocks are the $\frac{1}{2}v(v-1)$ unordered pairs of treatments, taken λ times each, and the parameters are

$$(v, \tfrac{1}{2}\lambda v(v-1), \lambda(v-1), 2, \lambda). \tag{10.1}$$

We have the trivial result

THEOREM 10.1

For every v and λ there is a unique balanced incomplete block design with $k = 2$, and its parameters are given by (10.1).

The existence problem for balanced incomplete block designs with $k = 2$ is therefore completely solved, so it is of little interest.

As we indicated in Section 1.2, we can impose further structure on these designs in the case where v is even, say $v = 2n$. We define a *one-factor* or *perfect matching* on a set of $2n$ objects to be a way of dividing the objects into n disjoint subsets of size 2. We discuss partitioning the blocks of the BIBD with parameters (10.1) into one-factors, the "resolvable design" problem for case $k = 2$. As there are n pairs in a one-factor and $\lambda n(2n-1)$ pairs in total, such a partition will consist of $\lambda(2n-1)$ one-factors.

Recall that a *graph* $G = G(V, E)$ consists of a finite set $V(G)$ of objects called *vertices* together with a set $E(G)$ of *edges*, which are unordered pairs of vertices. So the blocks of designs with $k = 2$ correspond to edges in a graph whose vertices are the treatments.

When discussing graphs, the edge containing x and y will simply be written xy or when no confusion is possible. The vertices x and y are called its *endpoints*, and we say this edge *joins* x to y; two vertices are *adjacent* if there is an edge joining them. The number of edges containing a vertex is called its *degree*. A graph is called *regular* if all its vertices have the same degree.

Given a set S of v vertices, the graph formed by joining all pairs of members of S is called the *complete* graph on S, and denoted K_S. We also write K_v to mean any complete graph with v vertices. Given a graph G, the set of all edges of $K_{V(G)}$ that are *not* in G will form a graph with the same vertex set $V(G)$; this new graph is called the *complement* of G, and written $\overline{G}$.

If G is a graph, it is possible to choose some of the vertices and some of the edges of G in such a way that these vertices and edges again form a graph, H say. H is then called a *subgraph* of G; one writes $H \leq G$. Clearly every graph G has itself and the 1-vertex graph (which we shall denote K_1) as subgraphs; we say H is a *proper* subgraph of G if it neither equals G nor K_1. A subgraph G of a graph H is called a *spanning* subgraph if $V(G) = V(H)$. Clearly any graph G is a spanning subgraph of $K_{V(G)}$.

We can now define matchings, one-factors and one-factorizations in graph-theoretic terms. A *matching* is any set of edges with no endpoint common to any two of them. A *one-factor* of a graph is the edge-set of a spanning subgraph that is regular of degree 1. So a one-factor is a matching that contains all the vertices; hence the term "perfect matching." A *one-factorization* of a general graph G is a way of partitioning $E(G)$ into one-factors of G. In the case $\lambda = 1$, finding a resolvable BIBD with block size 2 is the same as finding a one-factorization of a complete graph.

If a graph is to have a one-factorization, the number of vertices must be even (or else no one-factor can exist). Moreover the graph must be regular, and the degree will equal the number of factors in the factorization. These two necessary conditions are not sufficient. For example, consider the graph shown in Figure 10.1. The edge xy must belong to every one-factor, so no one-factorization is possible.

However, all even-order *complete* graphs have one-factorizations:

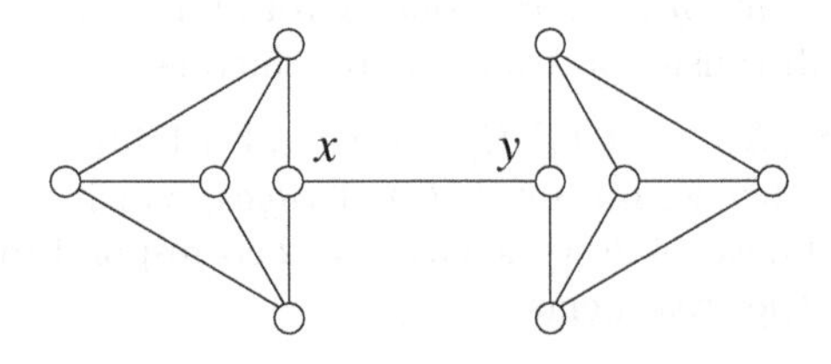

FIGURE 10.1: A regular graph without a one-factorization.

THEOREM 10.2

The complete graph K_{2n} has a one-factorization for all n.

Proof. We label the vertices of K_{2n} as $x_\infty, x_0, x_1, x_2, \cdots, x_{2n-2}$. The spanning subgraph F_i is defined to consist of the edges

$$x_\infty x_i,\ x_{i+1}x_{i-1},\ \cdots,\ x_{i+j}x_{i-j},\ \cdots,\ x_{i+n-1}x_{i-n+1} \tag{10.2}$$

where the subscripts other than ∞ are treated as integers modulo $2n-1$. Then F_i is a one-factor, as every vertex appears in the list (10.2) exactly once. We prove that $\{F_0, F_1, \cdots, F_{2n-2}\}$ forms a one-factorization. First, observe that every edge involving x_∞ arises precisely once: $x_\infty x_i$ is in F_i. If neither p nor q is ∞, then we can write $p+q = 2i$ in the arithmetic modulo $2n-1$, because either $p+q$ is even or $p+q+2n-1$ is even. Then $q = i-(p-i)$, and $x_p x_q$ is $x_{i+j}x_{i-j}$ in the case $j = p-i$. Since i is uniquely determined by p and q, this means that $x_p x_q$ belongs to precisely one of the F_i. So $\{F_0, F_1, \cdots, F_{2n-2}\}$ is the required one-factorization. □

This particular one-factorization of K_{2n} is denoted $\mathcal{G}K_{2n}$. It is common to write it with $\{\infty, 0, 1, 2, \cdots, (2n-2)\}$ replacing $\{x_\infty, x_0, x_1, x_2, \cdots, x_{2n-2}\}$.

COROLLARY 10.2.1

If $v = 2n$, the blocks of a design with parameters (10.1) *can be partitioned into $\lambda(2n-1)$ matchings.*

The *complete bipartite graph* on V_1 and V_2 has two disjoint sets of vertices, V_1 and V_2; two vertices are adjacent if and only if they lie in different sets. We write $K_{m,n}$ to mean a complete bipartite graph with m vertices in one set and n in the other.

If m and n are different, $K_{m,n}$ cannot have any one-factor, so we look at $K_{n,n}$.

THEOREM 10.3

One-factorizations of $K_{n,n}$ are equivalent to Latin squares of side n.

Proof. Suppose $L = (l_{ij})$ is a Latin square of side n, whose symbol-set is $v = \{1, 2, \ldots, n)$. We construct a one-factorization of the $K_{n,n}$ with vertex sets V_1 and V_2, where V_i is V with subscript i attached to each element. The k-th factor F_k consists of all the edges $\{i_1, (l_{ik})_2\}$ (reduced modulo n if necessary). The construction is easily reversed, so the existence of a one-factorization of $K_{n,n}$ implies the existence of a Latin square of side n. □

In particular, we could take

$$F_k = \{\{x_1(x+k)_2 : 1 \le x \le n\}\}$$

in the above construction. This one-factorization is called the *standard factorization* of $K_{n,n}$.

One could view K_{2n} as the union of three graphs, two disjoint copies of K_n and a copy of $K_{n,n}$. When n is even, this gives rise to a one-factorization in an obvious way: $n-1$ of the factors consist of the union of one factor from each of the K_n, and the other n factors are a one-factorization of the $K_{n,n}$. Such factorizations are called *twin* factorizations (see [2]). A slight generalization covers the case of odd n. We present here a particular case that is one of the standard one-factorizations.

First suppose n is even. Label the vertices of K_{2n} as

$$1_1, 2_1, \cdots, n_1, 1_2, 2_2, \cdots, n_2.$$

Write $P_{\alpha,1}, P_{\alpha,2}, \cdots, P_{\alpha,n-1}$ for the factors in $\mathcal{GK}_n$ with the symbols $x_1, x_2, \cdots, x_{n-2}, x_0, x_\infty$ replaced by $1_\alpha, 2_\alpha, \cdots, n_\alpha$ consistently, and write $H_n, H_{n+1}, \cdots, H_{2n-1}$ for the factors in the standard factorization of $K_{n,n}$ with $1, 2, \cdots, 2n$ replaced by $1_1, 2_1, \cdots, n_1, 1_2, \cdots, n_2$. If we define $H_i = P_{1,i} \cup P_{2,i}$ for $1 \le i \le n-1$, then $H_1, H_2, \cdots, H_{2n-1}$ make up a factorization denoted $\mathcal{GA}_{2n}$.

Now suppose n is odd. We use the same vertex set. Let $P_{\alpha,1}, P_{\alpha,2}, \cdots, P_{\alpha,n}$ be the factors of $\mathcal{GK}_{2n}$, where this time the vertices are $1_\alpha, 2_\alpha, \cdots, n_\alpha, \infty_\alpha$. We define

$$J_i = (P_{1,i} \cup P_{2,i} \cup \{i_1, i_2\}) \backslash \{\infty_1 i_1, \infty_2 i_2\}.$$

Then $J_1, J_2, \cdots, J_n$ are one-factors of K_{2n}, and they contain all the edges of the two copies of K_n together with the edges of the factor

$$1_1 1_2,\ 2_1 2_2,\ \cdots,\ n_1 n_2$$

of $K_{n,n}$. But these edges constitute one factor of the standard factorization. We write $J_{n+1}, J_{n+2}, \cdots, J_{2n-1}$ for the remaining factors in $K_{n,n}$, written in terms of our vertex set. Then $\{J_1, J_2, \cdots, J_{2n-1}\}$ is denoted $\mathcal{GA}_{2n}$. (For consistency, we shall always take J_{n+k} to consist of all the edges $i_1(i+k)_2$ in discussions of $\mathcal{GA}_{2n}$, where $i+k$ is reduced modulo n if necessary.)

THEOREM 10.4

$\mathcal{GA}_{2n}$, as defined above, is a one-factorization of K_{2n} for all n.

Exercises 10.1

10.1.1 Prove that the equations $vr = bk$, $\lambda(v-1) = r(k-1)$, $k = 2$ together imply $r = \lambda(v-1)$, $b = \frac{1}{2}\lambda v(v-1)$.

10.1.2 Prove that the graph shown in Figure 10.2 contains no one-factor.

10.1.3 Verify that if $m \neq n$ then $K_{m,n}$ has no one-factors.

10.1.4 What is the number of distinct one-factors in K_{2n}?

10.1.5 Idempotent Latin squares were discussed in Section 8.3. A *symmetric* idempotent Latin square A is a Latin square $A = (a_{ij})$ in which $a_{ii} = i$ for each i and $a_{ij} = a_{ji}$ for each i and j. Prove that the existence of a symmetric idempotent Latin square of side $2n-1$ is equivalent to the existence of a one-factorization of K_{2n}.

10.1.6 Find a one-factorization of K_8 that is different from $\mathcal{GK}_8$.

10.1.7 Suppose $\{F_1, F_2, \ldots, F_{2n-1}\}$ is a one-factorization of K_{2n}, $n > 1$. If F_i consists of the edges $a_{1i}b_{1i}$, $a_{2i}b_{2i}$, ..., $a_{ni}b_{ni}$. Write S_i for the set of $\frac{1}{2}n(n-1)$ quadruples $\{a_{xi}, b_{xi}, a_{yi}, b_{yi}\}$, where $1 \leq x < y \leq n$. Prove that the blocks $S_1 \cup S_2 \cup \cdots \cup S_{2n-1}$ constitute a 3-design. What are its parameters?

10.1.8 The nearest thing to a one-factor in K_{2n-1} is a set of $n-1$ edges that cover all but one vertex. Such a structure is called a *near-one-factor*. A set of near-one-factors that covers every edge precisely once is called a *near-one-factorization*.

(i) Prove that K_{2n-1} has a near-one-factorization for every n.

(ii) Prove that each vertex appears precisely once as an isolate in a near-one-factorization.

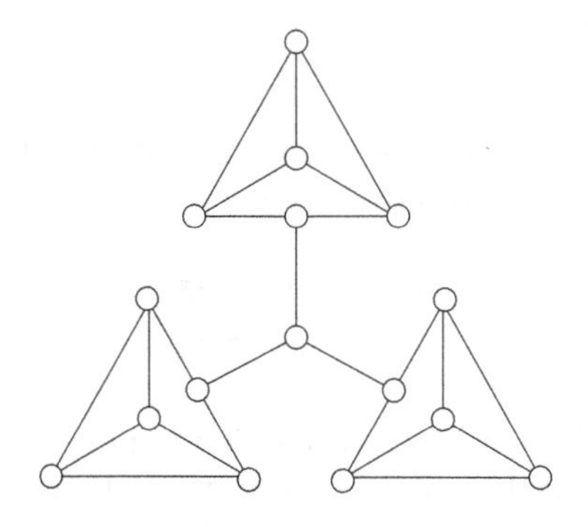

FIGURE 10.2: Graph for Exercise 10.1.2

10.1.9 Suppose $\mathcal{F}$ and $\mathcal{G}$ are near-one-factorizations of the graph K_{2n+1} with vertex set $\{0, 1, \cdots 2n\}$. Write F_i and G_i for the near-factors in $\mathcal{F}$ and $\mathcal{G}$ respectively that omit vertex i. Define $\mathcal{F}$ and $\mathcal{G}$ to be *disjoint* if F_i and G_i never have a common edge, for any i.

(i) Prove that there do not exist two disjoint near-one-factorizations of K_5.
Hint: Assume the first factor in the first factorization is $\{0\ 12\ 34\}$. *You will find that no pair is possible.*

(ii)* Prove that there can never be more than $2n - 1$ mutually disjoint near-one-factorizations of K_{2n+1}.

(iii) A set of $2n - 1$ mutually disjoint near-one-factorizations of K_{2n+1} is called *large*. Find a large set of mutually disjoint near-one-factorizations of K_7.

10.2 The Variability of One-Factorizations

It is obvious that there is only one one-factor in K_2, and it (trivially) constitutes the unique one-factorization. Similarly K_4 has just three one-factors, and together they form a factorization.

The situation concerning K_6 is more interesting. There are 15 one-factors of K_6, namely

01 23 45,	01 24 35,	01 25 34
02 13 45,	02 14 35,	02 15 34
03 12 45,	03 14 25,	03 15 24
04 12 35,	04 13 25,	04 15 23
05 12 34,	05 13 24,	05 14 23.

Consider the factor

$$F_1 = 01\,23\,45.$$

We enumerate the one-factorizations that include F_1. The only factors in the list that both contain 02 and have no edge in common with F_1 are

$$F_{21} = 02\,14\,35,$$
$$F_{22} = 02\,15\,34.$$

We now construct all one-factorizations that contain F_1 and F_{21}; those containing F_1 and F_{22} can be obtained from them by exchanging the symbols 4 and 5 throughout. But one easily sees that there is exactly one way of completing $\{F_1, F_{21}\}$ to a one-factorization, namely

03 15 24
04 13 25
05 12 34.

So there are exactly two one-factorizations containing F_1. It follows (by permuting the symbols $\{0, 1, 2, 3, 4, 5\}$) that there are exactly two one-factorizations containing any given one-factor.

Now let us count all the ordered pairs $(F, \mathcal{F})$ where F is a one-factor of K_6 and $\mathcal{F}$ is a one-factorization containing F. From what we have just said the total must be $2 \cdot 15$, that is, 30. On the other hand, suppose there are N one-factorizations. Since there are 5 factors in each, we have $5N = 30$, so $N = 6$; there are six one-factorizations of K_6.

The above calculations are simple in theory, but in practice they soon become unwieldy. There are 6240 one-factorizations of K_8 (see Exercise 10.2.1), 1,255,566,720 of K_{10} ([55]) and 252,282,619,805,368,320 of K_{12} ([48]). In any case, the importance of such numbers is dubious. It is more interesting to know about the existence of "essentially different" one-factorizations. As usual, the first step is to make a sensible definition of isomorphism of factorizations.

Suppose G is a graph with a one-factorization. Two one-factorizations $\mathcal{F}$ and $\mathcal{H}$ of G, say

$$\begin{aligned} \mathcal{F} &= \{F_1, F_2, \cdots, F_k\}, \\ \mathcal{H} &= \{H_1, H_2, \cdots, H_k\}, \end{aligned}$$

are called *isomorphic* if there exists a map ϕ from the vertex set of G onto itself such that

$$\{F_1\phi, F_2\phi, \cdots, F_k\phi\} = \{H_1, H_2, \cdots, H_k\},$$

where $F_i\phi$ is the set of all the edges $\{x\phi, y\phi\}$ such that $\{x, y\}$ is an edge in $\mathcal{F}$. In other words, when ϕ operates on one of the factors in $\mathcal{F}$, it produces one of the factors in $\mathcal{H}$. The map ϕ is called an *isomorphism.* This definition is consistent with the definition of isomorphism of designs.

In the case where $\mathcal{H}$ equals $\mathcal{F}$ itself, ϕ is called an *automorphism*; the automorphisms of a one-factorization $\mathcal{F}$ form a group, the *automorphism group* of $\mathcal{F}$, written $Aut(\mathcal{F})$.

Now that isomorphism is defined, we can more formally state our question about "essentially different" factorizations: What is the number of isomorphism classes (equivalence classes under isomorphism) of one-factorizations of K_{2n}? This is a very difficult question, and exact results are known only for $2n \leq 12$. There is a unique one-factorization of K_{2n}, up to isomorphism, for $2n = 2$, 4 or 6. There are exactly six for K_8; these were found by Safford (see [46]) and a full exposition is given in [173]. The six factorizations are shown in Figure 10.3. For K_{10} the number of one-factorizations up to isomorphism

$\mathcal{F}_1$

$A = 01\ 23\ 45\ 67$
$B = 02\ 13\ 46\ 57$
$C = 03\ 12\ 47\ 56$
$D = 04\ 15\ 26\ 37$
$E = 05\ 14\ 27\ 36$
$F = 06\ 17\ 24\ 35$
$G = 07\ 16\ 25\ 34$

$\mathcal{F}_2$

$A = 01\ 23\ 45\ 67$
$B = 02\ 13\ 46\ 57$
$C = 03\ 12\ 47\ 56$
$D = 04\ 15\ 26\ 37$
$E = 05\ 14\ 27\ 36$
$F = 06\ 17\ 25\ 34$
$G = 07\ 16\ 24\ 35$

$\mathcal{F}_3$

$A = 01\ 23\ 45\ 67$
$B = 02\ 13\ 46\ 57$
$C = 03\ 12\ 47\ 56$
$D = 04\ 16\ 25\ 37$
$E = 05\ 17\ 26\ 34$
$F = 06\ 14\ 27\ 35$
$G = 07\ 15\ 24\ 36$

$\mathcal{F}_4$

$A = 01\ 23\ 45\ 67$
$B = 02\ 13\ 46\ 57$
$C = 03\ 12\ 47\ 56$
$D = 04\ 16\ 27\ 35$
$E = 05\ 17\ 26\ 34$
$F = 06\ 14\ 25\ 37$
$G = 07\ 15\ 24\ 36$

$\mathcal{F}_5$

$A = 01\ 23\ 45\ 67$
$B = 02\ 13\ 46\ 57$
$C = 03\ 14\ 27\ 56$
$D = 04\ 16\ 25\ 37$
$E = 05\ 17\ 26\ 34$
$F = 06\ 12\ 35\ 47$
$G = 07\ 15\ 24\ 36$

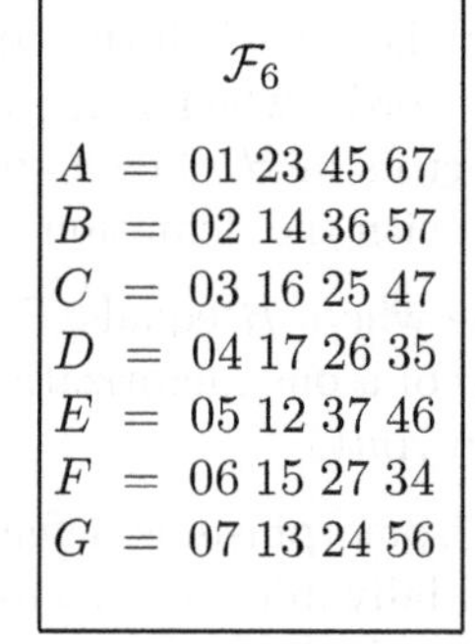

$\mathcal{F}_6$

$A = 01\ 23\ 45\ 67$
$B = 02\ 14\ 36\ 57$
$C = 03\ 16\ 25\ 47$
$D = 04\ 17\ 26\ 35$
$E = 05\ 12\ 37\ 46$
$F = 06\ 15\ 27\ 34$
$G = 07\ 13\ 24\ 56$

FIGURE 10.3: One-factorizations of K_8.

is 396; they were listed in [55] (see also [56]) and are reprinted in [170]. There are 526,915,620 isomorphism classes of one-factorizations of K_{12}; see [48].

One of the ideas that has been used in discriminating between nonisomorphic one-factorizations is the division structure. A *d-division* in a one-factorization is a set of d factors whose union is disconnected. For the one-factorizations of K_8 shown in Figure 10.3, the 3-divisions and maximal 2-divisions (those that cannot be extended to 3-divisions) in the six factorizations of K_8 are listed in Figure 10.4. $\mathcal{F}_1$ has 3-divisions, while $\mathcal{F}_5$ does not; this shows that $\mathcal{F}_1$ and $\mathcal{F}_5$ cannot be isomorphic. Similarly, $\mathcal{F}_2$ has three 3-divisions, while $\mathcal{F}_4$ has only one, so they are not isomorphic. The six factorizations have different division structures, so one can test for isomorphism by checking divisions, a much shorter procedure than checking all possible isomorphisms. For larger complete graphs, nonisomorphic factorizations often have the same division structure, but it is still a useful tool in testing for isomorphism.

	3-divisions	maximal 2-divisions
$\mathcal{F}_1$	ABC ADE AFG BDF BEG CDG CEF	
$\mathcal{F}_2$	ABC ADE AFG	BD BE CD CE
$\mathcal{F}_3$	ABC	CD CF DF EG
$\mathcal{F}_4$	ABC	DE DF DG EF EG FG
$\mathcal{F}_5$	$--$	AB CF EG
$\mathcal{F}_6$	$--$	$--$

FIGURE 10.4: Divisions in one-factorizations of K_8.

Another tool is the automorphism group; one-factorizations with nonisomorphic automorphism groups cannot themselves be isomorphic. Automorphisms and automorphism groups of one-factorizations have also been used to prove that the number of isomorphism classes of one-factorizations of K_{2n} tends to infinity as n does. This result was proven independently in [2] and [101]; the proofs are quite similar, but [101] also includes another proof using quasigroups.

Suppose $\mathcal{F} = \{F_1, F_2, \ldots, F_{2n-1}\}$ is a one-factorization of the K_{2n} with vertex set S. Suppose T is a subset of S; write H for the complete graph with vertex set T. It may happen that for every i, $H \cap F_i$ consists either only of edges or only of vertices; in other words each $H \cap F_i$ either is a one-factor of H or contains no edges. That is, the graphs $H \cap F_i$ (with isolated vertices deleted) form a one-factorization of H. In that case, we call it a *subfactorization* of $\mathcal{F}$. For convenience we usually reorder the F_i so that the factors contributing to the subfactorization occur first in the list.

Say $\mathcal{F}$ is a one-factorization of K_{2n} that contains a one-factorization $\mathcal{E}$ of K_{2s} as a subfactorization. (We call it a subfactorization *of order* $2s$.) Without loss of generality, we may assume the copies of K_{2s} and K_{2n} to be based on $\{0, 1, \cdots, 2s-1\}$ and $\{0, 1, \cdots, 2n-1\}$, and assume F_i contains E_i (plus isolated vertices) for $1 \leq i \leq 2s-1$. If $k \geq 2s$, then each of $0, 1, \cdots, 2s-1$ must appear in a different edge of F_k, so F_k contains at least $2s$ edges. Therefore $2n \geq 4s$.

Cruse [43] proved that the converse also holds. So:

THEOREM 10.5

There exists a subfactorization of K_{2s} in K_{2n} if and only if $2n \geq 4s$. □

In [171], subfactorizations are used to prove that the number of one-factorizations goes to infinity. In fact:

THEOREM 10.6 [171]

There are at least

$$\frac{6^{(n-t)n} \cdot 420n \cdot 2^5}{(5n+2t+1)!}$$

nonisomorphic one-factorizations of $K_{5n+2t+1}$, when $n > 10$, n is odd and $0 \leq t \leq n$.

Exercises 10.2

10.2.1 Suppose $\mathcal{F}$ is a one-factorization of K_{2n}.

(i) Show that if the automorphism group of $\mathcal{F}$ has order g, then there are $\frac{(2n)!}{g}$ one-factorizations isomorphic to $\mathcal{F}$.

(ii) Find the automorphism groups of the six one-factorizations $\mathcal{F}_1$, $\mathcal{F}_2$, $\mathcal{F}_3$, $\mathcal{F}_4$, $\mathcal{F}_5$, $\mathcal{F}_6$ shown in Figure 10.3; verify that their orders are 1344, 64, 16, 96, 24 and 42, respectively.

(iii) Hence prove that there are 6240 one-factorizations of K_8.

10.2.2* Prove that the number of one-factorizations of K_{2n} is divisible by

$$\frac{(2n-3)!}{(n-2)!2^{n-2}}.$$

10.2.3 Prove that a one-factorization containing a d-division must have at least $2(d+1)$ vertices when d is odd, and at least $2(d+2)$ vertices when d is even.

10.2.4 A d-division is *proper* if its factors are not part of any $(d+1)$-division. Write $D_d(\mathcal{F})$ for the number of proper d-divisions of a one-factorization $\mathcal{F}$. What are $D_3(\mathcal{F}_i)$ and $D_2(\mathcal{F}_i)$ for the six one-factorizations in Figure 10.3?

10.2.5 Find a one-factorization of K_8 containing a subfactorization of order 4.

10.2.6 Find a one-factorization of K_{10} containing a subfactorization of order 4.

10.2.7 We say the union of three one-factors contains the *triangle abc* if each of the factors contains one of the edges ac, bc and ca. Consider triangles in the factorization $\mathcal{F}_6$ of K_8.

(i) Show that the union of any three factors in $\mathcal{F}_6$ contains either exactly one or exactly two triangles.

(ii) If the union contains exactly one triangle, call the three points of the triangle a *triad*. Show that $\mathcal{F}_6$ contains precisely 14 triads, namely

$$025,\ 027,\ 034,\ 036,\ 047,\ 056,\ 236,$$
$$237,\ 245,\ 246,\ 345,\ 357,\ 467,\ 567.$$

(iii) Observing that the structure of the set of triads is preserved by automorphism, construct the automorphism group of $\mathcal{F}_6$.

10.2.8* Write V for the set of all edges of K_6. $\mathcal{S}$ is the set of triples of members of V that form a triangle (that is, triples ab, bc, ca where a, b and c are vertices of K_6), and $\mathcal{T}$ is the set of all one-factors of K_6 (considered as triples of edges). A design has treatment-set V and block-set $\mathcal{S} \cup \mathcal{T}$. Show that it is a BIBD, and find its parameters.

10.3 Starters

The factorization of Theorem 10.2 is easy to visualize in graphical terms. Suppose the vertices of K_{2n} are arranged in a circle with one vertex in the center. This arrangement is shown in Figure 10.5, with the vertices labeled $\{\infty, 0, 1, \cdots, 2n-2\}$. The one-factor F_0 is shown in that figure. F_1 is constructed by rotating F_0 through one-$(2n-2)$th part of a revolution, clockwise, and subsequent factors are obtained in the same way. It is easy to see that the $n-1$ chords separate pairs of vertices that differ (modulo $2n-1$) by different numbers, and that on rotation the one edge that represents distance d will generate all edges with distance d.

To generalize this construction we make the following definition. A *starter* in an Abelian group G of order $2n-1$ is an ordered partition of the nonzero members of G into 2-sets $\{x_1, y_1\}, \{x_2, y_2\}, \cdots, \{x_{n-1}, y_{n-1}\}$ with the property that the $2n-2$ differences $\pm(x_1-y_1), \pm(x_2-y_2), \cdots, \pm(x_{n-1}-y_{n-1})$ are all different and therefore contain every nonzero element of G precisely once. A starter is a special type of system of supplementary difference sets.

It is clear that the set F_0 used in the proof of Theorem 10.2 is a starter. On the other hand, suppose any other starter F had been used instead of F_0. We construct a set of $2n-1$ factors by systematically adding elements of G to F (F itself is obtained by adding the zero element). This process is *developing* F in G as defined in Section 5.1.

For convenience, we order the resulting factors so that the pair $\{\infty, i\}$ occurs in the ith factor. Now consider any unordered pair $\{x, y\}$ where neither x nor y is the symbol ∞. Suppose $x-y=d$. There will be exactly one pair $\{x_j, y_j\}$ in the starter for which $x_j - y_j = \pm d$; for convenience, assume $x_j - y_j = d$. Then

$$x = x_j + (x - x_j),\ y = y_j + (x - x_j),$$

so (x, y) will occur in the factor formed by adding $x - x_j$ to the starter. So every possible pair occurs at least once in the list of factors. On the other hand, the whole list of factors will contain only $n(2n-1)$ pairs, n in each factor, and if any one of the $n(2n-1)$ possible pairs occurred more than once, there would be too many pairs. So each pair occurs precisely once, and we have constructed a one-factorization. We have proven:

THEOREM 10.7

If there is a starter S in a group G of order $2n-1$, then the $2n-1$ factors $\{(\infty, x)\} \cup (S+x)$, where x ranges through G, form a one-factorization of K_{2n}.

(Following Section 5.1, we call this the one-factorization *generated by S*.)

The type of starter we used in Theorem 10.2, consisting of all the pairs $\{i, -i\}$ where i ranges through the nonzero elements of some group, is called a *patterned starter*; $\mathcal{GK}_{2n}$ is the factorization generated by the patterned starter in the cyclic group Z_{2n-1}. The one-factorization of K_{2n} constructed from the patterned starter in G is called the *patterned factorization* of G; when G is not named, it is assumed to be Z_{2n-1}, so "the patterned factorization" often means "$\mathcal{GK}_{2n}$."

Different starters may give rise to nonisomorphic one-factorizations, or they

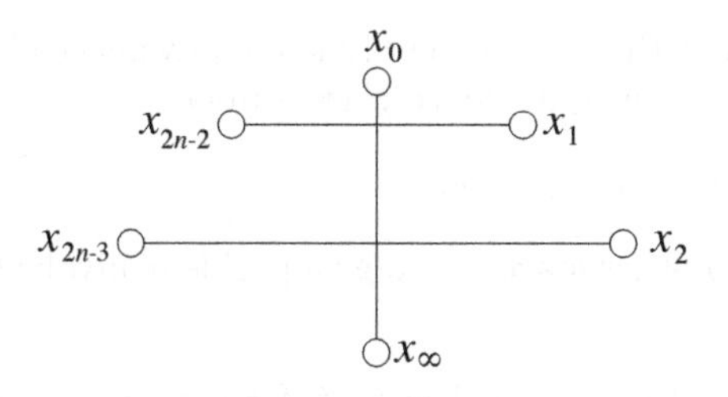

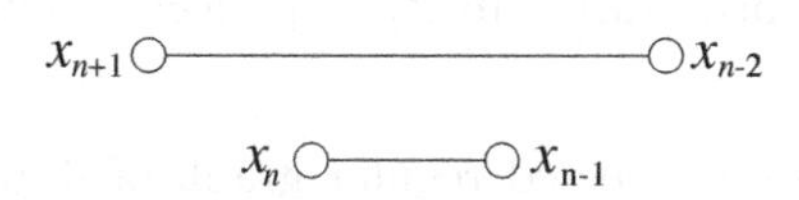

FIGURE 10.5: The patterned starter

may be isomorphic. Starters with additional properties are used in combinatorial design constructions, and for this reason several families of starters have been constructed. One important construction is due to Mullin and Nemeth [109]; the Mullin–Nemeth starters were generalized by Dinitz [47]. Another significant result is the recursive construction due to Horton [73].

If S is a starter in an Abelian group G, then the set of pairs $-S$, constructed by replacing each pair $\{x, y\}$ in S by $\{-x, -y\}$, is also a starter. More generally, if the group G has a multiplication operation defined on it—for example, if G is the set of integers modulo $2n - 1$ or if G is the additive group of any integral domain or field—one may define kS by

$$kS = \{(kx, ky) : \{x, y\} \in S\}.$$

The set kS is not always a starter, but in some cases it will be; in those cases, k is called a *multiplier* of S. For example, if S is a starter in the integers modulo $2n - 1$, and if $\gcd(k, 2n - 1) = 1$, then kS is a starter. In this case the two starters S and kS are called *equivalent.*

Exercises 10.3

10.3.1* Prove the existence of exactly one starter in the cyclic group of order 3, exactly one starter in the cyclic group of order 5, and exactly three starters in the cyclic group of order 7.

10.3.2 Suppose S is a starter in Z_{2n-1}.

(i) Prove that kS is a starter if and only if $\gcd(k, 2n - 1) = 1$.

(ii) Starters S and T are called *equivalent* if $T = kS$ for some S. Prove that the patterned starter is not equivalent to any other starter.

(iii) Prove that there are precisely two equivalence classes of starters in Z_7, and three equivalence classes in G_9.

10.3.3* Find all starters in $Z_3 \times Z_3$.

10.3.4 Prove that no other starter in a group G is equivalent to the patterned starter in G.

10.3.5 A one-factorization is called *perfect* if no two of its factors constitute a 2-division (see Section 10.1). Prove that the one-factorization of K_{p+1} from the patterned starter in Z_p is perfect if and only if p is an odd prime.

10.3.6 Prove that every connected regular graph of degree 3 on 8 vertices is a union of three factors in the factorization $\mathcal{F}_5$ of K_8, as presented in Figure 10.10.3. Hence, prove that every regular graph of degree 4 on 8 vertices has a one-factorization.

Chapter 11

Applications of One-Factorizations

11.1 An Application to Finite Projective Planes

As a somewhat unexpected application, we shall use the one-factorizations of K_6 to prove the uniqueness of the projective plane $PG(2,4)$.

We first return to the study of ovals in projective planes (see Section 4.1). Recall that any 2-arc in a $PG(2,n)$ has at most $n+2$ points; those with $n+1$ and $n+2$ points are called *type I* and *type II* ovals respectively. We shall define an arc to be *maximal* if it is not a proper subset of any (larger) arc.

LEMMA 11.1

If a maximal 2*-arc in a* $PG(2,n)$ *has* m *points, then*

$$n \leq \binom{m-1}{2}. \tag{11.1}$$

Proof. Suppose c is an m-point 2-arc in a $PG(2,n)$. If $m = n+2$, then (11.1) is true. So assume $m < n+2$. Then any point p of c lies on at least one tangent to c. Suppose l is a tangent at p. There are $m-1$ points other than p on c, so there are exactly $\binom{m-1}{2}$ secants to c that do not contain p. If $n > \binom{m-1}{2}$, there must be at least one point of l, other than p, that does not lie on any of these secants; call it q. Clearly, q cannot lie on any secant through p, either. So no line through q contains more than one point of c, and $c \cup \{q\}$ is also a 2-arc. So c was not maximal. It follows that for any maximal 2-arc the assumption that $n > \binom{m-1}{2}$ must be false. $\square$

We now apply Lemma 11.1 in the case $n = 4$. Since

$$4 \leq \binom{m-1}{2} \Rightarrow m \geq 5,$$

a maximal arc in a $PG(2,4)$ must contain at least 5 points. But if one contains 5 points, it is a type I oval, and can be extended to a type II oval on 6 points, since $n = 4$, which is even (see Corollary 4.1.2). So any maximal arc in $PG(2,4)$ must contain 6 points (6 is the maximum, since $6 = n+2$). Therefore any $PG(2,4)$ contains a 6-point 2-arc.

Suppose c is any 6-point 2-arc in a $PG(2,4)$. Select any point p not on c. Every line through p meets c in 0 or 2 points. So the set of lines through p determines a partition of c into three pairs of points. In other words, it determines a one-factor of the K_6 based on the points of c. Call this one-factor $F(p)$.

If p and q are distinct points not on c, then $F(p)$ and $F(q)$ have an edge in common if and only if the line pq is a secant to c; and $F(p)$ and $F(q)$ cannot have more than one common edge. So the 15 points not lying on c correspond to different one-factors of K_6; as we found in Section 10.2 that there are exactly 15 one-factors, each occurs precisely once. Moreover, if l is an external line to c, the five one-factors corresponding to the points on l are disjoint, so they form a one-factorization of K_6. One-factorizations corresponding to the distinct external lines have exactly one common factor, corresponding to the point of intersection of the two lines. Since a 6-point arc has no tangents and 15 secants, it has six external lines, corresponding precisely to the six one-factorizations of K_6.

We now observe that these structural facts are sufficient to determine $PG(2,4)$ completely.

THEOREM 11.2

All planes $PG(2,4)$ are isomorphic.

Proof. Any plane $PG(2,4)$ must contain a maximal 2-arc c of 6 points. Label its points 0, 1, 2, 3, 4, 5. There will be 15 secants to c; write l_{ij} for the secant containing i and j. Given two secants l_{ij} and l_{pq}, where $\{i,j,p,q\}$ are all different, label the points of intersection $l_{ij} \cap l_{pq}$ with the one-factor of K_6 containing the edges ij and pq. The third secant through the point will necessarily be labeled with the third edge of the factor. An elementary counting argument shows that all 15 exterior points will be labeled in this way, and the membership of all the secants is determined.

Finally, there will be six external lines to c. These must correspond to the six one-factorizations; the members of a line are labeled with the factors in one of the factorizations.

The structure is completely determined by these considerations, up to permutation of the point and line labels. So the plane is determined up to isomorphism. □

The theorem actually tells us more about $PG(2,4)$: Given any two ovals in the plane, there must be an automorphism that carries one into the other. These considerations apply only to $PG(2,4)$; when eight points are allowed, the variability and number of one-factorizations are already too great for useful application to the isomorphism problem.

11.2 Tournament Applications of One-Factorizations

Suppose several baseball teams play against each other in a league. The competition can be represented by a graph with the teams as vertices and with an edge xy representing a game between teams x and y. We shall refer to any such league where precisely two participants meet in each game as a *tournament*. (The word "tournament" is also used for the directed graphs derived from this model by directing the edge from winner to loser, but we shall not consider the results of games, just the scheduling.)

Sometimes multiple edges will be necessary; sometimes two teams do not meet. The particular case where every pair of teams plays exactly once is called a *round robin tournament*, and the underlying graph is complete.

A very common situation is when several matches must be played simultaneously. In the extreme case, when every team must compete at once, the set of games held at the one time is called a *round*. Clearly the games that form a round form a one-factor in the underlying graph. If a round robin tournament for $2n$ teams is to be played in the minimum number of sessions, we require a one-factorization of K_{2n}. If there are $2n-1$ teams, the relevant structure is a near-one-factorization of K_{2n-1}. In each case the factorization is called the *schedule* of the tournament.

In some cases the order of the rounds is important. If an ordering is applied to the set of one-factors in a one-factorization, the one-factorization will be called *ordered*. The tournament schedule is then the ordered one-factorization.

In many sports a team owns, or regularly plays in, one specific stadium or arena. We shall refer to this as the team's "home field." When the game is played at a team's home field, we refer to that team as the "home team" and the other as the "away team." Often the home team is at an advantage; and more importantly, the home team may receive a greater share of the admission charges. So it is usual for home and away teams to be designated in each match. We use the term *home-and-away schedule* (or just *schedule*) to refer to a round robin tournament schedule in which one team in each game is labeled the home team and one the away team. Since this could be represented by applying a direction to the edges in each factors, this is called *orienting* the edges in the factors, and the one-factorization is called *oriented.* A home-and-away schedule is equivalent to an oriented one-factorization. It is very common to conduct a double round robin, in which every team plays every other team twice. If $2K_v$ represents the multigraph on v vertices with precisely two edges joining every pair of distinct vertices, a double round robin corresponds to an oriented one-factorization of $2K_{2n}$. If the two matches for each pair of teams are arranged so that the home team in one is the away team in the other, we shall say the schedule and the corresponding oriented one-factorization of $2K_{2n}$ are *balanced.*

For various reasons one often prefers a schedule in which runs of successive away games and runs of successive home games do not occur (although there are exceptions; an East Coast baseball team, for example, might want to make a tour of the West, and play several away games in succession). We shall define a *break* in a schedule to be a pair of successive rounds in which a given team is at home, or away, in both rounds. A schedule is *ideal* for a team if it contains no break for that team. Oriented factorizations are called *ideal* for a vertex if and only if the corresponding schedules are ideal for the corresponding team.

THEOREM 11.3 [176]

Any schedule for $2n$ *teams is ideal for at most two teams.*

Proof. For a given team x, define a vector v_x to have j-th component $v_{xj} = 1$ if x is home in round j and $v_{xj} = 0$ if x is away in round j. If the schedule is ideal for team x, then v_x consists of alternating zeroes and ones, so there are only two possible vectors v_x for such teams. But $v_x \neq v_y$ when $x \neq y$; the vectors must differ in position j, where x plays y in round j. So the schedule can be ideal for at most two teams. □

The following theorem shows that the theoretical best-possible case can be attained.

THEOREM 11.4 [176]

There is an oriented one-factorization of K_{2n} *with exactly* $2n - 2$ *breaks.*

Proof. We orient the one-factorization $\mathcal{P} = \{P_1, P_2, \cdots, P_{2n-1}\}$ based on $\{\infty\} \cup Z_{2n-1}$, defined by

$$P_k = \{\{\infty, k\}\} \cup \{\{k+i, k-i\} : 1 \leq i \leq n-1\}. \tag{11.2}$$

Edge $\{\infty, k\}$ is oriented with ∞ at home when k is even and k at home when

k is odd. Edge $\{k+i, k-i\}$ is oriented with $k-i$ at home when i is even and $k+i$ at home when i is odd.

It is clear that ∞ has no breaks. For team x, where x is in Z_{2n-1}, we can write $x = k + (x - k) = k - (k - x)$. The way in which x occurs in the representation (11.2) will be as x when $k = x$, as $k + (x - k)$ when $1 \leq x - k \leq n - 1$, and as $k - (k - x)$ otherwise. The rounds other than P_x where x is at home are the rounds k where $x-k$ is odd and $1 \leq x-k \leq n-1$, and the rounds k where $k - x$ is even and $1 \leq k - x \leq n - 1$. It is easy to check that factors P_{2j-1} and P_{2j} form a break for symbols $2j - 1$ and $2j$, and that these are the only breaks. □

Suppose two teams share the same home ground. (In practice this occurs quite frequently. The cost of maintaining a football stadium, for example, is high, and two teams will often share the expenses.) Then those two teams cannot both be "home" in the same round (except for the round in which they play each other). To handle the most demanding case, one needs a home-and-away schedule with the following property: There is a way of pairing the teams such that only one member of each pair is at home in each round. We shall refer to this as property (P). Schedules with this property are discussed in [13], where an example for four teams is given in order to construct a more complicated type of tournament.

THEOREM 11.5 [166]

Given any one-factorization of K_{2n}, there exists a way of pairing the vertices and a way of orienting the edges such that the resulting home-and-away schedule has property (P) *with respect to the given pairing.*

Proof. Select one factor at random; call it F_1. The pairs that play against each other in this round will constitute the pairing referred to in property (P). No matter how home teams are allocated in the round corresponding to F_1, only one member of each pair will be at home.

Now select any other round. Let the one-factor associated with this round be F_2. Consider the union of F_1 and F_2. In each of the cycles comprising it, select one of the points at random; the teams corresponding to it and to every second point as you go around the cycle from it are the home teams in the round under discussion.

Exactly one home team has been selected in every edge of F_2. So there is exactly one home team in every match of the new round. So (P) is satisfied. □

Suppose a double round robin tournament is to be played with the sort of home ground sharing that we have been using. This is easily managed when the tournament consists of two copies of an ordinary round robin. First $2n-1$ rounds are played with home teams allocated as in the proof of Theorem 11.5, and then the competition is repeated with home-and-away teams exchanged. However, not all double round robins are of this type. The underlying factorization is a one-factorization of $2K_{2n}$, the multigraph on $2n$ vertices with exactly two edges joining each pair of distinct points. There are three types of factorization possible, namely two copies of a one-factorization of K_{2n}, copies of two different one-factorizations of K_{2n}, and factorizations that cannot be decomposed into two one-factorizations of K_{2n}. Factorizations of this third type are called *indecomposable.*

Each of the three possibilities can occur, even in as small an example as $2K_6$; in fact, up to isomorphism, there is exactly one factorization of each kind in that case. They are shown in Figure 11.1, in the order given above; factors are written as columns. Each factorization has been written so that, if the left-hand team in each match is the home team, then the pairing 1-2, 3-4, 5-6 satisfies (P), and each pair xy occurs once in a game with x at home and once with y. Let us call this a *proper orientation* of the factorization with regard to the pairing; Figure 11.1 shows that every one-factorization of $2K_6$ has a proper orientation. In general there is no known example of a one-factorization of $2K_{2n}$ with no proper orientation, although it has not been shown that none exists, even in the case of $2K_8$.

A number of other problems have been studied concerning scheduling and concerning the sharing of facilities; one particular problem concerns a competition with a junior and a senior league, where many clubs field teams in both leagues; naturally, teams from the same club share a stadium. The interested reader should consult, for example, [13, 145, 177, 178].

Problems can arise in the allocation of games even without home-and-away considerations. Suppose one wishes to run a round robin tournament with $2n$

```
12 13 14 15 16 21 31 41 51 61
34 52 62 42 32 43 25 26 24 23
56 46 35 63 54 65 64 53 36 45

12 13 14 51 61 21 31 41 15 16
34 52 62 24 23 43 26 25 32 42
56 46 35 36 45 65 54 63 64 53

12 21 31 13 14 41 51 15 16 61
34 35 24 62 52 26 23 42 32 25
56 64 65 45 36 53 46 63 54 43
```

FIGURE 11.1: The one-factorizations of $2K_6$, properly oriented.

teams. One technique is to play round 1 by pairing teams at random. The matches in round 2 are also arbitrary, but one demands "compatibility": No two teams play if they already met in round 1. If this process is continued, will it always be successful or are there cases where this premature assignment of matches, without checking to see whether the tournament can be completed, will result in an impossible situation? In fact, such unsatisfactory results can occur. The smallest case occurs when $2n = 6$; the rounds

$$14\ 25\ 36 \quad 15\ 26\ 34 \quad 16\ 24\ 35 \tag{11.3}$$

cannot be completed.

Exercises 11.2

11.2.1 Using the proof of Theorem 11.3, prove that there is a schedule for $2n-1$ teams that is ideal for every team, in the sense that no team plays two home or two away games in consecutive rounds.

11.2.2 Prove the existence of a double round robin in which home matches are allocated in such a way that two teams have no breaks and every other team has exactly one break.

11.2.3 A softball league plays in two conferences, each of four teams. Games are scheduled on Monday, Tuesday, Thursday and Friday nights, two games per night, plus an opening week of one game per night. No team may play twice per night, nor on consecutive nights. Teams meet other teams in the same conference three times and teams in the other conference twice. Construct a suitable schedule. [145]

11.2.4 Is there a one-factorization of $2K_{2n}$, for any n, which has no proper orientation for any pairing?

11.2.5* Verify that the partial tournament shown in (11.3) cannot be completed.

11.2.6 If n is odd, $n > 1$, prove that there exists a partial tournament for $2n$ players, of n rounds, that cannot be completed.

11.3 Tournaments Balanced for Carryover

In some sports there is a carryover effect from round to round. For example, if team x plays against a very strong team, then x may perform poorly in its next round, either because its members are demoralized or because it was weakened in the preceding round. To overcome this bias, Russell [121] proposed the idea of a tournament that is balanced for carryover.

Consider a tournament whose schedule is the ordered one-factorization $\mathcal{F}$,

$$\mathcal{F} = \{F_0, F_1, \cdots, F_{2n-2}\},$$

of the K_{2n} with vertex set S. It will be convenient to write x_i for the team that opposes x in round i, so that

$$F_i = \{(x, x_i) : x \in S\}.$$

(Of course, each edge is written twice in this representation.) We define the *predecessor* of y at round j to be that x such that

$$x_j = y_{j+1},$$

and denote it by $p_j(y)$; moreover $x = p_{2n-2}(y)$ means $x_{2n-2} = y_1$. In other words, "x precedes y at round j means there is a z such that xz is a match in round j and yz is a match in round $j+1$. The set of predecessors of y is $P(y)$,

$$P(y) = \{p_j(y) : 0 \le j \le 2n-2\}.$$

The ordered one-factorization $\mathcal{F}$ is *balanced for carryover* if $P(y) = S \backslash \{y\}$ for all $y \in S$.

THEOREM 11.6 [121]

There is a tournament schedule that is balanced for carryover whenever the number of teams is a power of 2.

Proof (after [78]). Say the number of teams is $2n = 2^m$, for some positive integer m. Select a primitive element ξ in the field $GF(2n)$. We take $S = GF(2n)$, so

$$S = \{0, 1, \xi, \cdots, \xi^{2n-2}\}.$$

We define the factor F_0 to consist of all the pairs $(z, z+1)$ where $z \in S$, and form F_i by multiplying every member of F_0 by ξ^i:

$$F_i = \{(\xi^i z, \xi^i(z+1)) : z \in S\}.$$

If $x = \xi^i z$ then

$$x_i = \xi^i(z+1) = x + \xi^i.$$

Say $p_j(y) = x$. Then

$$\begin{aligned} x_j &= y_{j+1}, \\ x + \xi^j &= y + \xi^{j+1}, \\ x &= y + \xi^j(1+\xi). \end{aligned}$$

So

$$P(y) = \{y + \xi^j(1+\xi) : 0 \le j \le 2n-2\}.$$

As j varies from 0 to $2n-2$, ξ^j ranges through the nonzerozero elements of S, and so does $\xi^j(1+\xi)$. (Certainly $1+\xi$ is nonzero, as otherwise $\xi = 1$, and 1 is not primitive). So

$$\begin{aligned} \{\xi^j(1+\xi)\} &= S\backslash\{0\}, \\ P(y) &= S\backslash\{y\}, \end{aligned}$$

and $\mathcal{F}$ is balanced for carryover. □

As an example, the schedule for eight teams is shown in Figure 11.2. (The power ξ^i is represented by the integer $i+1$.) A complete search shows that there is no example of order 6 or 10, and the only examples of order 8 are the one shown and the one derived from it by reversing the order of the factors.

The construction just given is a special case of the following idea. Consider orderings of the field $GF(q)$. An ordering $(0 = a_0, a_1, \cdots, a_{q-1})$ is called *balanced* if the partial sums $S_1, S_2, \cdots, S_{q-1}$ are all different, where

$$S_t = \sum_{i=1}^{t} a_i.$$

Given a balanced ordering of $GF(q)$, write a_q for the (unique) element of the field that does not arise as one of $\{S_1, S_2, \cdots, S_{q-1}\}$. Define

$$b_{i,r} = a_i + a_q + S_{r-1}. \tag{11.4}$$

THEOREM 11.7 [121]

If $(a_0, a_1, \cdots, a_{q-1})$ is a balanced ordering of $GF(q)$, where $q = 2^m$, then

$$F_r = \{\{a_i, b_{i,r}\}\}$$

01 26 34 57
02 37 45 61
03 41 56 72
04 52 67 13
05 63 71 24
06 74 12 35
07 15 23 46

FIGURE 11.2: A schedule balanced for carryover.

(where $b_{i,r}$ is defined as in (11.4)*) is a one-factor of K_q, and*

$$\{F_1, F_2, \cdots, F_{q-1}\}$$

is a one-factorization and is balanced for carryover.

Verifying this theorem is left to the reader. Notice that if ξ is a primitive element in $GF(q)$ then $(0, 1, \xi, \xi^2, \cdots)$ is a balanced ordering — if $S_i = S_j$ when $1 \leq i < j \leq q-1$ then

$$\xi^i + \xi^{i+1} + \cdots + \xi^{j-1} = 0,$$

so

$$1 + \xi + \cdots + \xi^{j-i-1} = 0;$$

multiplying both sides by $1 - \xi$ yields $\xi^{j-i} = 1$, which is impossible. So Theorem 11.6 is a special case of Theorem 11.7.

Russell [121] conjectured that schedules balanced for carryover exist only when the order is a power of 2. However, Anderson [4] exhibits examples for 20 and 22 teams. They come from starters:

$$\{1,5\}\ \{2,16\}\ \{3,11\}\ \{4,14\}\ \{6,7\}\ \{8,13\}\ \{9,15\}\ \{10,12\}$$

in Z_{19} and

$$\{1,4\}\ \{2,9\}\ \{3,18\}\ \{5,13\}\ \{6,11\},\ \{7,17\}\ \{8,14\}\ \{10,12\}\ \{15,16\}$$

in Z_{21}.[1]

A further restriction on schedules was proposed in [14]. We say that x is the *k-th predecessor* of y at round j, and write $x = p_{j,k}(y)$, if $x_j = y_{k+j}$ (addition of subscripts being carried out modulo $2n-1$). Write

$$P_k(y) = \{p_{j,k}(y) : 0 \leq j \leq 2n-2\}.$$

Then the schedule is *balanced at level* k if $P_k(y) = S \backslash \{y\}$, and it is *totally balanced* if it is balanced at every level.

It is easy to check that the construction of Theorem 11.6 is totally balanced. However, not all balanced orderings of $GF(2^m)$ yield totally balanced factorizations. Bonn [21] has generated all balanced orderings of $GF(16)$. He found many that are balanced at levels 1 and 14 only, one (and its reverse) that is totally balanced, and two different tournaments that are balanced at levels 1, 3, 6, 9, 12 and 14 but no others. The reason for the latter pattern is a mystery.

[1] These examples were constructed by Norman Finizio, by computer.

It is not difficult to prove

THEOREM 11.8 [14]

If there exists a totally balanced tournament of order $2n$, then there exists a set of $n-1$ mutually orthogonal Latin squares of order $2n$.

This theorem gives some weight to a weaker version of Russell's conjecture, that a totally balanced schedule of order $2n$ can exist only when $2n$ is a power of 2.

Exercises 11.3

11.3.1* Suppose F_1, F_2 and F_3 are three edge-disjoint one-factors of $2K_6$. Show that there is a pair $\{x, y\}$ such that x precedes y at round 1 and also at round 2. (Consequently there is no schedule for six teams that is balanced for carryover.)

11.3.2 Prove Theorem 11.7.

11.3.3 Prove that if a one-factorization of K_{2n} is balanced at level i, then it is balanced at level $2n-1-i$.

11.3.4 (i) Suppose the schedule $\mathcal{F}$ is balanced for carryover. Define an array $A^0 = (a_{i,j})$ of order $2n$ by

$$\begin{aligned} a_{i,j} &= j_i,\ 0 \le i \le 2n-2, \\ a_{2n-1,j} &= j. \end{aligned}$$

Prove that A^0 is a Latin square.

(ii) If $1 \le k \le 2n-2$, define $A^k = (a^k_{i,j})$ by

$$\begin{aligned} a^k_{i,j} &= (j_i)_{i-k},\ 0 \le i \le 2n-2, \\ a^k_{2n-1,j} &= j. \end{aligned}$$

Prove that if $\mathcal{F}$ is totally balanced then $A^1, A^2, \cdots, A^{2n-2}$ are Latin squares.

(iii) Prove Theorem 11.8.

Chapter 12

Steiner Triple Systems

12.1 Construction of Triple Systems

The first block designs to be studied in detail were BIBDs with block size 3. These are also called *triple systems*, and the words "block" and "triple" are used interchangeably.

In 1844 Woolhouse [181] asked about block designs with $\lambda = 1$; in 1845 [182] he reported a lack of solutions, and in 1846 [183] he simplified the problem to the case of triple systems, and gave some further comments the following year [184]. Meanwhile Kirkman solved the existence problem for triple systems in a lecture to the Royal Society; this was published in 1847 [88]. In 1853 Steiner [140] discussed t-designs with $k = t+1$ and $\lambda = 1$; when $t = 2$ these are triple systems with $\lambda = 1$, so they are called *Steiner triple systems*.

Suppose a triple system on v treatments exists. Then from (1.1) and (1.2) we have $3b = vr$, $\lambda(v-1) = 2r$. Therefore, $r = \lambda(v-1)/2$ and $b = \lambda v(v-1)/6$. So the parameters λ and v are sufficient to specify all the parameters of the design. We shall denote a triple system with parameters λ and v by $T(\lambda, v)$. (Another common notation for such a system is $S_\lambda(2, 3, v)$.)

Let us consider Steiner triple systems in particular. Then $\lambda = 1$, $r = (v-1)/2$, and $b = v(v-1)/6$. Since r and b must be integers, v must be odd and 6 must divide $v(v-1)$. It follows that $v \equiv 1$ or $3 \pmod 6$. Similar calculations can be carried out for larger λ. It is clear that the conditions depend on the primacy of λ to 6. It is easy to prove:

LEMMA 12.1

If there is a $T(\lambda, v)$, then: $(\lambda, 6) = 1$ implies $v \equiv 1$ or $3 \pmod 6$; $(\lambda, 6) = 2$ implies $v \equiv 0$ or $1 \pmod 3$; $(\lambda, 6) = 3$ implies $v \equiv 1 \pmod 2$.

We shall now prove that these necessary conditions are sufficient (with the obvious exception that v cannot equal 2 in the case $(\lambda, 6) = 6$) by giving elementary constructions using Latin squares. The proofs for $\lambda = 1$ were by

Bose [22] and Skolem [134], but the inherent simplicity of their methods was pointed out by Lindner (see, for example, [100]). The constructions for higher λ are taken from [142].

THEOREM 12.2

There is a $T(\lambda, v)$ for all positive integers v and λ not excluded by Lemma 12.1.

Proof. We first prove the theorem for $\lambda \leq 6$.

Suppose we have proven the existence of a $T(\lambda, v)$. Then, by taking s copies of each block in that design, we can construct a $T(s\lambda, v)$. So if we can find a $T(1, v)$ for all $v \equiv 1$ or $3 \pmod 6$, this will prove the existence of a $T(\lambda, v)$ for all λ when $v \equiv 1$ or $3 \pmod 6$. In particular, we shall have a $T(2, v)$ in those cases; if we also obtain a $T(2, v)$ when $v \equiv 0$ or $4 \pmod 6$, we shall have a $T(2, v)$ for all $v \equiv 0$ or $1 \pmod 3$, so we can get a $T(\lambda, v)$ for all $v \equiv 0$ or $1 \pmod 3$ and for all even λ. By similar arguments we see that it is sufficient to construct the following designs: $T(1, v)$ for all $v \equiv 1, 3 \pmod 6$; $T(2, v)$ for all $v \equiv 0, 4 \pmod 6$; $T(3, v)$ for all $v \equiv 5 \pmod 6$; $T(6, v)$ for all $v \equiv 2 \pmod 6), v \neq 2$.

(i) It is convenient to start with the case $\lambda = 2$. To construct a $T(2, 3n)$ and a $T(2, 3n + 1)$, we shall need an idempotent Latin square of side n. This is impossible when $n = 2$, so we first observe that there exist a $T(2, 6)$ and a $T(2, 7)$. Suitable sets of triples are:

$T(2,6)$: 123, 124, 135, 146, 156, 236, 245, 256, 345, 346

$T(2,7)$: 123, 124, 135, 146, 157, 167, 236, 247, 256, 257, 345, 347, 367, 456.

Now suppose $n \neq 2$. Let $A = (a_{ij})$ be an idempotent Latin square of order n. The existence of such a square follows from (8.4), Exercise 8.3.1, and Theorem 8.12 (and a much easier proof is given in Exercise 12.1.2). We construct a triple system based on the symbol set

$$S = \{1^1, 2^1, \ldots, n^1, 1^2, 2^2, \ldots, n^2, 1^3, 2^3, \ldots, n^3\}.$$

Consider all the triples

$$\{x^1, y^1, a_{xy}^2\},\ 1 \leq x \leq n,\ 1 \leq y \leq n,\ x \neq y.$$

If x and z are any two numbers in the range $\{1, 2, \ldots, n\}$, then $a_{xy} = z$ for exactly one value of y and $a_{yx} = z$ for exactly one value of y. Moreover, if $z \neq x$, the y-values are also not equal to x. So the triples listed will among them contain every unordered pair of the form $\{x^1, z^2\}$ with $1 \leq x \leq n, 1 \leq$

$z \le n$, and $x \ne z$, exactly twice each. Moreover, every pair x^1, y^1 with $x \ne y$ occurs exactly twice. To avoid "triples" with repeated elements, we omit the case $x \ne y$. Then the triples

$$\left.\begin{array}{l}\{x^1, y^1, a^2_{xy}\}\\ \{x^2, y^2, a^3_{xy}\}\\ \{x^3, y^3, a^1_{xy}\}\end{array}\right\} : 1 \le x \le n, 1 \le y \le n \qquad (12.1)$$

would form a $T(2, 3n)$ based on S, except for the fact that no pair $\{x^i, x^j\}$ with $i \ne j$ is represented. If we append

$$\{x^1, x^2, x^3\}, 1 \le x \le n, \text{ twice each} \qquad (12.2)$$

we have a $T(2, 3n)$.

To construct a $T(2, 3n+1)$ we use the treatments $1^1, 2^1, \dots, n^3$ again, together with a symbol ∞. The triples (12.1) may be used, but they do not contain any copies of the pairs x^i, x^j with $i \ne j$, nor the pairs x^i, ∞. The triples

$$\left.\begin{array}{l}\{x^1, x^2, x^3\}\\ \{x^1, x^2, \infty\}\\ \{x^2, x^3, \infty\}\\ \{x^3, x^1, \infty\}\end{array}\right\} : 1 \le x \le n, \qquad (12.3)$$

together with (12.1), form a $T(2, 3n+1)$.

(ii) We now consider the case $\lambda = 1$. Suppose the Latin square A used in part (i) is symmetric. Then, for example,

$$\{x^1, y^1, a^2_{xy}\} = \{y^1, x^1, a^2_{yx}\},$$

and the triples in (12.1) and (12.2) consist of a certain set of triples, each one taken twice. So the $T(2, 3n)$ is two copies of a $T(1, 3n)$ joined together. It follows that the existence of a symmetric idempotent Latin square L of side n implies the existence of a $T(1, 3n)$.

If n is odd, say $n = 2k - 1$, such a square L is easy to construct. One example is $L = (l_{ij})$, where

$$l_{ij} \equiv k(i+j) \pmod{2k-1}.$$

(In other words, L is obtained by "back-circulating" its first row; the first row has been chosen in such a way as to give the diagonal elements in the proper order.) So there is a $T(1, 3n)$ whenever n is odd, which is to say that there is a $T(1, v)$ when $v \equiv 3 \pmod 6$.

As one would expect, no symmetric and idempotent Latin square exists when n is even (see Exercise 12.1.3), so some modification is needed to handle the case of $T(1, v)$ with $v \equiv 1 \pmod 6$. A slightly different symmetric Latin square must be used.

We let $M = (m_{ij})$ be a Latin square of side $2t$ that is symmetric and has diagonal $(1, 2, \ldots, k, 1, 2, \ldots, t)$. We shall call M a symmetric *semi-idempotent* Latin square. Such a square exists for every even side $2t$. One example has first row $1, t+1, 2, t+2, \ldots, t, 2t$, and again the later rows are formed by back-circulating.

We define a $T(1, 6k+1)$ based on the symbols $\{x^i : 1 \leq i \leq 3,\ 1 \leq x \leq 2k\}$ and a symbol ∞, by the triples

$$\begin{array}{c}
\{x^1, x^2, x^3\} \ : \ 1 \leq x \leq t \\
\left.\begin{array}{c}\{\infty, x^1, (x-t)^2\} \\ \{\infty, x^2, (x-t)^3\} \\ \{\infty, x^3, (x-t)^1\}\end{array}\right\} \ : \ k+1 \leq x \leq 2t \\
\left.\begin{array}{c}\{x^1, y^1, m_{xy}^2\} \\ \{x^2, y^2, m_{xy}^3\} \\ \{x^3, y^3, m_{xy}^1\}\end{array}\right\} \ : \ 1 \leq x < y \leq 2t.
\end{array} \tag{12.4}$$

The verification is straightforward (see Exercise 12.1.7).

(iii) The case of $T(3, v)$ for $v \equiv 5 \pmod 6$ is the easiest of the constructions. One simply takes all the triples

$$\{\{y, x+y, 2x+y\} : 0 < x < \tfrac{1}{2}v,\ 0 \leq y < v\}$$

where all the additions are reduced modulo v.

(iv) For $\lambda = 6$ we construct a $T(6, 6k+2)$ based on the symbols $1^1, 2^1, \ldots, n^1, 1^2, 2^2, \ldots, n^2, 1^3, 2^3, \ldots, n^3, \infty^1, \infty^2$, where $n = 2k$. The general approach is to take three copies of a $T(2, 3n)$ and modify the triples to allow for the symbols ∞^1 and ∞^2.

Suppose A is an idempotent Latin square of side n, $n \neq 2$. If we take each of the triples (12.1) three times, we have covered all pairs of symbols exactly six times except that the pairs of type x^i, x^j and x^i, ∞^j are completely missing. If one were to take all the triples

$$\{\{\infty^i, x^1, x^2\}, \{\infty^i, x^2, x^3\}, \{\infty^i, x^3, x^1\} : 1 \leq x \leq n, i = 1, 2\} \tag{12.5}$$

three times each, all pairs would be covered *except* $\{\infty^1, \infty^2\}$. So we take all the triples (12.5) *except for those with* $x = 1$ and append them to three copies of the list (12.1). Then it is necessary to find a set of triples that contain three copies of all the pairs in the triples

$$\begin{array}{c}
\{\infty^1, 1^1, 1^2\},\ \{\infty^1, 1^2, 1^3\},\ \{\infty^1, 1^3, 1^1\}, \\
\{\infty^2, 1^1, 1^2\},\ \{\infty^2, 1^2, 1^3\},\ \{\infty^2, 1^3, 1^1\},
\end{array}$$

as well as six copies of the pair $\{\infty^1, \infty^2\}$. What is required is in fact a $T(6, 5)$ on the symbols $1^1, 1^2, 1^3, \infty^1$, and ∞^2. Such a system exists, from part (iii).

To summarize, the following triples are used:

$$\left.\begin{array}{l}\{x^1,y^1,a^2_{xy}\}\\ \{x^2,y^2,a^3_{xy}\}\\ \{x^3,y^3,a^1_{xy}\}\end{array}\right\} : 1 \le x \le n, 1 \le y \le n, x \ne y, \text{ each triple taken thrice}$$

$$\left.\begin{array}{l}\{\infty^1,x^1,x^2\},\ \{\infty^2,x^1,x^2\}\\ \{\infty^1,x^2,x^3\},\ \{\infty^2,x^2,x^3\}\\ \{\infty^1,x^3,x^1\},\ \{\infty^2,x^3,x^1\}\end{array}\right\} : 2 \le x \le n, \text{ each triple taken thrice}$$

$$\left.\begin{array}{lll}\{\infty^1,\infty^2,1^1\}, & \{\infty^1,\infty^2,1^2\}, & \{\infty^1,\infty^2,1^3\},\\ \{\infty^1,1^1,1^2\}, & \{\infty^1,1^2,1^3\}, & \{\infty^1,1^3,1^1\},\\ \{\infty^2,1^1,1^2\}, & \{\infty^2,1^2,1^3\}, & \{\infty^2,1^3,1^1\}\end{array}\right\} : \text{ each taken twice}$$

$$\{1^1,1^2,1^3\}.$$

This covers all the cases with $v \equiv 2 \pmod 6$ except $v = 8$. A $T(6,8)$ is easily constructed. Suppose B is the set of all blocks of a $T(5,7)$; for example, take five copies of the $T(1,7)$ based on $S = \{1,2,3,4,5,6,7\}$. Append to B the 21 blocks formed by taking the 21 unordered pairs of members of S and appending the element ∞ to each. These blocks form a $T(6,8)$.

(v) Finally, suppose a $T(\lambda,v)$ exists. Take the blocks of this design together with s copies of each block of a $T(6,v)$. The result is a $T(\lambda+6s,v)$. Thus the results for $\lambda \le 6$ imply all the results of the theorem. □

The assumption that n is even was not used in part (iv), so that part of our proof provides a construction for a $T(6,v)$ when $v \equiv 5 \pmod 6$. This was not needed, since one can double the $T(3,v)$ constructed in part (iii). However, the proof cannot be simplified by omitting this duplication. Moreover, the two $T(6,v)$'s constructed are not isomorphic. (Proof of this last remark is left as an exercise.)

Exercises 12.1

12.1.1 Verify that in a $T(1,v)$, $v \equiv 1$ or $3 \pmod 6$.

12.1.2 If n is odd, define $L_n = (l_{ij})$ by $l_{ij} = 2i - j \pmod n$. If n is even, $n > 2$, define T_n to be the Latin square of order n derived from L_{n-1} by replacing the $(1,2)$, $(2,3)$, ..., $(n-2,n-1)$ and $(n-1,1)$ elements by n and then appending a last row and column that make the square Latin; the last column is $(n-1,1,2,\ldots,n-2,n)$ and the last row is $(n-2,n-1,1,\ldots,n-3,n)$. Prove that the L_n and T_n are idempotent Latin squares.

12.1.3* Suppose L is a symmetric Latin square of side n, and that the symbol x occurs k times on the diagonal.

(i) How many times does x occur above the diagonal?

(ii) Prove that $k \equiv n \pmod 2$.

(iii) If L is also idempotent, prove that n is odd.

12.1.4 Use the Latin square

1	4	2	5	3
4	2	5	3	1
2	5	3	1	4
5	3	1	4	2
3	1	4	2	5

to construct a $T(1, 15)$.

12.1.5 Use the Latin square

1	4	2	3
4	2	3	1
2	3	1	4
3	1	4	2

to construct a $T(1, 13)$.

12.1.6 Suppose A is a symmetric Latin square of side n based on $\{1, 2, \ldots, n\}$, and B is derived from A by adding n to each entry. Prove that

$$\begin{matrix} A & B \\ B & A \end{matrix}$$

is a symmetric Latin square of side $2n$. Use this to prove the existence of symmetric semi-idempotent Latin squares of all even sides.

12.1.7 Show that the triples listed in (12.4) do in fact form a $T(1, 6k + 1)$.

12.1.8 Prove[1] that there is $PB(v; \{4, 3\}; 1)$ whenever $v \equiv 3$ or $4 \pmod 6$.

12.1.9* Suppose a $T(v - 3, v - 1)$ exists. Show that there is a $T(v - 2, v)$ with the blocks of the $T(v - 3, v - 1)$ as some of its blocks.

12.1.10 Use the triples 123, 135, 014 to generate, cyclically $\pmod 7$, a set of 21 triples. Show that it is *not* possible to partition these 21 triples into three $T(1, 7)$s.

12.2 Subsystems

By a *subsystem* of a triple system is meant a subset of the set of blocks that forms a triple system on some subset of the set of treatments. Although these can be discussed for any λ, we shall consider only the case $\lambda = 1$.

[1]See also Exercises 2.1.4 and 13.1.2.

THEOREM 12.3

Suppose there exist a Steiner triple system $\mathcal{A}$ with v_1 treatments, and a Steiner triple system $\mathcal{B}$ with v_2 treatments that contains a subsystem $\mathcal{C}$ on v_3 treatments $\{1, 2, \ldots, v_3\}$. Then there exists a Steiner triple system $\mathcal{D}$ on $v = v_3 + v_1(v_2 - v_3)$ treatments that contains v_1 subsystems on v_2 treatments each, one subsystem on v_1 treatments and one subsystem on v_3 treatments.

Proof. Write $s = v_2 - v_3$. We construct a $T(1, v)$ based on the v-set

$$S_0 \cup S_1 \cup, \ldots, S_{v_1},$$

where

$$\begin{aligned} S_0 &= \{m_1, m_2, \ldots, m_{v_3}\}, \\ S_1 &= \{n_{11}, n_{12}, \ldots, n_{1s}\}, \\ S_2 &= \{n_{21}, n_{22}, \ldots, n_{2s}\}, \\ &\cdots \\ S_{v_1} &= \{n_{v_1 1}, n_{v_1 2}, \ldots, n_{v_1 s}\}. \end{aligned}$$

The blocks are constructed according to the following rules.

(i) Associate the elements of S_0 with the treatments of $\mathcal{C}$ and form the triple $\{m_i, m_j, m_k\}$ if and only if $\{i, j, k\}$ is a triple of $\mathcal{C}$. This gives $\frac{1}{6}v_3(v_3 - 1)$ triples, which we take as blocks of $\mathcal{D}$.

(ii) For each i, associate $S_0 \cup S_i$ with the treatments of $\mathcal{B}$ in such a way that the elements of S_0 are always associated with the treatments of $\mathcal{C}$, just as they were in (i), and form the triples on $S_0 \cup S_i$ that correspond to the triples of $\mathcal{B}$. Some of these contain three members of S_0, and they are precisely the triples that were constructed in (i); these can be ignored. There remain

$$\frac{v_2(v_2 - 1) - v_3(v_3 - 1)}{6}$$

triples, each containing at most one element of S_0; we take these triples as blocks of $\mathcal{D}$.

(iii) Finally, consider the given system $\mathcal{A}$ on the treatments $\{1, \ldots, v_1\}$; if $\{j, k, l\}$ is a triple of this system, adjoin to $\mathcal{D}$ all triples $\{n_{jx}, n_{ky}, n_{lz}\}$ where x, y, and z satisfy

$$x + y + z \equiv 0 \pmod{s}, 1 \leq x, y, z \leq s.$$

In solving this congruence, x and y can be chosen in s ways each; z is then uniquely determined. Hence each of the $\frac{1}{6}v_1(v_1 - 1)$ blocks of A leads to s^2 additional blocks of D.

From (i) we get a subsystem isomorphic to $\mathcal{C}$, on the v_3 elements of S_0. From (ii) we get altogether v_1 subsystems, each isomorphic to $\mathcal{B}$, on the v_2 elements of $S_0 \cup S_i$ for each i; each of these subsystems contains that of (i) as a subsystem of itself. The blocks constructed in (iii), containing only $n_{1s}, n_{2s}, \ldots, n_{v_1 s}$, form a subsystem isomorphic to $\mathcal{A}$.

Obviously, $\mathcal{D}$ has $v_3 + v_1(v_2 - v_3)$ treatments and constant block size 3. We check on balance and verify that any two treatments mutually belong to exactly one triple of $\mathcal{D}$. Any triple containing two elements of S_0 in fact contains three elements of S_0, uniquely determined by (i); a triple containing one element of S_0 and one of $S_i, i \neq 0$, contains another element of S_i, uniquely determined by (ii) for each i; a triple containing two elements of $S_i, i \neq 0$, contains either an element of S_0 or a third element of the same S_i, uniquely determined by (ii); a triple containing n_{jx} and $n_{ky}, j \neq k$, contains also n_{lz}, uniquely determined by (iii). So $\lambda = 1$ for $\mathcal{D}$, and $\mathcal{D}$ is a $T(1, v)$. □

As an example, we construct a $T(1, 15)$ using $v_1 = 3$, $v_2 = 7$, and $v_3 = 3$. We have $s = 4$. The constituent subsystems are:

A, on 3 treatments, with block $\{1, 2, 3\}$;

B, on 7 treatments, with blocks $\{1, 2, 3\}, \{1, 4, 5\}, \{1, 6, 7\}, \{2, 4, 6\}, \{2, 5, 7\}, \{3, 4, 7\}, \{3, 5, 6\}$;

C, on 3 treatments, with block $\{1, 2, 3\}$.

Then

$$\begin{aligned}
S_0 &= \{m_1, m_2, m_3\}, \\
S_1 &= \{n_{11}, n_{12}, n_{13}, n_{14}\}, \\
S_2 &= \{n_{21}, n_{22}, n_{23}, n_{24}\}, \\
S_3 &= \{n_{31}, n_{32}, n_{33}, n_{34}\}.
\end{aligned}$$

The block of type (i) is $\{m_1, m_2, m_3\}$. To form the blocks of type (ii) from $S_0 \cup S_1$, we set up the correspondence

$$(1, 2, 3, 4, 5, 6, 7) \leftrightarrow (m_1, m_2, m_3, n_{11}, n_{12}, n_{13}, n_{14});$$

the blocks of the $T(1, 7)$ are $\{m_1, m_2, m_3\}$, $\{m_1, n_{11}, n_{12}\}$, $\{m_1, n_{13}, n_{14}\}$, and so on. The first block is deleted, and the remaining blocks are

$$\begin{array}{lll}
\{m_1, n_{11}, n_{12}\}, & \{m_1, n_{13}, n_{14}\}, & \{m_2, n_{11}, n_{13}\}, \\
\{m_2, n_{12}, n_{14}\}, & \{m_3, n_{11}, n_{14}\}, & \{m_3, n_{12}, n_{13}\}.
\end{array}$$

Replacing S_1 by S_2 and then by S_3 we obtain the blocks

$$\begin{array}{lll}
\{m_1, n_{21}, n_{22}\}, & \{m_1, n_{23}, n_{24}\}, & \{m_2, n_{21}, n_{23}\}, \\
\{m_2, n_{22}, n_{24}\}, & \{m_3, n_{21}, n_{24}\}, & \{m_3, n_{22}, n_{23}\}, \\
\{m_1, n_{31}, n_{32}\}, & \{m_1, n_{33}, n_{34}\}, & \{m_2, n_{31}, n_{33}\}, \\
\{m_2, n_{32}, n_{34}\}, & \{m_3, n_{31}, n_{34}\}, & \{m_3, n_{32}, n_{33}\}.
\end{array}$$

Finally the system on v_1 treatments has just one block $\{1, 2, 3\}$; so we take all the blocks of the form $\{n_{1x}, n_{2y}, n_{3z}\}$, where $x + y + z \equiv 0 \pmod 4$, as the blocks of type (iii). If we choose x and y, then z is uniquely determined, so these blocks are

$$\begin{array}{llll} \{n_{11}, n_{21}, n_{32}\}, & \{n_{11}, n_{22}, n_{31}\}, & \{n_{11}, n_{23}, n_{34}\}, & \{n_{11}, n_{24}, n_{33}\}, \\ \{n_{12}, n_{21}, n_{31}\}, & \{n_{12}, n_{22}, n_{34}\}, & \{n_{12}, n_{23}, n_{33}\}, & \{n_{12}, n_{24}, n_{32}\}, \\ \{n_{13}, n_{21}, n_{34}\}, & \{n_{13}, n_{22}, n_{33}\}, & \{n_{13}, n_{23}, n_{32}\}, & \{n_{13}, n_{24}, n_{31}\}, \\ \{n_{14}, n_{21}, n_{33}\}, & \{n_{14}, n_{22}, n_{32}\}, & \{n_{14}, n_{23}, n_{31}\}, & \{n_{14}, n_{24}, n_{34}\}. \end{array}$$

These 35 blocks form the required $T(1, 15)$.

THEOREM 12.4

Given a $T(1, v)$, there is a $T(1, 2v+1)$ with the original system as a subsystem.

Proof. Suppose the v treatments of the $T(1, v)$ are $\{1, 2, \ldots, v\}$. Select a one-factorization of the K_{v+1} with vertices $\{v+1, v+2, \ldots, 2v+1\}$; say that the factors are $M_1, M_2, \ldots, M_v$. (Such a factorization exists, since v is necessarily odd.) The $T(1, 2v+1)$ has as its triples all the triples in the $T(1, v)$ together with all the triples of the form $\{i, j, k\}$, where $1 \le i \le v$ and $\{j, k\}$ is an edge in M_i. It is easy to see that the design is balanced: If i and j are in the range $1 \le i \le v, 1 \le j \le v$, then they occur together in one triple from the original system and no other; if $1 \le i \le v$ and $j > v$, j occurs in exactly one edge of M_i, so i and j occur together in exactly one block derived from M_i; if both $j > v$ and $k > v$, then $\{j, k\}$ is an edge in exactly one M_i and $\{i, j, k\}$ is the unique triple containing j and k. □

Theorem 12.4 can be applied even in quite small cases. For example, using the $T(1, 3)$ with one block $\{1, 2, 3\}$ and the one-factorization

$$M_1 = 45, 67 \quad M_2 = 46, 57 \quad M_3 = 47, 56,$$

we obtain the $T(1, 7)$ with blocks

$$\{1, 2, 3\}, \{1, 4, 5\}, \{1, 6, 7\}, \{2, 4, 6\}, \{2, 5, 7\}, \{3, 4, 7\}, \{3, 5, 6\}.$$

There is a corresponding theorem in which we embed a $T(1, v)$ in a $T(1, 2v-7)$. We first need a preliminary result on factorizations, which is proven in [137].

LEMMA 12.5

Given a set of $2n$ treatments, $n \ge 4$, one can find $2n$ 3-sets and $2n - 7$ one-factors that between them contain every pair of treatments precisely once.

THEOREM 12.6

If there is a $T(1, v)$, then there is a $T(1, 2v+7)$ with the original $T(1, v)$ as a subsystem.

Proof. Suppose S is a $T(1, v)$ on the v symbols $\{1, 2, \ldots, v\}$ and T is a collection of 3-sets and one-factors as outlined in Lemma 12.5 on the $2n = v+7$ symbols $v+1, v+2, \ldots, 2v+7$, into factors $M_1, M_2, \ldots, M_v$ and triples. The triples for the new system are:

the triples of S;

the triples of T;

the triples $\{\{x, y, z\} : \{y, z\} \in M_x, 1 \le x \le v\}$. □

Exercises 12.2

12.2.1* Use Theorem 12.3, in the case where $v_3 = 1$, to construct two $T(1, 19)$ designs, using the two equations

$$19 = 1 + 9(3-1), 19 = 1 + 3(7-1).$$

Are the designs isomorphic?

12.2.2 Construct a $T(1, 15)$, using a $T(1, 7)$ and Theorem 12.4.

12.2.3 Prove that no $T(1, 13)$ can contain a subsystem $T(1, 9)$.

12.2.4 Prove that if $u \equiv 1$ or $3 \pmod 6$, then there exists a positive integer v, $v \equiv 1$ or $3 \pmod 6$, such that $u = 2v+1$ or $u = 2v+7$. Hence show that assuming Theorems 12.4 and 12.6 and the existence of a $T(1, 1)$ and a $T(1, 3)$, one can prove the existence of Steiner triple systems of all possible orders without using Theorem 12.2.

12.2.5* Find a $T(1, 13)$ with a subsystem $T(1, 3)$, using Theorem 12.6.

12.2.6 (i) Consider the complete graph K_{6n} based on the vertices x^i, where $1 \le x \le 2n$ and $0 \le i \le 2$. The graph G is formed from K_{6n} by deleting all the edges $x^i x^j$, where $i \ne j$. Select any one-factorization $\{F_1^i, F_2^i, \ldots, F_{2n-1}^i\}$ of the K_{2n} on vertices $1^i, 2^i, \ldots, (2n)^i$; define one-factors

$$H_j^i = \{x^{i+1}(x+j)^{i+2} : 1 \le x \le 2n\} \cup F_j^i$$

for $0 \le i \le 2, 1 \le j \le 2n-1$. (Additions are reduced modulo $2n$, when necessary.) Prove that the H_j^i form a one-factorization of G.

(ii) Use the factorization above, in the case $6n = v + 3$, to prove that when $v \equiv 3 \pmod 6$, any $T(1, v)$ can be embedded as a subsystem of a $T(1, 2v+3)$.

12.3 Simple Triple Systems

A balanced incomplete block design is called *simple* if it has no repeated block. Simple designs are sometimes preferred in geometrical and statistical design theory. For this reason it is interesting to know when simple designs exist. We restrict ourselves to the one case of simple triple systems $T(2, v)$. (The designs with $\lambda = 1$ are necessarily simple.) The constructions come from [143].

We use a special design that we call a *skew transversal square*: This is an idempotent Latin square $L = (l_{ij})$ with the property that $l_{ij} = l_{ji}$ is never true unless $i = j$.

LEMMA 12.7

There is a skew transversal square of side n for every positive n other than 2 and 3.

Proof. We first define a family of idempotent Latin squares $L_n = (l_{ij})$ of all orders other than 2. If n is odd, $l_{ij} \equiv 2i - j \pmod{n}$. If n is even, $n \neq 2$, we construct L_n from L_{n-1} by replacing the $(1,2), (2,3), \ldots, (n-1,1)$ entries by n and appending last row $(n-2, n-1, 1, 2, \ldots, n-3, n)$ and last column $(n-1, 1, 2, \ldots, n-2, n)$. Clearly L_n is an idempotent Latin square. When n is odd, L_n will be skew unless $2i - j \equiv 2j - i \pmod{n}$, and L_{n+1} will also be skew. So we have skew transversal squares of every side n except $n = 2$ and $n = 3m$ or $3m + 1$, m odd.

Side 4 and side 9 are covered by examples in Figure 12.1.

In the remaining odd cases, say $n = 3m$. Write $n = pq$, where p is a power of 3 and q is prime to 3. From the first paragraph of the proof, we can find

$$\begin{bmatrix} 1 & 4 & 2 & 3 \\ 3 & 2 & 4 & 1 \\ 4 & 1 & 3 & 2 \\ 2 & 3 & 1 & 4 \end{bmatrix} \quad \begin{bmatrix} 1 & 8 & 7 & 6 & 9 & 3 & 5 & 4 & 2 \\ 4 & 2 & 9 & 8 & 3 & 7 & 1 & 5 & 6 \\ 5 & 6 & 3 & 2 & 7 & 9 & 4 & 1 & 8 \\ 3 & 5 & 6 & 4 & 8 & 2 & 9 & 7 & 1 \\ 2 & 9 & 4 & 7 & 5 & 1 & 8 & 6 & 3 \\ 7 & 1 & 9 & 5 & 2 & 6 & 3 & 9 & 4 \\ 9 & 4 & 2 & 1 & 6 & 8 & 7 & 3 & 5 \\ 6 & 3 & 5 & 9 & 1 & 4 & 2 & 8 & 7 \\ 8 & 7 & 1 & 3 & 4 & 5 & 6 & 2 & 9 \end{bmatrix}$$

FIGURE 12.1: Skew transversal squares of sides 4 and 9.

a skew transversal square S of side q. Select a symmetric idempotent Latin square R of side p. (As p is odd, such a square was constructed in part (ii) of the proof of Theorem 12.2.) Write $T^{(k)}$ to mean T with every entry increased by k and then reduced modulo q to the usual range $\{1, 2, \ldots, q\}$. Then the required skew transversal square of side pq is a $p \times p$ array of $q \times q$ blocks; the i-th diagonal block is $S+(i-1)q$, and if $i \neq j$ the (i, j) block is $S_k+(r_{ij}-1)q$, where $k \in \{0, 1, 2\}$ and $k \equiv i-j \pmod 3$. This is obviously a skew transversal square.

The remaining even cases $n = 3m + 1$ are easily solved from the preceding construction. In the square of side $3m$ the diagonals of the $(1, 2), (2, 3), \ldots, (p, 1)$ blocks are replaced by the symbol $3m + 1$, and the deleted entries are transported to a new last row and column just as in the construction of L_n for even n. □

THEOREM 12.8 [143]

There is a simple $T(2, v)$ whenever $v \equiv 0$ or $1 \pmod 3$.

Proof. We divide the proof into three parts.

(i) *Case* $v \equiv 1 \pmod 3$. We construct a simple $S(2, v)$ on the $3n$ symbols x^i, where $1 \leq x \leq n$ and $0 \leq i \leq 2$, and a symbol ∞. If $n \geq 4$, select a skew transversal square $L = (l_{ij})$ of side n. Then (with superscripts reduced modulo 3 when necessary) the triples

$$\begin{array}{ll} \{x^0, x^1, x^2\} & : 1 \leq x \leq n; \\ \{x^i, x^{i+1}, \infty\} & : 1 \leq x \leq n; 0 \leq i \leq 2; \\ \{x^i, y^i, {l_{xy}}^{i+1}\} & : 1 \leq x \leq n, 1 \leq y \leq n, x \neq y; 0 \leq i \leq 2 \end{array} \tag{12.6}$$

form a simple $S_2(2, 3, 3n+1)$. To handle the case $n = 2$, we exhibit the triples:

$$123, 145, 167, 246, 257, 347, 356, 124, 137, 156, 235, 267, 346, 357$$

form a simple $S_2(2, 3, 7)$. For $n = 3$ we could also exhibit the triples. However, writing $n = 3$ and

$$L = (l_{ij}) = L_3 = \begin{bmatrix} 1 & 3 & 2 \\ 3 & 2 & 1 \\ 2 & 1 & 3 \end{bmatrix}$$

we have the following interesting formulation:

$$\begin{array}{ll} \{x^0, x^1, x^2\} & : 1 \leq x \leq n; \\ \{x^i, x^{i+1}, \infty\} & : 1 \leq x \leq n; 0 \leq i \leq 2; \\ \{x^i, y^i, {l_{xy}}^{i+1}\} & : 1 \leq x < y \leq n; 0 \leq i \leq 2; \\ \{x^i, y^i, {l_{xy}}^{i+1}\} & : 1 \leq y < x \leq n; 0 \leq i \leq 2. \end{array}$$

This construction could be used whenever there is a symmetric idempotent Latin square of order n; but such designs can obviously exist only when n is odd, so we would still need formula 12.6 for the cases $n = 1 \pmod 6$.

(ii) *Case* $v \equiv 0 \pmod 6$. When $v = 6n$ we use symbols x^i for $1 \le x \le n$ and $0 \le i \le 5$, and any idempotent Latin square $L = (l_{ij})$ of side n. Elements of L will be superscripted in the obvious way. The triples are:

$$\{x^1, x^3, x^5\} \;\; : 1 \le x \le n;$$

$$\left.\begin{matrix}\{x^i, x^{i+1}, x^{i+3}\} \\ \{x^i, x^{i+3}, x^{i+4}\}\end{matrix}\right\} : 1 \le x \le n; 0 \le i \le 2;$$

$$\{x^i, x^{i+4}, x^{i+5}\} \; \} \; : 1 \le x \le n; i = 0, 2, 4;$$

$$\left.\begin{matrix}\{x^i, y^i, {l_{xy}}^{i+2}\},\ \{x^i, y^i, {l_{xy}}^{i+3}\} \\ \{x^i, y^{i+1}, {l_{xy}}^{i+3}\},\ \{x^i, y^{i+1}, {l_{xy}}^{i+2}\} \\ \{x^{i+1}, y^i, {l_{xy}}^{i+3}\},\ \{x^{i+1}, y^i, {l_{xy}}^{i+2}\} \\ \{x^{i+1}, y^{i+1}, {l_{xy}}^{i+2}\},\ \{x^{i+1}, y^{i+1}, {l_{xy}}^{i+3}\}\end{matrix}\right\} 1 \le x < y \le n; i = 0, 2, 4.$$

The case $n = 2, v = 12$ must be treated separately. If $T(a, b, c, d)$ denotes the set of four possible triples with elements in $\{a, b, c, d\}$, then

$$\begin{matrix}T(j, j+3, j+6, j+9) \; : \; j = 0, 1, 2; \\ \{\{3, 3y+1, 3z+2\} : x, y, z \in \{0, 1, 2, 3\}, x+y+z \text{ odd}\}\end{matrix}$$

form a simple $T(2, 12)$ on $\{0, 1, \ldots, 11\}$.

(iii) *Case* $v \equiv 3 \pmod 6$. We use symbols x^i for $1 \le x \le n$ and $i = 1, 2, 3$, and a symmetric idempotent Latin square $L = (l_{ij})$ of side n. Elements of L will again be superscripted. The triples are

$$\{x^1, x^2, x^3\},\ \{x^1, x^2, (x+1)^3\} \;\; : \; 1 \le x \le n;$$

$$\left.\begin{matrix}\{x^1, y^1 {l_{xy}}^2\},\ \{x^1, y^1, (l_{xy}+1)^3\} \\ \{x^2, y^2, {l_{xy}}^3\},\ \{x^2, y^2, {l_{xy}}^1\} \\ \{x^{i+1}, y^i, {l_{xy}}^{i+3}\},\ \{x^{i+1}, y^i, {l_{xy}}^{i+2}\} \\ \{x^3, y^3, {l_{xy}}^1\},\ \{x^3, y^3, (l_{xy}+1)^2\}\end{matrix}\right\} 1 \le x < y \le n;$$

where additions and subtractions are carried out modulo n. □

12.4 Cyclic Triple Systems

A *difference triple* modulo v is a set of three integers $\{x, y, z\}$ such that either $x + y + z$ is congruent to zero or else one of x, y and z is congruent to the sum of the other two (mod v). A *difference partition* of a set S is a partition of S into difference triples. Some samples are shown in Figure 12.2.

Difference partitions were introduced by Heffter [70] because of their relationship to a certain type of Steiner triple system. If a design $\mathcal{D}$ is based on Z_v, we say $\mathcal{D}$ is *cyclic* if increasing every element of a block by 1 (mod v) always results in another block. For example, any design generated from initial blocks in Z_v (not including an infinity element) is cyclic, and similarly partial circulation modulo v yields a cyclic design (see Section 5.4).

For the remainder of this section, S_v denotes the subset $\{1, 2, \ldots, (v-1)/2\}$ of Z_v and $S'_v = S_v \backslash \{v/3\}$.

LEMMA 12.9

If $v \equiv 1 \pmod 6$, and there is a difference partition of S_v, then there is a cyclic Steiner triple system of order v. If $v \equiv 3 \pmod 6$, and there is a difference partition of S'_v, then there is a Steiner triple system of order v.

Proof. Suppose $v \equiv 1 \pmod 6$ and $\mathcal{P}$ is a difference partition of S_v. If $\{x, y, z\}$ is a member of $\mathcal{P}$ with the property that some member is congruent to the sum of the other two, suppose without loss of generality that $z = x + y$, so that every triple satisfies

$$z \equiv \pm(x + y) \pmod v.$$

With the triple $\{x, y, z\}$ we associate the triple $\{0, x, x + y\}$, and write

$$\mathcal{B} = \{\{0, x, x + y\} : \{x, y, z\} \in P\}\}.$$

Then it is easy to see that the triples $\{j, x+j, t+j\}$, where addition is modulo v, j ranges through $\{0, 1, \ldots, v-1\}$ and $\{0, x, t\}$ ranges through $\mathcal{B}$, form a triple system $T(1, v)$. (Either $t = z$ or $t = -z$.) The set $\mathcal{B}$ is a set of initial blocks, in the sense of Section 5.4.

In the case $v = 3 \pmod 6$, say $v = 3k$, the same construction is used, and the blocks $\{j, k + j, 2k + j\}$ are also included for every j and k such that $0 \le j < k$.

In both cases the triple system is obviously cyclic. □

It is a straightforward matter to prove that no difference partition exists for $v = 9$, and there is no cyclic triple system of that order (see Exercise 12.4.1).

It is easy to construct difference partitions of small orders. As an example we consider the case of $v = 15$. The set S'_{15} is $\{1, 2, 3, 4, 6, 7\}$. The difference partition must contain a triple that includes 1. It might have the form $\{1, x, x + 1\}$ or $\{1, x, 15 - (x + 1)\}$ or $\{x, 15 + 1 - x, 1\}$. Since no element can

$v = 19$	$\{1,5,6\}\{2,8,9\}\{3,4,7\}$
$v = 27$	$\{1,12,13\}\{2,5,7\}\{3,8,11\}\{4,6,10\}$
$v = 45$	$\{1,11,12\}\{1,17,19\}\{3,20,22\}\{4,10,14\}$ $\{5,8,13\}\{6,18,21\}\{7,9,16\}$
$v = 63$	$\{1,15,16\}\{2,27,29\}\{3,25,28\}\{4,14,18\}\{5,26,31\}$ $\{6,17,23\}\{7,13,20\}\{8,11,19\}\{9,24,30\}\{10,12,22\}$

FIGURE 12.2: Some small difference partitions.

be greater than 7, the only possibility is $\{1, x, x+1\}$ with $x + 1 \leq 7$; since 5 is not in S'_{15}, x must equal 2, 3, or 6. So the candidates are

$$\{\{1,2,3\},\{4,6,7\}\}, \quad \{\{1,3,4\},\{2,6,7\}\}, \quad \{\{1,6,7\},\{2,3,4\}\}.$$

By inspection the only case in which both triples are difference triples is the second one:

$$\{\{1,3,4\},\{2,6,7\}\}.$$

The cases $v = 7$ and $v = 13$ are left as exercises. Figure 12.2 shows difference partitions for $v = 19$, 27, 45 and 63.

THEOREM 12.10 [113]

There is a difference partition of S_v for every $v \equiv 1 \,(\mathrm{mod}\ 6)$ and a difference partition of S'_v for every $v \equiv 3 \,(\mathrm{mod}\ 6)$ except $v = 9$.

Proof. Figure 12.3 gives constructions for difference partitions of S_v whenever $v \equiv 1 \,(\mathrm{mod} 6)$, $v \neq 7, 13, 19$, and of S'_v whenever $v \equiv 3 \,(\mathrm{mod} 6)$, $v \neq 19, 15, 27, 45, 63$. Together with the partitions in Figure 12.2 and the examples for $v = 15$ above and for $v = 7$ and 13 in Exercise 2, this proves the theorem. □

COROLLARY 12.10.1

There is a cyclic triple system of every possible order except 9.

Exercises 12.4

12.4.1* Prove that there is no difference partition of S'_9.

12.4.2* Find difference partitions of S_v when $v = 7$ and $v = 13$.

$v = 18s + 1, s \geq 2$	$\{3r + 1, 4s - r + 1, 4s + 2r + 2\}$, $r = 0, \ldots, s - 1$ $\{3r + 2, 8s - r, 8s + 2r + 2\}$ $r = 0, \ldots, s - 1$ $\{3r + 3, 6s - 2r - 1, 6s + r + 2\}$, $r = 0, \ldots, s - 1$ $\{3s, 3s + 1, 6s + 1\}$
$v = 18s + 3, s \geq 1$	$\{3r + 1, 8s - r + 1, 8s + 2r + 2\}$ $r = 0, \ldots, s - 1$ $\{3r + 2, 4s - r, 4s + 2r + 2\}$ $r = 0, \ldots, s - 1$ $\{3r + 3, 6s - 2r - 1, 6s + r + 2\}$ $r = 0, \ldots, s - 1$
$v = 18s + 7, s \geq 1$	$\{3r + 1, 8s - r + 1, 8s + 2r + 2\}$ $r = 0, \ldots, s - 1$ $\{3r + 2, 6s - 2r + 1, 6s + r + 3\}$, $r = 0, \ldots, s - 1$ $\{3r + 3, 4s - r + 1, 4s + 2r + 4\}$, $r = 0, \ldots, s - 1$ $\{3s + 1, 4s + 2, 7s + 3\}$
$v = 18s + 9, s \geq 4$	$\{3r + 1, 4s - r + 3, 4s + 2r + 4\}$, $r = 0, \ldots, s$ $\{3r + 2, 8s - r + 2, 8s + 2r + 4\}$, $r = 2, \ldots, s - 2$ $\{3r + 3, 6s - 2r + 1, 6s + r + 4\}$, $r = 1, \ldots, s - 2$ $\{2, 8s + 3, 8s + 5\}, \{3, 8s + 1, 8s + 4\}$, $\{5, 8s + 2, 8s + 7\}, \{3s, 7s + 3, 8s + 6\}$, $\{3s - 1, 3s + 2, 6s + 1\}$
$v = 18s + 13, s \geq 1$	$\{3r + 2, 6s - 2r + 3, 6s + r + 5\}$, $r = 0, \ldots, s - 1$ $\{3r + 3, 8s - r + 5, 8s + 2r + s\}$, $r = 0, \ldots, s - 1$ $\{3r + 1, 4s - r + 3, 4s + 2r + 4\}$, $r = 0, \ldots, s$ $\{3s + 2, 7s + 5, 8s + 6\}$
$V = 18s + 15, s \geq 1$	$\{3r + 1, 4s - r + 3, 4s + 2r + 4\}$, $r = 0, \ldots, s$ $\{3r + 2, 8s - r + 6, 8s + 2r + s\}$, $r = 0, \ldots, s$ $\{3r + 3, 6s - 2r + 3, 6s + r + 6\}$, $r = 0, \ldots, s - 1$

FIGURE 12.3: Families of difference partitions.

12.4.3 Verify that the sets shown in Figures 12.2 and 12.3 are difference partitions.

12.5 Large Sets and Related Designs

By a *large set* of designs with block size k, based on a support set S, is meant a set of designs whose block-sets together include all k-subsets of S. We shall consider only large sets with $k = 3$ where the constituent designs have $\lambda = 1$. Obviously such sets need only be considered in cases where $v \equiv 1$ or $3 \pmod 6$.

The number of blocks in a $T(1, v)$ is $\frac{1}{6}v(v-1)$. The total number of 3-sets that can be chosen from v objects is $\frac{1}{6}v(v-1)(v-2)$, precisely enough for $v-2$ triple systems. A partition of the set of all such triples into $v-2$ parts, each of which is a $T(1, v)$, is called a *large set of disjoint triple systems.*

There is no large set of disjoint triple systems for $v = 7$. However, it has been shown ([103, 152]; see also [82]) that there is a large set for every other order congruent to 1 or $3 \pmod 6$.

There is a unique $T(1, 9)$ up to isomorphism. Any realization of it may be described as follows. Write the nine treatments in a 3×3 array. Then the blocks are the rows of the array, the columns, the diagonals and the back-diagonals. The design is resolvable, and these four sets of blocks are the parallel classes. For example, if the array is

1	2	3
4	5	6
7	8	9

then the parallel classes are

123 456 789 (rows) 147 258 369 (columns)
159 267 348 (diagonals) 168 249 357 (back-diagonals).

This representation makes it easy to display a large set; seven 3×3 arrays suffice. In fact, even less space is needed. There are precisely two large sets on 9 elements, up to isomorphism [11]. In the first set, the triple systems are written on the support set $\{\infty, \infty'\} \cup Z_7$, and are represented as

∞	∞'	i
$1+i$	$2+i$	$3+i$
$4+i$	$5+i$	$6+i$

where $i = 0, 1, \ldots, 6$ and addition is carried out modulo 7. The second large set is based on $\{\infty, x, y\} \cup Z_6$; one system is represented by

$$\begin{array}{|ccc|}\hline \infty & x & y \\ 0 & 2 & 4 \\ 3 & 1 & 5 \\ \hline\end{array}$$

together with the six systems

$$\begin{array}{|ccc|}\hline \infty & 2i & 1+2i \\ x & 2+2i & 5+2i \\ 3+2i & y & 4+2i \\ \hline\end{array}\ , \quad \begin{array}{|ccc|}\hline \infty & 1+2i & 2+2i \\ x & 3+2i & 2i \\ 4+2i & y & 5+2i \\ \hline\end{array}$$

where $i = 0, 1, 2$ and addition is carried out modulo 6.

A large set of triple systems on seven symbols would be a set of five triple systems whose block-sets are disjoint. While this is impossible, there exists a set of three disjoint triple systems. One example is

$$\begin{array}{ccc} 012 & 013 & 014 \\ 034 & 025 & 026 \\ 056 & 046 & 035 \\ 135 & 124 & 125 \\ 146 & 156 & 136 \\ 236 & 234 & 246 \\ 245 & 356 & 345. \end{array}$$

Two designs are called *almost disjoint* if they have precisely one block in common. Mutually almost disjoint triple systems are often called MAD systems.

Lindner and Rosa [102] showed that, when $v \neq 7$, a large set of n MAD systems on v treatments can exist only when $n = v - 1, v$ or $v + 1$, except that there exists a set of 15 systems of order 13 (which they exhibited). They proved:

THEOREM 12.11 [102]

There is a large set of v MAD systems on v treatments whenever $v \equiv 1$ or $3 \pmod 6$.

Proof. Select a 3-$(v + 1, 4, 1)$ design $\mathcal{D}$ (recall from Section 2.4 that Hanani [65] showed that one exists for all $v \equiv 1$ or $3 \pmod 6$). Suppose the support set is $S = \{0, 1, \ldots, v\}$. Write $\mathcal{E}_x$ for the set of all quadruples of the design that contain x; delete x from each of them; then replace 0 by x each time it occurs. This is a $T(1, v)$; call it $\mathcal{B}_x$.

We show that $\mathcal{B}_1, \mathcal{B}_2, \ldots \mathcal{B}_v$ form the required large set. First, every triple is covered; if abc is any triple of nonzero elements of S then there is a unique block $abcd$ in $\mathcal{D}$; if $d = 0$ then abc is in $\mathcal{B}_a$ (and in $\mathcal{B}_b$ and $\mathcal{B}_c$), and otherwise abc is in $\mathcal{B}_d$. Second, if x and y are distinct and nonzero then x, y and 0 lie in a unique block of $\mathcal{D}$, say $0xyz$, and xyz is the unique common block in $\mathcal{B}_x$ and $\mathcal{B}_y$. □

LEMMA 12.12

Suppose we have a large set of $T(1, v)$'s consisting of $v - 1$ designs. There must be $v(v-1)/6$ triples that occur in precisely two of the triple systems, while the others occur precisely once. The triples that occur twice form a $T(1, v)$.

Proof. Consider any two of the elements. They occur together in $v - 1$ blocks, one per triple system. As there are $v - 2$ different triples containing two given elements, there is precisely one repetition. There are $v(v-1)/2$ pairs of elements, and each repeated block contains three pairs, so there must be $v(v-1)/6$ repeated blocks. As each pair occurs in exactly one repetition, the repeated blocks form a $T(1, v)$, and they contain no duplicates. This is enough to prove the theorem. □

THEOREM 12.13

There is no large set of $v - 1$ MAD systems on v treatments.

Proof. Suppose a set of $v - 1$ MAD systems were to exist at order v. From Lemma 12.12, exactly $\frac{1}{6}v(v-1)$ triples occur in two of the systems. Each represents the intersection of one pair of systems, and there are $\frac{1}{2}(v-1)(v-2)$ such pairs. Equating these, $\frac{1}{6}v = \frac{1}{2}(v-2)$, or $v = 3$. So there are no cases of $v - 1$ MAD systems other than the trivial one. □

There is a set of $v+1$ MAD systems when $v = 13$ or 15 [57], but no such set when $v = 9$ [102]; the existence of sets of $v + 1$ for other orders is undecided.

So we come to order 7. Theorem 12.11 provides a set of seven MAD systems, but we can do more.

It is easy to count all $T(1, 7)$s. Suppose the systems are based on $\{0, 1, 2, 3,$

$4, 5, 6\}$. There are six systems that contain the triple 012, namely

$$\begin{array}{cccccc}
012 & 012 & 012 & 012 & 012 & 012 \\
034 & 034 & 035 & 035 & 036 & 036 \\
056 & 056 & 046 & 046 & 045 & 045 \\
135 & 136 & 134 & 136 & 134 & 135 \\
146 & 145 & 156 & 145 & 156 & 146 \\
236 & 235 & 236 & 234 & 235 & 234 \\
245 & 246 & 245 & 256 & 246 & 256
\end{array}.$$

There are 35 possible triples, and each system contains 7 of them, so the number of systems is $6 \times 35/7 = 30$. Of course, they are all isomorphic.

The first, fourth and fifth systems listed above have no common blocks except 012. If we go through the remaining 24 systems we find that this pattern can be continued, yielding a set of 15 MAD systems:

$$\begin{array}{ccccccccccccccc}
012 & 012 & 012 & 013 & 013 & 013 & 014 & 014 & 014 & 015 & 015 & 015 & 016 & 016 & 016 \\
034 & 035 & 036 & 024 & 025 & 026 & 023 & 035 & 036 & 023 & 024 & 026 & 023 & 035 & 034 \\
056 & 046 & 045 & 056 & 046 & 045 & 056 & 026 & 025 & 046 & 036 & 034 & 045 & 024 & 025 \\
135 & 136 & 134 & 126 & 124 & 125 & 136 & 123 & 135 & 126 & 123 & 124 & 134 & 123 & 135 \\
146 & 145 & 156 & 145 & 156 & 146 & 134 & 146 & 136 & 125 & 156 & 126 & 125 & 145 & 124 \\
236 & 234 & 235 & 235 & 326 & 234 & 345 & 346 & 234 & 245 & 256 & 235 & 356 & 346 & 236 \\
245 & 256 & 246 & 346 & 345 & 356 & 356 & 345 & 456 & 246 & 245 & 456 & 246 & 256 & 456
\end{array}.$$

If these 15 triple systems are deleted from the complete collection of 30, the remaining 15 systems also form a MAD set.

There is another design whose blocks are the 35 triples on seven objects, each occurring precisely once. The blocks are arranged in seven rounds, such that:

(i) each round contains the same number of blocks;

(ii) each block occurs exactly once in the design;

(iii) each treatment occurs either twice or three times in each round;

(iv) no two treatments occur together in two or more blocks in any round.

A design satisfying (i) through (iv) is called a *triad design* on seven points.

It follows immediately that each round of such a design contains five blocks; one treatment occurs in three of the blocks and the others in two each. If we refer to the treatment of frequency 3 as the *focus* of the round, then each treatment is the focus of exactly one round. We shall label the treatments 1, 2, 3, 4, 5, 6, 7, and assume that treatment i is the focus of round i.

From property (iv), round i is of the form

$$ix_1^iy_1^i,\ ix_2^iy_2^i,\ ix_3^iy_3^i,\ x_1^ix_2^ix_3^i,\ y_1^iy_2^iy_3^i,$$

where $\{i, x_1^i, y_1^i, x_2^i, y_2^i, x_3^i, y_3^i\}$ is some permutation of $\{1, 2, 3, 4, 5, 6, 7\}$. The first three triples in round i are associated with the near-one-factor

$$N_i = \{i \ \ x_1^i y_1^i \ \ x_2^i y_2^i \ \ x_3^i y_3^i\}.$$

We exhibit one example of a triad design on 7 points:

127	316	451	532	674
231	427	562	643	715
342	531	673	754	126
453	642	714	165	237
564	753	125	276	341
675	164	236	317	452
712	275	347	321	563

This is a highly structured example. As can be seen on inspection, the rounds are formed by developing the first round modulo 7. The seven near-one-factors N_i associated with the rounds form the near-one-factorization

1 27 36 45
2 31 47 56
3 42 51 67
4 53 62 71
5 64 73 12
6 75 14 23
7 16 25 34

and the other blocks form two disjoint $T(1, 7)$s (the vertical lines in the presentation of the triad design display these triple systems).

Not all triad designs have this degree of structure. There are in fact six isomorphism classes of triad designs of order 7, as was shown in [79]. Examples are shown in Figure 12.4.

Exercises 12.5

12.5.1 In each of the following cases write down a pair of $T(1, 7)$s, based on the set $S = \{1, 2, 3, 4, 5, 6, 7\}$, that satisfy the given condition:

(i) the 14 triples are all distinct;

(ii) precisely *one* of the triples is repeated;

(iii) precisely *three* of the triples are repeated;

(iv) there are seven triples, all repeated.

12.5.2 Verify that there is no large set of triple systems on 7 points.

12.5.3* Prove that the set of all triples on a v-set forms a $T(v - 2, v)$. Use this and Exercise 12.5.2 to show that there is no set of four pairwise disjoint triple systems on 7 points.

123	145	167	257	346	123	145	167	256	347
214	235	267	156	347	214	235	267	157	346
315	324	367	147	256	315	324	367	146	257
413	426	457	127	356	413	426	457	127	356
512	537	546	136	247	512	537	546	136	247
614	623	657	137	245	615	623	647	137	245
715	723	746	126	345	714	723	756	126	345
123	145	167	247	356	123	145	167	256	347
214	235	267	156	347	214	235	267	136	457
314	326	357	125	467	314	326	357	127	456
417	423	456	136	257	417	426	435	125	367
517	526	534	146	237	517	524	536	146	237
612	634	657	137	245	612	634	657	135	247
712	736	745	246	135	713	725	746	156	234
123	145	167	247	356	123	145	167	356	247
214	235	267	137	456	214	236	257	137	456
315	326	347	146	257	315	327	346	126	457
413	426	457	125	367	417	426	435	125	367
516	524	537	127	346	516	524	537	134	267
613	625	647	157	234	613	625	647	157	234
714	723	756	126	345	712	734	756	146	235

FIGURE 12.4: All triad designs on seven treatments.

12.5.4* Consider the 15 triple systems in a MAD set on seven objects; label them $T_1, T_2, \ldots, T_{15}$. Consider the design constructed as follows:

- its support set is $\{1, 2, \ldots, 15\}$;
- its blocks are labeled with the triples on $\{1, 2, 3, 4, 5, 6, 7\}$;
- object i belongs to block abc if and only if, among the 15 triple systems, block abc belongs to system T_i.

Show that this is a balanced incomplete block design. What are its parameters?

12.5.5 In the original application of triad designs, it was required that the blocks be ordered so that no treatment occurs twice in the same position in any round. Show that this condition can be met by any triad design on 7 points.

Chapter 13

Kirkman Triple Systems and Generalizations

13.1 Kirkman Triple Systems

Recall from Section 6.2 that the first mention of resolvable designs is the question posed by the Reverend Thomas Kirkman [90] in 1850: "Fifteen young ladies in a school walk out three abreast for seven days in succession: it is required to arrange them daily, so that no two shall walk twice abreast." Kirkman [91] later showed that a solution exists, and gave another single answer by four of the readers of the *Lady's and Gentleman's Diary.*

In Section 6.2 a *parallel class* of blocks in a design was defined to be a set of blocks that between them contain every element precisely once. A design is called *resolvable* when its blocks admit of a partition into parallel classes. So Kirkman was in fact asking whether there is a resolvable Steiner triple system on 15 treatments.

More generally, Kirkman asked: For which v does there exist a resolvable $T(1, v)$? We know that a $T(1, v)$ must have $v \equiv 1$ or $3 \pmod 6$. Moreover, for resolvability, the block size 3 must divide v. So a necessary condition is $v \equiv 3 \pmod 6$. We shall show that this condition is in fact sufficient, a fact that was finally proven by Ray-Chaudhury and Wilson [118] some 121 years after Kirkman originally posed the problem, although Anstice [5, 6] found an infinite family of solutions (for $v = 12n + 3$, where $6n + 1$ is prime) as early as 1852.

A resolvable Steiner triple system is called a *Kirkman triple system*, and a Kirkman triple system on v points is denoted $KTS(v)$.

A parallel class in which each block has three elements is also called a *triangle factor*. This terminology reflects the fact that parallel classes are factors in the underlying graph of the design; if no pair appears in blocks of two triangle factors, they are called *edge-disjoint*.

Suppose $v = 6m+3$ for some integer m. Then the parameters of a $KTS(v)$

$KTS(9)$:

1 2 3	1 4 7	1 5 9	1 6 8
4 5 6	2 5 8	2 6 7	2 4 9
7 8 9	3 6 9	3 4 8	3 5 7

$KTS(15)$:

7 8 9	1 810	2 811	4 6 9	5 710	2 3 9	1 5 9
2 610	3 711	1 412	3 812	4 813	1 611	3 410
4 511	5 612	6 713	2 513	3 614	5 814	2 712
1 313	2 414	3 515	1 714	1 215	4 715	6 815
121415	91315	91014	101115	911 2	101213	111314

$KTS(21)$:

1 2 3	4 5 6	7 8 9	10 11 12	13 14 15
4 7 13	7 10 16	10 13 19	1 13 16	4 16 19
5 8 14	8 11 17	11 14 20	2 14 17	5 17 20
6 9 15	9 12 18	12 15 21	3 15 18	6 18 21
10 17 21	3 13 20	2 6 16	5 9 19	1 8 12
11 18 19	1 14 21	3 4 17	6 7 20	2 9 10
12 16 20	2 15 19	1 5 18	4 8 21	3 7 11
16 17 18	19 20 21	10 18 20	11 16 21	12 17 19
1 7 19	1 4 10	2 13 21	3 14 19	1 15 20
2 8 20	2 5 11	3 5 16	1 6 17	2 4 18
3 9 21	3 6 12	6 8 19	4 9 20	5 7 21
4 11 15	7 14 18	1 9 11	2 7 12	3 8 10
5 12 13	8 15 16	4 12 14	5 10 15	6 11 13
6 10 14	9 13 17	7 15 17	8 13 18	9 14 16

FIGURE 13.1: Small Kirkman triple systems.

are

$$(6m + 3, (2m + 1)(3m + 1), 3m + 1, 3, 1).$$

There are $2m + 1$ blocks in each parallel class, so there are $3m + 1$ parallel classes.

It is instructive to have some examples of Kirkman triple systems, and some small examples will also be useful later in this section. For this reason we present Kirkman triple systems on $v = 9, 15$, and 21 treatments in Figure 13.1. (The case of 15 treatments is of course a solution to Kirkman's original problem.)

Another interesting small case is $v = 39$. The treatments are represented by the integers modulo 39. Each parallel class consists of the 13 blocks

$$\{x + i, y + i, z + i\} : i = 0, 3, 6, 9, 12, 15, 18, 21, 24, 27, 30, 33, 36\},$$

where $\{x, y, z\}$ is a starter block for the parallel class. The starter blocks are

$$\begin{array}{llll} \{0,1,2\}, & \{3,9,27\}, & \{4,10,28\}, & \{5,11,29\}, \\ \{6,15,18\}, & \{7,16,19\}, & \{8,17,20\}, & \{12,31,38\}, \\ \{13,32,36\}, & \{14,30,37\}, & \{21,25,35\}, & \{22,26,33\}, \\ \{23,24,34\}, & \{12,32,37\}, & \{13,30,0\}, & \{14,31,36\}, \\ \{21,26,34\}, & \{22,24,35\}, & \{23,25,33\}. & \end{array}$$

LEMMA 13.1

Suppose there exists a $PB(v; K; 1)$, and suppose there exists a $KTS(2k + 1)$ for every $k \in K$. Then there is a $KTS(2v + 1)$.

Proof. Let us write V for the set $\{1, 2, \ldots, v\}$; we shall assume that the $PB(v; K; 1)$ has treatment set V. If X is any set, we write X_i for the set resulting from subscripting every element of X with an i. In this notation we exhibit a $KTS(2v + 1)$ with treatment set $\{\infty\} \cup V_1 \cup V_2$.

Select any block B of the $PB(v; K; 1)$; suppose $|B| = k$. Then there exists a $KTS(2k+1)$; take its treatment set to be $\{\infty\} \cup B_1 \cup B_2$, and permute the treatments if necessary so as to ensure that $\{\infty, x_1, x_2\}$ is a triple for every $x \in B$. We denote the resulting $KTS(2k + 1)$ by $\mathcal{K}(B)$, and write $\mathcal{T}(B)$ for the set of triples of $\mathcal{K}(B)$. Then we claim that the union of the $\mathcal{T}(B)$, over all blocks B of the $PB(v; K; 1)$, is the required $KTS(v)$.

It is easy to verify that the triples form a $T(1, 2v + 1)$. The pair $\{\infty, x_1\}$ appears in the triple $\{\infty, x_1, x_2\}$ and no other, and the same is true of $\{x_1, x_2\}$. (The triple $\{\infty, x_1, x_2\}$ will arise in more than one of the sets $\mathcal{T}(B)$, but since we took the union of the block sets, these multiple occurrences are not counted multiple times.) If $x \neq y$, then $\{x_i, y_j\}$ will arise in precisely one member of $\mathcal{T}(B)$, namely when B is the unique block containing $\{x, y\}$ in the $PB(v; K; 1)$.

Now suppose x is any member of V. Say s is the number of blocks of the pairwise balanced design that contain x; write $B_{x1}, B_{x2}, \ldots, B_{xs}$ for those blocks of the pairwise balanced design that contain x. Write $\mathcal{P}_{xi}$ for the parallel class containing the triple $\{\infty, x_1, x_2\}$ in $\mathcal{K}(B_{xi})$, and $\mathcal{Q}_x$ for the union of the P_{xi} for $1 \leq i \leq s$ (with the triple $\{\infty, x_1, x_2\}$ only counted once, of course). Every member y of V, other than x, will belong to exactly one of the B_{xi}, so y_1 and y_2 will belong to exactly one block in $\mathcal{Q}_x$. Clearly, ∞, x_1 and x_2 occur exactly once. So $\mathcal{Q}_x$ is a parallel class. The sets $\mathcal{Q}_x$ and $\mathcal{Q}_y$ have no common triple when $x \neq y$, and there are exactly v of these sets—precisely the number of parallel classes in a $KTS(2v+1)$. So the sets $\mathcal{Q}_x$ partition the blocks of the $T(1, 2v+1)$ into parallel classes, and we have a Kirkman triple system. □

In particular, suppose $K = \{4, 7, 10, 19\}$. Then $\{2k+1 : k \in K\} = \{9, 15, 21, 39\}$, and we know that Kirkman triple systems of those four orders exist. From the above lemma, any $PB(v; K; 1)$ gives rise to a $KTS(2v+1)$. In particular, if v ranges through all positive integers congruent to $1 \pmod 3$, $2v+1$ will range through all positive integers congruent to $3 \pmod 6$. Therefore, we investigate the existence of $PB(v; \{4, 7, 10, 19\}; 1)$.

LEMMA 13.2

There is a $PB(v; \{4, 7, 10, 19\}; 1)$ *when* $v = 1, 4, 7, 10, 13, 16, 19, 22, 25, 28, 31, 34, 37$ *or* 40.

Proof. The case $v = 1$ is trivial and $v = 4$, 7, 10, 19 are realized by one-block designs. From Corollary 8.10.1 we obtain a $PB(28; \{4, 7\}; 1)$, a $PB(37; \{4, 10\}; 1)$ and a $PB(40; \{4, 10\}; 1)$. When $v = 13$, 16 or 25, there are balanced incomplete block designs with $k = 4$ and $\lambda = 1$; the cases $v = 13$ and $v = 16$ are the projective and affine planes of parameters 3 and 4, respectively, while a $(25, 50, 8, 4, 1)$-design was exhibited in Section 5.4. (We could also have used this approach for $v = 28$, since a suitable block design was constructed in Section 6.2.)

When $v = 22$, we use a Kirkman triple system on 15 treatments; suppose its treatment set is $\{1, 2, \ldots, 15\}$. Let $\mathcal{F}_1, \mathcal{F}_2, \ldots, \mathcal{F}_7$ be the seven parallel classes in a resolution of the design. We define a pairwise balanced design with treatment set $\{1, 2, \ldots, 15, t_1, t_2, \ldots, t_7\}$ as follows. Each block in $\mathcal{F}_i$ is extended to a 4-set by adding treatment t_i to it. Then block $\{t_1, t_2, t_3, t_4, t_5, t_6, t_7\}$ is appended to the design. The result is a $PB(22; \{4, 7\}; 1)$ with exactly one 7-block. A similar construction works for $v = 31$ (see Exercise 13.1.1). When $v = 34$, we first construct a $PB(28; \{3, 4\}; 1)$ whose treatments are the ordered pairs (x, y), where x is an integer modulo 3 and y is an integer modulo 9, together with ∞. We write

$$\begin{array}{ll} B_1 = \{00, 12, 22, 23\}, & B_2 = \{00, 13, 15\}, \\ B_3 = \{00, 14, 18\}, & B_4 = \{00, 03, 06, \infty\}. \end{array}$$

Then we define

$$B_{ijk} = B_k + ij = \{xy + ij : xy \in B_k\}$$

(where $xy + ij = (x + i, y + j)$, except for the rule $\infty + ij = \infty$), and

$$\mathcal{B}_k = \{B_{ijk} : 0 \le i \le 2, 0 \le j \le 8\}.$$

The blocks $\mathcal{B}_1 \cup \mathcal{B}_2 \cup \mathcal{B}_3 \cup \mathcal{B}_4$ form the $PB(28; \{3, 4\}; 1)$. Moreover, if we define

$$\begin{aligned} \mathcal{S}_h &= \{B_{ij2} : j - 1 \equiv h \pmod 3\} \\ \mathcal{T}_h &= \{B_{ij3} : j \equiv h \pmod 3\}, \end{aligned}$$

then clearly the $\mathcal{S}_h$ and the $\mathcal{T}_h$ partition the blocks of size 3 into six parallel classes. We define six new treatments $s_1, s_2, s_3, t_1, t_2, t_3$, and we append s_h to all blocks in $\mathcal{S}_h$, and t_h to all blocks in $\mathcal{T}_h$. Then a block $\{\infty, s_1, s_2, s_3, t_1, t_2, t_3\}$ is added. We have a $PB(34; \{4, 7, 10\}; 1)$. □

In Section 8.4 we used simple group divisible designs or SGDDs. Recall that an SGDD $(X, \mathcal{G}, \mathcal{A})$ consists of a set of treatments X and sets of subsets of X called groups ($\mathcal{G}$) and blocks ($\mathcal{A}$) such that the groups are a partition of X and such that any two treatments belong to precisely one group or precisely one block, but not both. We shall sometimes refer to the SGDD as a $GD(v; G; A)$, where G and A are the sets of cardinalities of members of $\mathcal{G}$ and $\mathcal{A}$, respectively. (If G or A has only one element, we write that number, omitting set brackets.) If $\mathcal{G} \cup \mathcal{A}$ is interpreted as a set of blocks, we have a pairwise balanced design: If there exists a $GD(v; G; A)$, then there exists a $PB(v; G \cup A; 1)$.

To investigate the existence of $PB(v; \{4, 7, 10, 19\}; 1)$, we shall use two specific group divisible designs repeatedly, a $GD(12; 3; 4)$ and a $GD(15; 3; 4)$. From Theorem 8.10, the existence of two orthogonal Latin squares of side 3 implies the existence of a transversal design $TD(4, 3)$, which is a $GD(12; 3; 4)$. If the technique of puncturing a balanced incomplete block design (as described in Exercise 8.4.4) is applied to a $(16, 20, 5, 4, 1)$-design, we obtain a $GD(15; 3; 4)$. The same technique applied to a $(28, 63, 9, 4, 1)$-design yields a $GD(27; 3; 4)$.

We also need the following lemma, which is a special case of a more general theorem (see Exercise 13.1.4).

LEMMA 13.3

Let w be a fixed positive integer. Say there exists a $GD(v; G; A)$ and, for every k belonging to A, there exists a $GD(wk; w; A_k)$, for some set A_k. Then there exists a $GD(wv; G^*; A^*)$, where

$$\begin{aligned} G^* &= \{wg : g \in G_i\} \\ A^* &= \bigcup_{k \in A} A_k. \end{aligned}$$

Proof. Suppose the original SGDD is $(X, \mathcal{G}, \mathcal{A})$. For each x in X, define $S_x = \{x_1, x_2, \ldots, x_w\}$. If B is a block of size k in the SGDD, construct a $GD(wk; w; A_k)$ whose treatments are the elements $x_1, x_2, \ldots, x_w$ for every x in B and whose groups are $\{S_x : x \in B\}$. Write $\mathcal{A}(B)$ for the set of blocks of this design. Now consider the design $(X^*, \mathcal{G}^*, \mathcal{A}^*)$,

$$X^* = \bigcup_{x \in X} S_x;$$

if H is any member of $\mathcal{G}$ then $H^* = \bigcup_{x \in H} S_x$ and

$$\begin{aligned} \mathcal{G}^* &= \{H^* : H \in \mathcal{G}\}; \\ \mathcal{A}^* &= \bigcup_{B \in \mathcal{A}} \mathcal{A}(B). \end{aligned}$$

It is easy to verify that $(X^*, \mathcal{G}^*, \mathcal{A}^*)$ is a simple group divisible design with the required parameters. □

COROLLARY 13.3.1

If there exists a $GD(v; g; 4)$, then there exists a $GD(3v; 3g; 4)$ and a $PB(3v + 1; \{3g + 1, 4\}; 1)$.

Proof. We take $w = 3$, $G = \{g\}$, and $A = \{4\}$. The condition "for every k belonging to A there exists a $GD(wk; w; A_k)$" now reduces to "there exists a $GD(12; 3; A_4)$." We have constructed a $GD(12; 3; 4)$. So the lemma applies with $A_4 = \{4\}$, and we have a $GD(3v; 3g; 4)$. If an additional treatment ∞ is added to each group and these augmented groups are treated as blocks, we have a $PB(3v + 1; \{3g + 1, 4\}; 1)$. □

COROLLARY 13.3.2

There is a $PB(v; \{4, 7, 10, 19\}; 1)$ when $v = 43, 46, 79$ or 82.

Proof. We apply Corollary 13.3.1 to designs of the following kinds: $GD(14; 2; 4)$, $GD(15; 3; 4)$, $GD(26; 2; 4)$ and $GD(27; 3; 4)$.

To construct a $GD(14;2;4)$, take as treatments the set of integers modulo 14, as groups the sets $\{x, x+7\}$ and as blocks

$$\{i, 2+i, 5+i, 6+i\} : 0 \le i \le 13.$$

The construction of a $GD(26;2;4)$ is analogous (see Exercise 13.1.5). The other two designs were constructed earlier in this section. □

THEOREM 13.4

There exists a $PB(v; \{4,7,10,19\}; 1)$ *whenever* $v \equiv 1 \pmod 3$.

Proof. The theorem is true when $v \le 46$ or $v = 79$ or 82, from Lemma 13.2 and Corollary 13.3.2. So suppose v is not one of those values.

We first observe that we can write

$$v = 12m + 3t + 1,$$

where m and t are integers such that $0 \le t \le m$ and such that there exist three pairwise orthogonal Latin squares of side m. From Theorem 9.7 it is sufficient to require that $m \ge 4$, $m \ne 6$, and $m \ne 10$. This is easy whenever $v \ge 133$, since we can select $m \ge 11$. For $49 \le v \le 58$, we use $m = 4, 0 \le t \le 3$. For $61 \le v \le 76$, we use $m = 5, 0 \le t \le 5$. For $85 \le v \le 106$, we use $m = 7, 0 \le t \le 7$. The cases $109 \le v \le 130$ are handled by $m = 9, 0 \le t \le 7$.

We now proceed by induction. Suppose $v \equiv 1 \pmod 3$ and suppose there is a $PB(v'; \{4,7,10,19\}; 1)$ whenever $1 \le v' < v$ and $v' \equiv 1 \pmod 3$. If $v \le 49$ or $v = 79$ or 82, there is nothing to prove. So suppose that

$$v = 12m + 3t + 1,$$

where $0 \le t \le m$ and three pairwise orthogonal Latin squares of side m exist. From Theorem 8.10 we can construct a $GD(5m; m; 5)$. From one of the groups of this design, delete $m-t$ elements. The result is a $GD(4m+t; \{m,t\}; \{4,5\})$ with just one block of size t.

Now apply Lemma 13.3 in the case $w = 3$. There exist group divisible designs $GD(12;3;4)$ and $GD(15;3;4)$. So we can construct a $GD(12m + 3t; \{3m, 3t\}; 4)$. Adding one new point to each group we obtain a $PB(12m + 3t + 1; \{3m+1; 3t+1; 4\}; 1)$.

Finally, we know that there are simple pairwise balanced designs on $3m+1$ and $3t+1$ treatments with block sizes $\{4,7,10,19\}$, by the induction hypothesis. We replace each block of those sizes by the appropriate pairwise balanced design (the construction of Theorem 2.1). The result is a $PB(v; \{4,7,10,19\}; 1)$. □

COROLLARY 13.4.1

There is a Kirkman triple system on v treatments for all $v \equiv 3 \pmod 6$.

Exercises 13.1

13.1.1 (i) Suppose there exists a Kirkman triple system with $6n+3$ treatments, and consequently $3n+1$ parallel classes. Prove that there exists a $PB(9n+4; \{4, 3n+1\}; 1)$ with exactly one block of size $3n+1$. Hence prove that there is a $PB(31; \{4, 10\}; 1)$.

(ii) Given that there exists a Kirkman triple system with $v = 27$, prove that there exists a balanced incomplete block design with parameters $(40, 130, 13, 4, 1)$.

(iii) Assuming further the existence of a Kirkman triple system with $v = 81$, prove the existence of a balanced incomplete block design with parameters $(121, 1210, 40, 4, 1)$.

13.1.2* Use the existence of Kirkman triple systems to show[1] that a pairwise balanced design $PB(v; \{3, 4\}; 1)$ exists whenever $v \equiv 0$ or $1 \pmod 6$, $v > 6$.

13.1.3* For what values of v does there exist a $PB(v; \{4, 3\}; 1)$ that contains at least one block of size 4 and at least one block of size 3?

13.1.4 Suppose $(X, \mathcal{G}, \mathcal{A})$ is an SGDD. With each treatment x associate a positive integer weight w_x, and select a collection of pairwise disjoint sets $\{S_x : x \in X\}$ where S_x has size w_x. Given a block $B \in \mathcal{A}$, define

$$\begin{aligned} S(B) &= \bigcup_{x \in B} S_x, \\ \mathcal{G}(B) &= \{S_x : x \in B\}, \end{aligned}$$

and for each B select a collection $\mathcal{A}(B)$ of subsets of $S(B)$ such that $(S(B), \mathcal{G}(B), \mathcal{A}(B))$ is an SGDD. If G is any member of $\mathcal{G}$, then define

$$G^* = \bigcup_{x \in G} S_x.$$

If $X^* = \bigcup_{x \in X}$, $\mathcal{G}^* = \{G : G \in \mathcal{G}\}$, $\mathcal{A}^* = \bigcup_{B \in \mathcal{A}} \mathcal{A}(B)$, prove that $(X^*, \mathcal{G}^*, \mathcal{A}^*)$ is an SGDD.

13.1.5 A design is constructed using as treatments the set of integers modulo 26, groups $\{x, x+13\}$ and blocks

$$\begin{aligned} \{i, 6+i, 8+i, 9+i\} &: 0 \le i \le 25 \\ \{i, 4+i, 11+i, 16+i\} &: 0 \le i \le 25. \end{aligned}$$

[1]See also Exercises 2.1.4 and 12.1.8.

Prove that the design is a $GD(26; 2; 4)$.

13.2 Kirkman Packings and Coverings

In Kirkman's schoolgirl problem, what if the class size is not an odd multiple of 3? In that case one can ask for a set of daily walks in which as many pairs walk together as is possible without repetitions (called a *Kirkman packing design*), or that every pair walks together at least once and the number of repetitions is minimized (a *Kirkman covering design*). When the number of schoolgirls is not divisible by 3, one row of 2 or 4 girls is allowed. Another solution in the case of $v \equiv 1 (\mathrm{mod}\, 3)$ is to allow one girl to walk alone; each day's walk is called a *near-triangle-factor*, and one can discuss packings and coverings with these factors. Similarly, in the case of $v \equiv 2 (\mathrm{mod}\, 3)$, packings with one row of five girls have been considered.

In those cases where a Kirkman triple system cannot exist, we would like to find the maximum number of rounds in a Kirkman packing. Writing $KP(v, r)$ for a Kirkman packing with r rounds (or *of length* r), we would like to determine $r(v)$, the largest value of r such that a $KP(v, r)$ exists. If $b(v)$ denotes the number of unordered pairs of objects covered by a round of a $KP(v, r)$ then the number of pairs covered by the design is $rb(v)$; clearly, this cannot exceed $\frac{1}{2}v(v-1)$, so

$$r(v) \leq M(v) = \left\lfloor \frac{v(v-1)}{2b(v)} \right\rfloor. \tag{13.1}$$

A Kirkman packing that attains this upper bound $M(v)$ will be called *optimal*. One question is: For what values of v does an optimal Kirkman packing exist?

First, suppose the number of schoolgirls is an even multiple of 3. Then the nearest one can come to the exact solution is a design in which each pair will occur exactly once, except that, given a schoolgirl x, there will exist a schoolgirl x' such that the pair xx' never occurs.

Resolvable packing designs called *Nearly Kirkman triple systems* were introduced by Kotzig and Rosa [93]. The same idea was invented independently by Irving [80]. A *Nearly Kirkman triple system* $NKTS(v)$ is a way of selecting from a v-set S one one-factor and $\frac{1}{2}v - 1$ triangle factors that together contain every pair of elements precisely once. If the one-factor is deleted, the triangle factors form a $KP(v, \frac{1}{2}v - 1)$. When $v = 6m$, $b(v) = 6m$ also (there are $2m$ triples, and each covers three pairs). So

$$M(v) = \left\lfloor \frac{v(v-1)}{2v} \right\rfloor = \tfrac{1}{2}v - 1.$$

Thus Nearly Kirkman triple systems provide optimal Kirkman packing designs for the case $v \equiv 0 \pmod 6$ (simply delete the one-factor).

No $NKTS(6)$ or $NKTS(12)$ exists [93]. When $v = 6$, obviously only one round is possible, and $r(6) = 1$. For $v = 12$, a hand computation shows that five rounds cannot be constructed. On the other hand, a set of 4 rounds can be constructed from the affine plane on 16 points. Select one line; delete all points on this line and all lines in its parallel class. The remaining blocks form the design, with the rounds corresponding to the original parallel classes (see Figure 13.2). So $r(12) = 4$; there exists a $KP(12, 4)$.

1	2	3	4												
5	6	7	8												
9	10	11	12												
13	14	15	16												
1	5	9	13	1	7	12	14		1	5	9		1	7	12
2	6	10	14	2	8	11	13		2	6	10		2	8	11
3	7	11	15	3	5	10	16		3	7	11		3	5	10
4	8	12	16	4	6	9	15		4	8	12		4	6	9
								$\Rightarrow$							
1	6	11	16	1	8	10	15		1	6	11		1	8	10
2	5	12	15	2	7	9	16		2	5	12		2	7	9
3	8	9	14	3	6	12	13		3	8	9		3	6	12
4	7	10	13	4	5	11	14		4	7	10		4	5	11

FIGURE 13.2: Construction of $KP(12, 4)$.

It has been shown that an $NKTS(v)$ exists whenever v is a multiple of 6 greater than 12 (see for example [115]; a new proof appears in [116]).

If the number of schoolgirls is not a multiple of 3, the situation is more complicated. A Kirkman packing design (or *project design*—for the history of this name, see [32]) of order v and length r, or $KP(v, r)$, is a way of choosing sets called blocks from a set of v objects, and of partitioning the set of blocks into r subsets called rounds, so that: each object occurs exactly once per round; all blocks in each round are triples except for at most one, and that one can contain two or four objects; and each object-pair occurs in at most one block in the design.

When $v \equiv 2 \pmod 3$, say $v = 3n - 1$, there is one pair in each round, and $b(v) = 3(n-1) + 1 = v - 1$, so $M(v) = \lfloor \frac{v(v-1)}{2(v-1)} \rfloor = \lfloor \frac{v}{2} \rfloor$. If v is odd ($v \equiv 5 \pmod 6$), $M(v) = (v-1)/2$, while if v is even ($v \equiv 2 \pmod 6$), $M(v) = v/2$. The maximum is attained by deleting one point from a Kirkman triple system or a Nearly Kirkman triple system. In the cases $v = 5$ and 11, the maximums are 1 and 4 rounds respectively; one might hope for five rounds on 11 points, but no example exists. A $KP(11,4)$ can be constructed from a $KP(12,4)$ by deleting one point.

In the case $v \equiv 1 \pmod 3$, (13.1) yields $r(v) \leq \lfloor \frac{1}{2}v(v-1)/(v+2) \rfloor$. The following formulation gives a simpler bound that is slightly lower, although this has no practical importance because r must be an integer.

THEOREM 13.5

When $v \equiv 1 \pmod 3$,

$$r(v) \leq \lfloor (v-3)/2 \rfloor.$$

Proof. Suppose a $KP(v,r)$ exists, where $v \equiv 1 \pmod 3$. Each round contains one block of size 4. Write a_x and b_x respectively for the number of blocks of size 3 and of size 4 containing element x. Counting elements in blocks of size 4, $\sum b_x = 4r$; if $b_x \leq 1$ for all x (that is, x appears in at most one 4-block) then $4r \leq v$, and certainly $r \leq (v-3)/2$. So we can assume x was chosen so that $b_x = 2$. Then

$$\begin{aligned} a_x + b_x &= r; \\ 2a_x + 3b_x &\leq v - 1. \end{aligned}$$

So $b_x \leq v - 1 - 2r$. Therefore $2 \leq v - 1 - 2r$, and $r \leq (v-3)/2$. As r is integral, $r \leq \lfloor (v-3)/2 \rfloor$. □

The upper bound in Theorem 13.5 cannot be met for small orders. When $v = 7, 10, 13$ the bounds are respectively 2, 3 and 5; but the maximum rounds are 1, 1 and 4. However, the bound can be met for almost all values.

THEOREM 13.6 [32, 114, 28]

When $v \equiv 1 \pmod 3$, $r(v) = \lfloor (v-3)/2 \rfloor$ *except when* $v = 7, 10, 13$ *and possibly* $v = 19$.

One interesting class of designs is the set of Kirkman packings in which one treatment appears only in 4-blocks. These are *focused*, and the special element is the *focus*. Clearly a focused Kirkman packing on $3n+1$ treatments can have at most n rounds.

THEOREM 13.7

There is a focused $KP(3n+1, n)$ *whenever* $n \geq 4$.

Proof. Suppose $n \neq 6$. Select a pair A, B of orthogonal transversal squares of side n. We may assume A and B each have main diagonal $(1, 2 \ldots, n)$. Then round i consists of the quadruple

$$\{i\ (n+i)\ (2n+i)\ \infty\}$$

and the $n-1$ triples

$$\{j\ (n+a_{ij})\ (2n+b_{ij})\} : j = 1, 2, \ldots, n, j \neq i.$$

This leaves the case $n = 6, 3n+1 = 19$. A suitable design is

$\infty 12c$	$34d$	$56e$	$78f$	$90g$	abh
$\infty 9ad$	$57c$	$10h$	$2bg$	$38e$	$46f$
$\infty 7be$	$28h$	$49c$	$60d$	$1af$	$35g$
$\infty 50f$	$16g$	$8bd$	$3ac$	$47h$	$29e$
$\infty 48g$	$0ae$	$23f$	$59h$	$6bc$	$17d$
$\infty 36h$	$9bf$	$7ag$	$14e$	$25d$	$80c$

□

We write $\overline{r}(v)$ for the minimum possible number of factors in a Kirkman covering design $KCD(v)$.

THEOREM 13.8

(i) *For every* $v \equiv 2 \pmod 6$, $\overline{r}(v) = v/2$;
(ii) *For every* $v \equiv 5 \pmod 6$ *with* $v \geq 17$, $\overline{r}(v) = (v+1)/2$.

THEOREM 13.9

There is a Kirkman covering design $KCD(v)$ *for every* $v \equiv 1 (\bmod 3)$, $v \geq 4$, *and* $\overline{r}(v) = \lfloor (v-1)/2 \rfloor$.

Theorem 13.8 is not difficult. Constructions proving Theorem 13.9 are given in [116]. The following example, a $KCD(13)$, is given in [114].

2 7 9	5 6 11	8 12 13	1 3 4 10
2 5 12	4 6 9	1 7 13	3 10 8 11
2 6 10	3 7 12	9 11 13	1 4 5 8
1 2 11	3 6 13	5 7 10	4 8 9 12
2 4 13	6 7 8	10 11 12	1 5 3 9
2 3 8	4 7 11	1 6 12	5 9 10 13

The *excess* of a $KCD(v)$ is the multiset of pairs of objects that appear together more than once in the design. It is usually interpreted as a graph (with multiple edges allowed). For example, if pair ab appears in three blocks, there are two extra appearances, and the excess would have a 2-fold edge ab. Similarly, the *leave* of a $KPD(v)$ consists of those pairs that are not covered.

In the case $v \equiv 2 \pmod 3$ where the leaves of $KPD(v)$s and excesses of $KCD(v)$s are uniquely determined, the leave or excess of a $KPD(v)$ or $KCD(v)$ when $v \equiv 1 \pmod 3$ can take many forms. For example, here is a $KCD(19)$ with treatment set $\{1, 2, \ldots, 18\} \cup \{\infty\}$:

∞,7,14	6,3,5	2,16,12	8,17,11	1,18,10	9,4,15,13
∞,8,15	7,4,6	3,17,13	9,18,12	2,10,11	1,5,16,14
∞,9,16	8,5,7	4,18,14	1,10,13	3,11,12	2,6,17,15
∞,1,17	9,6,8	5,10,15	2,11,14	4,12,13	3,7,18,16
∞,2,18	1,7,9	6,11,16	3,12,15	5,13,14	4,8,10,17 .
∞,3,10	2,8,1	7,12,17	4,13,16	6,14,15	5,9,11,18
∞,4,11	3,9,2	8,13,18	5,14,17	7,15,16	6,1,12,10
∞,5,12	4,1,3	9,14,10	6,15,18	8,16,17	7,2,13,11
∞,6,13	5,2,4	1,15,11	7,16,10	9,17,18	8,3,14,12

The excess consists of nine disjoint double edges (two copies of $\{i, i+9\}$, for $i = 1, 2, \ldots, 9$).

Kirkman packings have applications in cryptography (see Section 16.2). For these, block size 2 is undesirable, so two variations have been proposed for packings on v treatments, where $v \equiv 2 \pmod 3$.

One variation is a design in which each round is composed of $(v-5)/3$ triples and one quintuple. Such a design with r rounds wil be denoted $KP(3, 5^*; v, r)$. Cao and Zhu [30] establish up to 22 unsettled values of $v \equiv 2 \pmod 3$, the maximum number $m(v)$ of factors possible in such a resolvable packing of v points.

THEOREM 13.10 [30]

Write $S = \{23, 29, 35, 59, 83, 89, 95, 101, 107, 113, 119, 125, 137, 149, 155, 173, 179, 185, 197, 203, 227\}$. *Then*

(i) $m(v) = 1$ *for* $v \in \{5, 8, 11, 14\}$, *and* $m(17) = 4$;

(ii) $m(v) = (v-8)/2$ *for* $v \equiv 2 \pmod 6$ *and* $v \geq 20$, $v \neq 26$;

(iii) $m(v) = (v-7)/2$ *for every* $v \equiv 5 \pmod 6$ *with* $v > 17$, *except possibly when* $v \in S$.

The only unsettled value when $v \equiv 2 \pmod 6$ is $v = 26$; in [30] it is shown that $8 \leq m(26) \leq 10$.

An interesting example is the case $v = 20$. Each factor consists of five triples and a quintuple, so 25 pairs are covered in each factor. Therefore

$$m(20) \le \left\lfloor \binom{20}{2}/25 \right\rfloor = 7.$$

Cao and Zhu establish that 7 factors cannot be achieved, and provide a solution with 6 factors, establishing $m(20) = 6$. The treatments are

$$\{1, 2, \ldots, 18\} \cup \{\infty_1, \infty_2\},$$

and the factors are

$$\begin{array}{cccccc}
5,6,13 & 15,11,\ 7 & 16,18,1 & 3,12,\infty_1 & 10,17,\infty_2 & 2,4,\ 8,\ 9,14 \\
6,1,14 & 16,12,\ 8 & 17,13,2 & 4,\ 7,\infty_1 & 11,18,\infty_2 & 3,5,\ 9,10,15 \\
1,2,15 & 17,\ 7,\ 9 & 18,14,3 & 5,\ 8,\infty_1 & 12,13,\infty_2 & 4,6,10,11,16 \\
2,3,16 & 18,\ 8,10 & 13,15,4 & 6,\ 9,\infty_1 & 7,14,\infty_2 & 5,1,11,12,17 \\
3,4,17 & 13,\ 9,11 & 14,16,5 & 1,10,\infty_1 & 8,15,\infty_2 & 6,2,12,\ 7,18 \\
4,5,18 & 14,10,12 & 15,17,6 & 2,11,\infty_1 & 9,16,\infty_2 & 1,3,\ 7,\ 8,13
\end{array}.$$

Another variant in the case $v \equiv 2 \pmod 3$ is a packing with precisely two blocks of size 4, denoted $KP(3, 4^{**}; v, r)$. In this case the maximum number of rounds is at most $\lfloor (v-5)/2 \rfloor$. Cao and Tang [29] showed that this can be realized provided $v \ge 32, v \equiv 2 \pmod 3$.

Exercises 13.2

13.2.1 A design has support set

$$(Z_{17} \times \{1, 2\}) \cup \{\infty_1, \infty_2\}.$$

One parallel class is

$$\{\{(i,1),(i,2)\} : i \in Z_{17}\} \cup \{\{\infty_1, \infty_2\}\}$$

and 12 others are formed by developing

$$\begin{array}{ccccccccc}
(2,1) & (7,2) & (8,2) & (12,1) & (10,2) & (13,2) & (3,1) & (12,2) & (0,2) \\
(11,1) & (14,2) & (4,2) & (7,1) & (6,2) & (15,2) & (0,1) & (9,1) & (11,2) \\
(13,1) & (8,1) & (3,2) & (10,1) & (14,1) & (4,1) & (15,1) & (16,1) & (1,1) \\
(16,2) & (1,2) & (5,2) & \infty_1 & (5,1) & (9,2) & (6,1) & (2,2) & \infty_2
\end{array}$$

modulo 17. Prove that the design is an $NKTS(36)$.

13.2.2 Write $S = \{1, 2, \ldots, 3n\}$ and $T = \{1, 2, \ldots, 6n\}$.

(i) Suppose the sets of triples $\{a_ib_ic_i : 1 \le i \le n\}$ and $\{d_ie_if_i : 1 \le i \le n\}$ are edge-disjoint triangle factors on S. Show that the following four factors are edge-disjoint triangle factors of T.

$$\begin{array}{l}\{a_i, b_i, c_i : 1 \le i \le n\} \cup \{(d+3n)_i, (e+3n)_i, (f+3n)_i : 1 \le i \le n\},\\ \{a_i, (b+3n)_i, (c+3n)_i : 1 \le i \le n\} \cup \{(d+3n)_i, e_i, f_i : 1 \le i \le n\},\\ \{(a+3n)_i, (b+3n)_i, c_i : 1 \le i \le n\} \cup \{d_i, e_i, (f+3n)_i : 1 \le i \le n\},\\ \{(a+3n)_i, b_i, (c+3n)_i : 1 \le i \le n\} \cup \{d_i, (e+3n)_i, f_i : 1 \le i \le n\}.\end{array}$$

(ii) Use part (i) to prove: If there exists a Kirkman triple system on $v = 12n + 9$ points, then there is a Nearly Kirkman triple system on $2v$ points.

13.2.3* Suppose there exist an $NKTS(v)$ and a pair of orthogonal Latin squares of side v. Construct an $NKTS(3v)$.

13.2.4 Show that there is no $KP(11, 5)$.

13.2.5* Verify the following results for small Kirkman packings: $r(7) = 1$, $r(10) = 1$.

13.2.6 Verify for Kirkman packings: $r(13) = 4$.
Hint: This is a lengthy brute-force calculation.

13.2.7 Prove Theorem 13.8.

13.2.8 Describe the leave of a $KPD(v)$ and the excess of a $KCD(v)$ when $v \equiv 2 \pmod 3$.

13.2.9 What is the excess of the $KCD(13)$ shown after Theorem 13.9?

13.2.10 Prove by direct calculation that no $KP(3, 4^{**}; 14, 5)$ exists.

13.2.11* Prove that a $KP(3, 4^{**}; v, r)$ with $v \ge 17$ can have at most $\lfloor (v-5)/2 \rfloor$ rounds.

13.2.12 Find an upper bound for the number of rounds in a $KP(3, 5^{*}; v, r)$.

Chapter 14

Hadamard Matrices

14.1 Basic Ideas

In 1893, Hadamard [59] addressed the problem of the maximum absolute value of the determinant of an $n \times n$ complex matrix H with all its entries on a unit circle. The maximum value is $\sqrt{n^n}$. Among real matrices, this value is attained if and only if H has every entry either 1 or -1, and satisfies

$$HH^T = nI. \tag{14.1}$$

Such matrices had appeared earlier [149], in the case where H is a power of 2, but they are now called *Hadamard* matrices.

Suppose H is an $n \times n$ matrix whose (i, r) entry is h_{ir}, where $h_{ir} = \pm 1$. The (i, r) entry of H^T will be h_{ri} so the (i, j) entry of HH^T is

$$\sum_{r=1}^{n} h_{ir} h_{jr}.$$

In particular, the (i, i) entry is

$$\sum_{r=1}^{n} h_{ir}^2.$$

If every entry of H is 1 or -1, then $h_{ir}^2 = 1$ for every i and r, so the (i, i) entry above equals n. Therefore, an $n \times n$ $(1, -1)$-matrix will be Hadamard if and only if

$$\sum_{r=1}^{n} h_{ir} h_{jr} = 0 \text{ when } i \neq j. \tag{14.2}$$

This observation can be expressed by defining a Hadamard matrix to be a square $(1, -1)$-matrix whose rows are pairwise orthogonal (as vectors).

Suppose H is a Hadamard matrix of side n. Then $HH^T = nI$ implies that H has inverse $n^{-1}H^T$. Since $H^{-1}H = I$ we have

$$H^T H = nI,$$

and H^T is also Hadamard. So the transpose of a Hadamard matrix is always Hadamard, and consequently the columns of a Hadamard matrix are orthogonal:

$$\sum_{r=1}^{n} h_{ri}h_{rj} = 0 \text{ when } i \neq j. \tag{14.3}$$

LEMMA 14.1

If H is a Hadamard matrix, and if K is obtained from H by negating some or all rows, negating some or all columns, permuting the rows, and permuting the columns, then K is Hadamard.

Proof. Assume H has order n. We consider the four types of transformation separately. In each case K has all entries 1 and -1. We verify equation (14.2).

(i) Suppose K is obtained by permuting rows—say, rows i and j of K were rows x and y of H. Then

$$\sum_{r=1}^{n} k_{ir}k_{jr} = \sum_{r=1}^{n} h_{xr}h_{yr} = 0.$$

(ii) Suppose row i of K is the negative of row i of H. Then

$$\begin{aligned}\sum_{r=1}^{n} k_{ir}k_{jr} &= \sum_{r=1}^{n} (-k_{ir})k_{jr} \\ &= -\sum_{r=1}^{n} h_{ir}k_{jr}.\end{aligned}$$

If row j of K equals row j of H, we have

$$-\sum_{r=1}^{n} h_{ir}k_{jr} = -\sum_{r=1}^{n} h_{ir}h_{jr} = -0 = 0,$$

while if row j of K is the negative of row j of H, then

$$-\sum_{r=1}^{n} h_{ir}k_{jr} = -\sum_{r=1}^{n} h_{ir}(-h_{jr}) = -\sum_{r=1}^{n} h_{ir}h_{jr} = 0.$$

In either case we have (14.2).

(iii) If K is obtained from H by permuting the columns, then $k_{i1}k_{j1} = h_{is}h_{js}$ for some s, $k_{i2}k_{j2} = h_{it}h_{jt}$ for some t, and so on. Therefore

$$\sum_{r=1}^{n} k_{ir}k_{jr} = h_{i1}h_{j1} + h_{i2}h_{j2} + \ldots + h_{in}h_{jn} = 0,$$

although the terms may arise in a different order.

(iv) Suppose K is formed by negating certain columns of H. Since

$$h_{ir}h_{jr} = (-h_{ir})(-h_{jr}),$$

$k_{ir}k_{jr} = h_{ir}h_{jr}$ whether column r has been negated or not, and

$$\sum_{r=1}^{n} k_{ir}k_{jr} = \sum_{r=1}^{n} h_{ir}h_{jr} = 0.$$

Now suppose K is obtained from H by a combination of the four operations. Then we can form a sequence of matrices $K_1, K_2, \ldots, K_m$, where K_1 is obtained from H by one operation (such as a row permutation), K_2 comes from performing one operation on K_1 (for example, column negation), and so on; $K = K_m$. Then

H Hadamard implies K_1 Hadamard;

K_1 Hadamard implies K_2 Hadamard;

. . . .

K_{m-1} Hadamard implies K Hadamard. □

We shall say that two Hadamard matrices are *equivalent* if one can be obtained from the other by a sequence of row and column permutations and negations. Since other equivalence relations on Hadamard matrices are also important, we shall refer to the relation we have just defined as "Hadamard equivalence" whenever any confusion might otherwise occur.

Given a Hadamard matrix of side n, suppose one negates every row in which the first entry is negative, and then negates every column in which the first element is now negative. The resulting matrix has first row and column entries all $+1$. Such a Hadamard matrix is called *normalized.*

It follows from Lemma 14.1 that every Hadamard matrix is equivalent to a normalized Hadamard matrix. This does not define a unique "normalized" form for a Hadamard matrix; it is possible to find more than one normalized matrix equivalent to a given Hadamard matrix.

As an example, let us consider normalized Hadamard matrices of side 4. Any such matrix must have the form

$$\begin{bmatrix} 1 & 1 & 1 & 1 \\ 1 & a & b & c \\ 1 & d & e & f \\ 1 & g & h & i \end{bmatrix}$$

where each of $a, b, \ldots, i$ is chosen from $\{1, -1\}$. Since rows 1 and 2 are orthogonal, we have

$$1 + a + b + c = 0;$$

the only possible solutions have one of $\{a, b, c\}$ equal to $+1$ and the others equal to -1. Similarly, rows 3 and 4 must contain exactly two entries -1 each. Now consider the orthogonality of the columns. A similar argument to the above proves that each of columns 2, 3, 4, contains exactly two entries -1. It follows that the matrix

$$A = \begin{bmatrix} a & b & c \\ d & e & f \\ g & h & i \end{bmatrix}$$

has one entry $+1$ per row and per column. So the entries $+1$ form a permutation matrix within A. There are six possible permutations, so there are six normalized Hadamard matrices; using the convenient notation that $+$ represents $+1$ and $-$ represents -1, they are

$$\begin{array}{cccccc} ++++ & ++++ & ++++ & ++++ & ++++ & ++++ \\ ++-- & ++-- & +-+- & +-+- & +--+ & +--+ \\ +-+- & +--+ & ++-- & +--+ & ++-- & +-+- \\ +--+ & +-+- & +--+ & ++-- & +-+- & ++-- \end{array}.$$

It is easy to see that these six matrices are equivalent—in fact, each one can be obtained from each other by row permutation alone. So there is exactly one equivalence class of Hadamard matrices of side 4. It is easy to construct Hadamard matrices of sides 1 and 2:

$$+ \qquad \begin{array}{cc} + & + \\ + & - \end{array}.$$

A natural question presents itself: For what sides n do Hadamard matrices exist? We start with a necessary condition.

THEOREM 14.2

If there is a Hadamard matrix of side n, then $n = 1$, $n = 2$, or n is a multiple of 4.

Proof. Suppose $n > 2$, and suppose there is a Hadamard matrix of side n. Then there is a Hadamard matrix of side n whose first row has every entry 1. By permuting columns we can make sure that the first $\frac{1}{2}n$ columns of the matrix have second entry 1 and the later columns have second entry -1. Similarly, we can permute columns so that the first three rows of the matrix look like

$$\begin{array}{cccc} +++\cdots & +++\cdots & +++\cdots & +++\cdots \\ +++\cdots & +++\cdots & ---\cdots & ---\cdots. \\ +++\cdots & ---\cdots & +++\cdots & ---\cdots \end{array}$$

Call the resulting matrix H.

Suppose H contains a columns that start $+++\ldots$, b columns that start $++-\ldots$, c columns that start $+-+\ldots$, and d columns that start $+--\ldots$. Then

$$a+b+c+d=n. \tag{14.4}$$

If H has (i,j) entry h_{ij}, then HH^T has $(1,2)$ entry

$$\sum_{r=1}^{n} h_{1r}h_{2r} = a+b-c-d,$$

so equation (14.2) yields

$$a+b-c-d=0. \tag{14.5}$$

Similarly, we get

$$\begin{aligned} a-b+c-d &= 0, & (14.6)\\ a-b-c-d &= 0, & (14.7) \end{aligned}$$

by considering the $(1,3)$ and $(2,3)$ entries, respectively, in HH^T. Adding equations (14.5) and (14.6) yields $d=a$; similarly, $c=a$ and $b=a$. So (14.4) reduces to

$$4a=n,$$

and n is a multiple of 4. □

No other restrictions on the side of a Hadamard matrix are known; in fact, it is conjectured that there is a Hadamard matrix of every side permitted by Theorem 14.2. The smallest side for which the existence of a Hadamard matrix is currently undecided is 668 (the previous "record-holder," 428, was constructed in [84]). We shall present a number of methods of construction in later sections.

Exercises 14.1

14.1.1* Prove that there are exactly 16 $(1,-1)$-matrices of size 2×2, and that exactly 8 of them are Hadamard.

14.1.2* How many $(1,-1)$ matrices are there of size 4×4? How many of them are Hadamard?

14.1.3 As a generalization of Hadamard matrices, we define a *c-matrix* of side n to be an $n \times n$ matrix with one entry 0 and all other entries 1 or -1 in each row, whose rows are pairwise orthogonal. Show that the transpose of a c-matrix is necessarily a c-matrix. Find examples for $n = 2$ and $n = 6$.

14.1.4 A is a Hadamard matrix of side n, and H is a Hadamard matrix of side $m + n$, where

$$H = \begin{bmatrix} A & B \\ C & D \end{bmatrix}$$

for some matrices B, C, D. Prove that $m \geq n$.

14.2 Hadamard Matrices and Block Designs

One reason for studying Hadamard matrices is their relationship to balanced incomplete block designs. This has also been the source of several results about Hadamard matrices, including one of the main constructions.

THEOREM 14.3

There exists a Hadamard matrix of side $4m$ *if and only if there exists a* $(4m-1, 2m-1, m-1)$*-design.*

Proof. Suppose H is a Hadamard matrix of side $4m$. Without loss of generality we may assume that H is normalized. Since H is Hadamard, we know that

$$\sum_{r=1}^{4m} h_{ir}h_{jr} = 0 \text{ if } j \neq i,$$

so $\sum_{r=1}^{4m} h_{jr} = 0$ if $j \neq 1$. Similarly, $\sum_{r=1}^{4m} h_{rj} = 0$ when $j \neq 1$. So the sum of entries in any row or any column, except the first, is zero.

Let A be the $(4m-1) \times (4m-1)$ matrix formed by removing the first row and column from H, so that

$$H = \begin{bmatrix} 1 & 1 & \cdots & 1 \\ 1 & & & \\ \vdots & & A & \\ 1 & & & \end{bmatrix}.$$

Then the sum of any row or any column of A must be -1, or equivalently,

$$AJ = JA = -J,$$

where as usual J is the $(4m-1) \times (4m-1)$ matrix with every entry $+1$.

Since $HH^T = 4mI$, we see that

$$AA^T = \begin{bmatrix} 4m-1 & -1 & \cdots & -1 \\ 1 & 4m-1 & \cdots & -1 \\ \vdots & \vdots & & \vdots \\ -1 & -1 & \cdots & 4m-1 \end{bmatrix} = 4mI - J.$$

Now consider the matrix $B = \frac{1}{2}(A + J)$. We have

$$\begin{aligned} BJ &= \tfrac{1}{2}(AJ + JJ) \\ &= \tfrac{1}{2}[-J + (4m-1)J] \\ &= (2m-1)J, \end{aligned}$$

and similarly

$$JB = (2m-1)J; \tag{14.8}$$

again

$$\begin{aligned} BB^T &= \tfrac{1}{4}(A + J)(A^T + J) \\ &= \tfrac{1}{4}(AA^T + JA^T + AJ + JJ) \\ &= \tfrac{1}{4}[(4mI - J) - J - J + (4m-1)J] \\ &= mI + (m-1)J. \end{aligned} \tag{14.9}$$

From Theorem 2.7, equations (14.8) and (14.9) show that B is the incidence matrix of a $(4m-1, 2m-1, m-1)$-design.

Conversely, say a $(4m-1, 2m-1, m-1)$-design exists. Let C be its incidence matrix, and set $A = 2C - J$. Form a matrix H by appending a new first row and column to A with every entry $+1$. It is easy to show that H is Hadamard. $\square$

The designs with $v = 4m-1$, $k = 2m-1$, and $\lambda = m-1$ are often called Hadamard 2-designs, as we said in Section 5.1. These are extremal symmetric balanced incomplete block designs, in the sense that they represent the nearest approach to a design with the same parameters as its complement. (The finite projective planes form the opposite extreme.)

In view of Theorems 5.7 and 5.8, we now know:

COROLLARY 14.3.1

If p^r is a prime power congruent to 3 (mod 4)*, then there is a Hadamard matrix of side $p^r + 1$. If p^r and $p^r + 2$ are both odd prime powers, then there is a Hadamard matrix of side $p^r(p^r + 2) + 1$.* □

The matrices of order p^r+1, constructed from Paley difference sets, are often called *Paley Hadamard matrices*. As an example we construct a Paley Hadamard matrix of order 8. The difference set $\{0, 1, 3\}$ can be used to construct a $(7, 3, 1)$-design. The incidence matrix of that design, and the corresponding Hadamard matrix, are shown in Figure 14.1.

$$\begin{bmatrix} 1 & 0 & 0 & 0 & 1 & 0 & 1 \\ 1 & 1 & 0 & 0 & 0 & 1 & 0 \\ 0 & 1 & 1 & 0 & 0 & 0 & 1 \\ 1 & 0 & 1 & 1 & 0 & 0 & 0 \\ 0 & 1 & 0 & 1 & 1 & 0 & 0 \\ 0 & 0 & 1 & 0 & 1 & 1 & 0 \\ 0 & 0 & 0 & 1 & 0 & 1 & 1 \end{bmatrix} \quad \begin{bmatrix} + & + & + & + & + & + & + & + \\ + & + & - & - & - & + & - & + \\ + & + & + & - & - & - & + & - \\ + & - & + & + & - & - & - & + \\ + & + & - & + & + & - & - & - \\ + & - & + & - & + & + & - & - \\ + & - & - & + & - & + & + & - \\ + & - & - & - & + & - & + & + \end{bmatrix}$$

FIGURE 14.1: Incidence matrix and Hadamard matrix.

THEOREM 14.4

The existence of a Hadamard matrix of side $4m$ implies the existence of balanced incomplete block designs of parameters:

(i) $(2m - 1, 4m - 2, 2m - 2, m - 1, m - 2)$;

(ii) $(2m, 4m - 2, 2m - 1, m, m - 1)$;

(iii) $(2m - 1, 4m - 2, 2m, m, m)$.

Proof. Suppose A is the incidence matrix of the $(4m - 1, 2m - 1, m - 1)$-design corresponding to the Hadamard matrix. Assume that the rows have been reordered, if necessary, so that all rows with first entry 1 precede all rows with first entry 0. Then

$$A = \begin{bmatrix} e & B \\ 0 & C \end{bmatrix}$$

where e is a column of $2m - 1$ entries 1, 0 is a column of $2m$ entries 0, and B

and C are $(0,1)$-matrices. Since $AA^T = mI + (m-1)J$ we have

$$mI + (m-1)J = \begin{bmatrix} ee^T + BB^T & BC^T \\ CB^T & CC^T \end{bmatrix} = \begin{bmatrix} J + BB^T & BC^T \\ CB^T & CC^T \end{bmatrix}$$

whence

$$BB^T = mI + (m-2)J, \qquad (14.10)$$
$$CC^T = mI + (m-1)J. \qquad (14.11)$$

Since A is the incidence matrix of a symmetric design, it also satisfies $A^T A = mI + (m-1)J$, so

$$mI + (m-1)J = \begin{bmatrix} e^T & 0 \\ B^T & C^T \end{bmatrix} \begin{bmatrix} e & B \\ 0 & C \end{bmatrix} = \begin{bmatrix} e^T e & e^T B \\ B^T e & B^T B + C^T C \end{bmatrix}.$$

Since $e^T B = (m-1)J$ we have

$$JB = (m-1)J. \qquad (14.12)$$

So every column of B has $m - 1$ entries 1. Therefore, $B^T B$ has diagonal $(m-1, m-1, \ldots, m-1)$. The equation

$$B^T B + C^T C = mI + (m-1)J$$

therefore implies that $C^T C$ has every diagonal entry $(2m-1) - (m-1) = m$, so

$$JC = mJ. \qquad (14.13)$$

By Theorem 2.7, equations (14.10) to (14.13) imply that B and C are the incidence matrices of designs with parameters (i) and (ii). Design (iii) is the complement of design (i). □

Exercises 14.2

14.2.1* Suppose B and C are the incidence matrices of designs with the parameters $(2m-1, 4m-2, 2m-2, m-1, m-2)$ and $(2m, 4m-2, 2m-1, m, m-1)$, respectively. Is

$$\begin{bmatrix} e & B \\ 0 & C \end{bmatrix}$$

necessarily the incidence matrix of a $(4m-1, 2m-1, m-1)$-design?

14.2.2 Suppose $\mathcal{D}$ is a $(4m-1, 2m-1, m-1)$-design on the treatment-set T, with blocks $B_1, B_2, \ldots, B_{4m-1}$. Let x be an object that does not belong to T. Define $8m-2$ blocks $B_{11}, B_{12}, \ldots, B_{1,4m-1}, B_{21}, B_{22}, \ldots, B_{2,4m-1}$ by

$$\begin{aligned} B_{1i} &= B_i \cup \{x\}, \\ B_{2i} &= T \backslash B_i. \end{aligned}$$

Prove that these blocks form a 3-design with support $T \cup \{x\}$. (This is called a *Hadamard 3-design.*)

14.2.3 A balanced incomplete block design is called *quasi-symmetric* if there are two integers μ and ν such that any two blocks intersect in μ elements, except that given any block there is exactly one other block that intersects it in ν elements.

(i) Prove that the design formed by taking two copies of a given symmetric balanced incomplete block design is a quasi-symmetric design.

(ii) Prove that the Hadamard 3-design on $4m$ points, when interpreted as a 2-design, is a $(4m, 8m-2, 4m-1, 2m, 2m-1)$-design and is quasi-symmetric.

(iii) Prove that constructions (i) and (ii) yield all quasi-symmetric designs.

14.3 Further Hadamard Matrix Constructions

Kronecker Product Constructions

To discuss the easiest of all Hadamard matrix constructions, we introduce the Kronecker product (or direct product) of two matrices. If A is of size $p \times q$ and B is of size $r \times s$, then $A \otimes B$ denotes the $pr \times qs$ matrix

$$\begin{bmatrix} a_{11}B & a_{12}B & \cdots & a_{1q}B \\ a_{21}B & a_{22}B & \cdots & a_{2q}B \\ & & \cdots & \\ a_{p1}B & a_{p2}B & \cdots & a_{pq}B \end{bmatrix}.$$

We also write $A \otimes B = [a_{ij}B]$ (square brackets denote the general element of a block matrix).

LEMMA 14.5

If A and B are any matrices, then

$$(A \otimes B)^T = A^T \otimes B^T. \tag{14.14}$$

If, further, C and D are any matrices such that the products AC and BD exist, then

$$(A \otimes B)(C \otimes D) = AC \otimes BD. \tag{14.15}$$

Proof. The transpose of any block matrix is formed by using the rule

$$\begin{bmatrix} X & Y \\ Z & W \end{bmatrix}^T = \begin{bmatrix} X^T & Y^T \\ Z^T & W^T \end{bmatrix}.$$

So $(A \otimes B)^T = [(a_{ij}B)^T] = [a_{ji}B^T]$. We have (14.14).

Now $[X_{ij}][Y_{ij}]$ has (i,j) block entry

$$\sum_k X_{ik}Y_{kj}$$

provided the blocks are conformable and the number of blocks per row in X equals the number of blocks per column in Y. So $(A \otimes B)(C \otimes D)$ has (i,j) block

$$\sum_k (a_{ik}B)(c_{kj}D) = \left(\sum_k a_{ik}c_{kj}\right) BD.$$

But $\sum_k a_{ik}c_{kj}$ is the (i,j) entry of AC. So

$$(A \otimes B)(C \otimes D) = AC \otimes BD.$$

□

THEOREM 14.6

If A and B are Hadamard matrices, then $A \otimes B$ is Hadamard.

Proof. Suppose A and B are of sides n and r, respectively. Then $AA^T = nI_n$ and $BB^T = rI_r$. So from Lemma 14.5,

$$\begin{aligned}(A \otimes B)(A \otimes B)^T &= (A \otimes B)(A^T \otimes B^T) \\ &= AA^T \otimes BB^T \\ &= nI_n \otimes rI_r,\end{aligned}$$

which obviously equals nrI_{nr}. The Kronecker product of $(1,-1)$ matrices is clearly a $(1,-1)$ matrix, so $A \otimes B$ is Hadamard. □

The most important consequence for our work of the discussion above is:

COROLLARY 14.6.1

If there are Hadamard matrices of sides n and r, then there is a Hadamard matrix of side nr.

For example, we can construct a matrix of order 2^k for any k by an inductive process. A is a Hadamard matrix of side 2 and B is a Hadamard matrix of side 2^{k-1}.

Williamson's Method

Corollary 14.6.1 tells us that a Hadamard matrix of side $2^k h$ can be constructed if one of side h is known. So there is particular interest in Hadamard matrices of side $h = 4n$, where n is odd.

To treat this case, John Williamson [179] considered the array

$$H = \begin{bmatrix} A & B & C & D \\ -B & A & -D & C \\ -C & D & A & -B \\ -D & -C & B & A \end{bmatrix} \tag{14.16}$$

which is derived from the quaternions. If A, B, C and D are replaced by square matrices of order n, H becomes a square matrix of order $4n$. One can attempt to choose A, B, C and D in such a way that H will be Hadamard.

If HH^T is considered as a block matrix with $n \times n$ blocks, the diagonal blocks each equal $AA^T + BB^T + CC^T + DD^T$. This must be $4nI$ for H to be Hadamard. The $(1,2)$ block is $BA^T - AB^T + DC^T - CD^T$. This will be zero if $AB^T = BA^T$ and $CD^T = DC^T$. Similar results hold for the other off-diagonal elements. So we have

THEOREM 14.7

If there exist square $(1,-1)$ *matrices* A, B, C, D *of order* n *that satisfy*

$$AA^T + BB^T + CC^T + DD^T = 4nI \tag{14.17}$$

and for every pair X, Y *of distinct matrices chosen from* A, B, C, D,

$$XY^T = YX^T \tag{14.18}$$

then the matrix H *of* (14.16) *is a Hadamard matrix of order* $4n$.

As an example, suppose the submatrices to be inserted were

$$A = \begin{bmatrix} + & + & + \\ + & + & + \\ + & + & + \end{bmatrix}, \ B = C = D = \begin{bmatrix} + & - & - \\ - & + & - \\ - & - & + \end{bmatrix}.$$

Then

$$AB^T = \begin{bmatrix} - & - & - \\ - & - & - \\ - & - & - \end{bmatrix} = BA^T$$

and similarly it may be verified that every other combination satisfies $XY^T = YX^T$. Moreover,

$$AA^T = \begin{bmatrix} 3 & 3 & 3 \\ 3 & 3 & 3 \\ 3 & 3 & 3 \end{bmatrix}, \; BB^T = CC^T = DD^T = \begin{bmatrix} 3 & -1 & -1 \\ -1 & 3 & -1 \\ -1 & -1 & 3 \end{bmatrix}$$

so $AA^T + BB^T + CC^T + DD^T = 12I = 4nI$. Therefore, these matrices satisfy the theorem. The Hadamard matrix is

$$\begin{bmatrix}
+ & + & + & + & - & - & + & - & - & + & - & - \\
+ & + & + & - & + & - & - & + & - & - & + & - \\
+ & + & + & - & - & + & - & - & + & - & - & + \\
- & + & + & + & + & + & - & + & + & + & - & - \\
+ & - & + & + & + & + & + & - & + & - & + & - \\
+ & + & - & + & + & + & + & + & - & - & - & + \\
- & + & + & + & - & - & + & + & + & - & + & + \\
+ & - & + & - & + & - & + & + & + & + & - & + \\
+ & + & - & - & - & + & + & + & + & + & + & - \\
- & + & + & - & + & + & + & - & - & + & + & + \\
+ & - & + & + & - & + & - & + & - & + & + & + \\
+ & + & - & + & + & - & - & - & + & + & + & +
\end{bmatrix}.$$

The main problem in this construction is finding the matrices A, B, C, D. If they are chosen to be symmetric, the condition (14.18) reduces to a requirement that the four matrices commute. If we further require that each matrix is circulant—the $(i+1, j+1)$ entry equals the (i, j) entry, with i and j being reduced modulo n if necessary—the four matrices will be polynomials in the matrix K,

$$K = \begin{bmatrix} 0 & 1 & 0 & 0 & \cdots & 0 \\ 0 & 0 & 1 & 0 & \cdots & 0 \\ 0 & 0 & 0 & 1 & \cdots & 0 \\ & & & & \cdots & \\ 0 & 0 & 0 & 0 & \cdots & 1 \\ 1 & 0 & 0 & 0 & \cdots & 0 \end{bmatrix}$$

so they will commute. A further advantage is that the four matrices are completely specified by their first rows.

Figure 14.2 is a table of suitable first rows for the matrices A, B, C, D for $n = 3, 5, 7$.

$$n = 3 : \begin{array}{c} +-- \\ +-- \\ +-- \\ +++ \end{array} \quad n = 5 : \begin{array}{c} ++--+ \\ +-++- \\ +---- \\ +---- \end{array} \quad n = 7 : \begin{array}{c} ++-++-+ \\ ++----+ \\ ++-++-+ \\ +-+++-- \end{array}$$

FIGURE 14.2: Initial rows for matrices used in Williamson's construction.

Exercises 14.3

14.3.1 Prove that the Kronecker product of matrices satisfies the distributive laws

$$\begin{aligned} (A + B) \otimes C &= (A \otimes C) + (B \otimes C) \\ A \otimes (B + C) &= (A \otimes B) + (A \otimes C) \end{aligned}$$

and the associative law

$$A \otimes (B \otimes C) = (A \otimes B) \otimes C.$$

14.3.2* Find all possible sets of symmetric circulant matrices $\{A, B, C, D\}$ of side 3 for use in Theorem 14.7; and similarly for side 5.

14.3.3* Find an 8×8 array based on the symbols $\pm A, \pm B, \pm C, \pm D, \pm E, \pm F, \pm G, \pm H$ such that each of A, B, C, D, E, F, G, H appears exactly once per row and per column, which would be suitable for use in a generalization of Theorem 14.7.

14.4 Regular Hadamard Matrices

A Hadamard matrix is called *regular* if every row contains the same number of elements $+1$.

THEOREM 14.8

There exists a regular Hadamard matrix of side $4t$ if and only if there is a symmetric balanced incomplete block design with parameters

$$(4t, 2t - t^{1/2}, t - t^{1/2}).$$

Proof. Suppose H is a regular Hadamard matrix of side $4t$, with h elements $+1$ per row. Then

$$\begin{aligned} HH^T &= 4tI, \\ HJ &= (2h-4t)J. \end{aligned}$$

Write $A = (H+J)/2$. A is a square $(0,1)$-matrix of side $4t$. Moreover,

$$\begin{aligned} AA^T &= \tfrac{1}{4}(HH^T + HJ + (HJ)^T + J^2) \\ &= \tfrac{1}{4}(4tI + 2(2h-4t)J + 4tJ \\ &= tI + (h-t)J \qquad (14.19) \\ AJ &= \frac{1}{2}(HJ + J^2) \\ &= hJ. \qquad (14.20) \end{aligned}$$

By Theorem 2.10, A is the incidence matrix of a symmetric balanced incomplete block design with

$$\begin{aligned} v &= b = 4t, \\ r &= k = h, \\ \lambda &= h - t. \end{aligned}$$

Then equation (1.2) yields

$$h(h-1) = (h-t)(4t-1),$$

whence

$$h^2 - 4th + t(4t-1) = 0,$$

which has solutions

$$\begin{aligned} h &= \tfrac{1}{2}(4t \pm \sqrt{4t}) \\ &= 2t \pm t^{\frac{1}{2}}. \end{aligned}$$

If the smaller root occurs, then A is the incidence matrix of the required design; otherwise, $J - A$ is the incidence matrix.

The converse is left as an exercise. □

This proof in fact tells us that a regular Hadamard matrix exists if and only if two designs, with parameters

$$(4t, 2t \pm t^{\frac{1}{2}}, t \pm t^{\frac{1}{2}})$$

exist; but the two are complements, so either they both exist or neither one does.

COROLLARY 14.8.1

If there is a regular Hadamard matrix of side n, then n is a perfect square.

Proof. If $n = 4t$, then both $2t - t^{\frac{1}{2}}$ and $4t$ are required to be integers, so $t^{\frac{1}{2}}$ must be an integer also. It remains to check the two trivial cases: $n = 1$ is possible, $n = 2$ is not. □

COROLLARY 14.8.2

If H is a regular Hadamard matrix, then each column of H contains the same number of entries $+1$ as do the rows of H. □

This follows from the fact that the dual of a symmetric design is a symmetric design with the same parameters.

Suppose H and K are regular Hadamard matrices. Then it is easy to see that $H \otimes K$ is regular, and we already know it is Hadamard. So we have

THEOREM 14.9

If there exist regular Hadamard matrices of sides n and r, then there exists a regular Hadamard matrix of side nr. □

THEOREM 14.10

Suppose there is a Hadamard matrix of side n. Then there is a regular Hadamard matrix of side n^2.

Proof. Let H be a Hadamard matrix of side n, and denote by H_j the matrix derived from H by negating columns in such a way that H_j has j-th row all $+1$. Now consider the $n^2 \times n^2$ matrix K, defined to be the $n \times n$ array of $n \times n$ blocks

$$K = [h_{ij}H_j].$$

Clearly, K is Hadamard. Any row of K contains one block of n consecutive entries either all $+1$ or all -1, and $n-1$ blocks of n entries, half of which are $+1$ and half -1. So any given row contains either $\frac{1}{2}n(n+1)$ or $\frac{1}{2}n(n-1)$ entries $+1$. If the row is of the former kind, negate it. The resulting matrix is a regular Hadamard matrix. □

These constructions give infinitely many regular Hadamard matrices, but they do not cover any of the cases of side $4t^2$ where t is odd, $t > 1$. These are difficult to construct. We present one example (from [160]) as Exercise

$$I = \begin{bmatrix} 1 & 0 & 0 \\ 0 & 1 & 0 \\ 0 & 0 & 1 \end{bmatrix} \quad J = \begin{bmatrix} 1 & 1 & 1 \\ 1 & 1 & 1 \\ 1 & 1 & 1 \end{bmatrix} \quad O = \begin{bmatrix} 0 & 0 & 0 \\ 0 & 0 & 0 \\ 0 & 0 & 0 \end{bmatrix}$$

$$K = \begin{bmatrix} 0 & 1 & 0 \\ 0 & 0 & 1 \\ 1 & 0 & 0 \end{bmatrix} \quad L = \begin{bmatrix} 0 & 0 & 1 \\ 1 & 0 & 0 \\ 0 & 1 & 0 \end{bmatrix}$$

$$A = \begin{bmatrix} 0 & J & J \\ J & 0 & J \\ J & J & 0 \end{bmatrix} \quad B = \begin{bmatrix} I & I & I \\ I & I & I \\ I & I & I \end{bmatrix}$$

$$C = \begin{bmatrix} I & I & I \\ K & K & K \\ L & L & L \end{bmatrix} \quad D = \begin{bmatrix} I & I & I \\ L & L & L \\ K & K & K \end{bmatrix} \quad E = \begin{bmatrix} I & L & K \\ L & K & I \\ K & I & L \end{bmatrix}$$

$$M = \begin{bmatrix} A & C & D & B \\ C^T & A & E^T & E \\ D^T & E & A & E^T \\ B & E^T & E & A \end{bmatrix}$$

FIGURE 14.3: Construction of a $(36, 15, 6)$-design.

14.4.3, a $(36, 15, 6)$ design (and consequently, a regular Hadamard matrix of side 36) that is constructed using block matrices.

In this section we have outlined a few results concerning one special class of Hadamard matrices, regular Hadamard matrices, that have attracted some interest. Other special classes have been studied. Some other special classes, namely skew-Hadamard matrices, cyclic Hadamard matrices and symmetric Hadamard matrices with constant diagonal (which have sometimes been called graphical Hadamard matrices), are presented in the exercises.

Exercises 14.4

14.4.1 A regular Hadamard matrix of side $4t^2$ is called *subregular* if it has $2t^2 - t$ entries $+1$ per row, and *superregular* if it has $2t^2 + t$. Prove that the direct product of two regular Hadamard matrices is superregular if the two constituent matrices are both superregular or both subregular, and is subregular if the two constituents are of different kinds.

14.4.2 (Completion of Theorem 14.8) Suppose A is the incidence matrix of a symmetric balanced incomplete block design with parameters

$$(4u^2, 2u^2 - u, u^2 - u).$$

Prove that $2A - J$ is a regular Hadamard matrix.

14.4.3 Verify that M, whose construction is outlined in Figure 14.3, is the incidence matrix of a $(36, 15, 6)$-design.

14.4.4* Prove that there exists a $(15, 7, 3)$-design whose incidence matrix is a 5×5 block matrix based on the matrices I, J, K, L, O of Figure 14.3.

14.4.5* A Hadamard matrix is called *graphical* if it is symmetric and has every diagonal entry equal to $+1$.

(i) Suppose H is a symmetric Hadamard matrix of side n. Prove that

$$H^2 = nI$$

and consequently that the eigenvalues of H are $n^{\frac{1}{2}}$ and $-n^{\frac{1}{2}}$ (with some multiplicities).

(ii) Prove that there are nonnegative integers a and b such that $a+b = n$ and $(a-b)n^{\frac{1}{2}}$ equals the trace of H.

(iii) Suppose further that every diagonal entry in H equals $+1$. Prove that n is a perfect square. (So the side of a graphical Hadamard matrix equals an integer square.)

14.4.6* Suppose B is a balanced incomplete block design with parameters

$$(2u^2 - u, 4u^2 - 1, 2u + 1, u, 1).$$

(i) Prove that any two distinct blocks of B have either no common element or one common element.

(ii) Order the blocks of B as $B_1, B_2, \ldots, B_{4u^2-1}$ in some way. Define a_{ij} by

$$a_{ij} = |B_i \cap B_j|$$

when $i \neq j$, and set $a_{ii} = 0$ for all i. Prove that $A = (a_{ij})$ is the incidence matrix of a symmetric balanced incomplete block design. What are its parameters?

(iii) Prove that there exists a graphical Hadamard matrix of side $4u^2$.

14.4.7* A *cyclic* Hadamard matrix of side n is a Hadamard matrix H that satisfies $h_{i+1,j+1} = h_{i,j}$ (with subscripts reduced modulo n when necessary).

(i) Prove that the side of cyclic Hadamard matrix is necessarily a perfect square.

(ii) Prove that there is a cyclic Hadamard matrix of side 4, but not one of side 16.

14.4.8 A *skew-Hadamard* matrix is a Hadamard matrix H with the properties that:

(a) every diagonal entry is $+1$;

(b) $H + H^T = 2I$.

(i) Prove that if there is a skew-Hadamard matrix $H = S - I$ of side n, then

$$\begin{bmatrix} S+I & S+I \\ S-I & S+I \end{bmatrix}$$

is a skew-Hadamard matrix of side $2n$. [180]

(ii) Prove that a Paley Hadamard matrix is skew-Hadamard.

(iii) Prove that there is a skew-Hadamard matrix of side 16.

(iv) A skew-Hadamard matrix is called *normalized* if its first row and its diagonal have every entry $+1$ and its first column has every entry -1 except the first. Prove that every skew-Hadamard matrix is equivalent to a normalized skew-Hadamard matrix.

14.4.9 See the definitions in Exercise 14.4.8. If M is a normalized skew-Hadamard matrix, the *core* of M is the matrix W obtained from M by deleting the first row and column and setting the diagonal equal to zero.

(i) Prove that the core of a normalized skew-Hadamard matrix of order k satisfies

$$WW^T = (k-1)I - J,\ WJ = 0,\ W^T = -W.$$

(ii) Suppose H is a normalized skew-Hadamard matrix of side n whose core is W. Write $S = H - I_n$ and $D = W + I_{n-1}$. If

$$K = S \otimes D + I_n \otimes J_{n-1},$$

prove that K is a Hadamard matrix of side $n(n-1)$.

14.4.10 A *conference matrix* C of side k is a symmetric $k \times k$ $(1, -1)$-matrix C having all diagonal entries 1 and satisfying

$$(C - I)^2 = (k-1)I.$$

A *normalized* conference matrix is one with all positive entries in its first row and column.

(i) Prove that a conference matrix has side congruent to $2 \pmod 4$.

(ii) Prove that every conference matrix is Hadamard-equivalent to a normalized conference matrix.

(iii) Suppose H is a Hadamard matrix of order $h > 1$, say $h = 2g$, and C is a conference matrix of order k. Define

$$\begin{aligned} U &= I_g \otimes \begin{bmatrix} 0 & 1 \\ -1 & 0 \end{bmatrix}, \\ M &= ((C - I) \otimes H) + (I_k \otimes UH). \end{aligned}$$

Show that M is a Hadamard matrix of order hk.

14.5 Equivalence

As we said in Section 14.1, two Hadamard matrices are called *equivalent* (or *Hadamard equivalent*) if one can be obtained from the other by some sequence of row and column permutations and negations. This is the same as saying that two Hadamard matrices are equivalent if one can be obtained from the other by premultiplication and postmultiplication by monomial matrices (signed permutation matrices). So the equivalence class of H is

$$\{PHQ : P, Q \text{ monomial}\}.$$

Another operation that clearly respects the Hadamard property is transposition. A Hadamard matrix and its transpose may have very different properties as matrices, so this operation is not usually counted as one of the basic equivalence relations. However, it is of special interest, and we say Hadamard matrices H and K are *transpose-equivalent* if H is equivalent either to K or to K^T.

We denote the number of equivalence classes of Hadamard matrices of side n by $h(n)$. It is easy to calculate the value of $h(n)$ for small n. In Section 14.1 we verified that $h(4) = 1$, and similarly one finds that

$$h(1) = h(2) = h(4) = h(8) = h(12) = 1.$$

(Only the last case requires very much calculation.) Exhaustive searches have been carried out (see [60, 61, 81, 86, 87]) that show

$$h(16) = 5, \ h(20) = 3, \ h(24) = 60, \ h(28) = 487.$$

These are the only exact results known. Four inequivalent Hadamard matrices of side 16 (taken from [60]) are shown in Figure 14.4. The matrix H_i^T is equivalent to H_i for $i = 1, 2, 3$; H_4^T is a suitable representative for the fifth equivalence class of Hadamard matrices of side 16. Given our current computing capability, it is most unlikely that $h(n)$ can be determined for $n \geq 32$. Hundreds of thousands of these matrices were found in a search restricted to very specialized matrices [99].

Precise determination of whether or not two given Hadamard matrices are equivalent is not easy, but it is possible to employ efficient computer programs. Perhaps the best approach uses the *graph* of a Hadamard matrix. If H is a Hadamard matrix of side n, then the graph $G(H)$ of H has $4n$ vertices $\{a_1, a_2, \ldots, a_n, \ b_1, b_2, \ldots, b_n, \ c_1, c_2, \ldots, c_n, \ d_1, d_2, \ldots, d_n\}$. Each of $a_1, a_2, \ldots, a_n, \ b_1, b_2, \ldots, b_n$ has a loop on it (an edge from the vertex to itself). The other edges are

$$\begin{aligned} a_i c_j \text{ and } b_i d_j \text{ if } h_{ij} &= \ \ 1; \\ a_i d_j \text{ and } b_i c_j \text{ if } h_{ij} &= -1. \end{aligned}$$

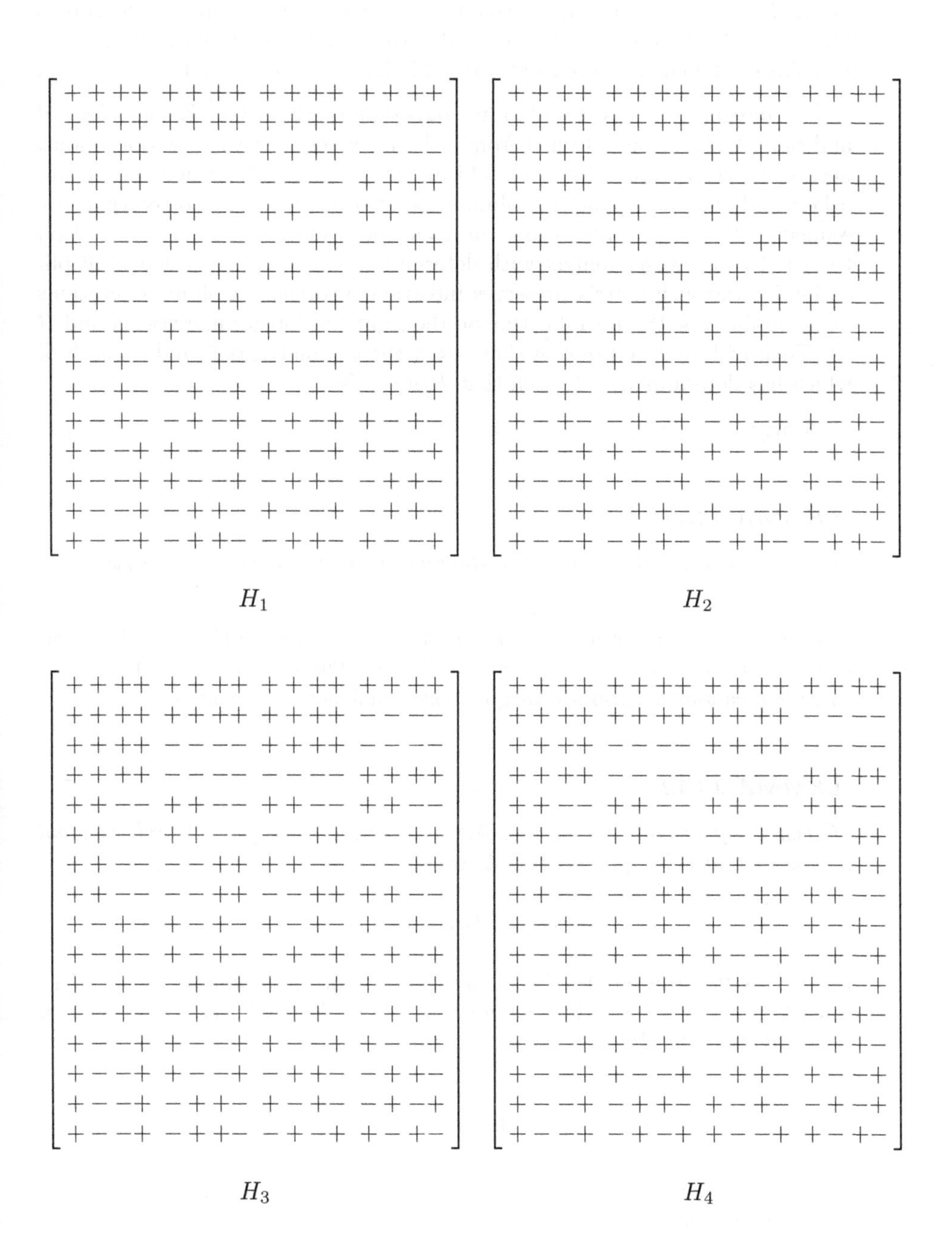

FIGURE 14.4: Representative Hadamard matrices of order 16.

Then H and K are equivalent Hadamard matrices if and only if the graphs $G(H)$ and $G(K)$ are isomorphic, under the usual definitions of graph isomorphism, and efficient programs are available to test for graph isomorphism.

Two integer matrices A and B are *integrally equivalent,* or $\mathbb{Z}$-equivalent, if and only if B can be obtained from A by a sequence of the row operations: negate a row, permute rows, or add an integer multiple of one row to another; and the corresponding column operations. It is easy to see that any sequence of these row operations on A can be achieved by multiplying A on the left by an integer matrix with determinant ± 1; conversely, left multiplication by any such matrix preserves integral equivalence. Column operations work similarly, with multiplication on the right. So integer matrices A and B are $\mathbb{Z}$-equivalent if and only if there exist *integer* matrices P and Q, each of which has determinant ± 1, such that $PAQ = B$.

Clearly:

LEMMA 14.11

Two Hadamard matrices that are Hadamard equivalent must be $\mathbb{Z}$-equivalent.

Therefore we can obtain a lower bound on the number of Hadamard equivalence classes by counting $\mathbb{Z}$-equivalence classes. The significance of this is that there is a canonical form for $\mathbb{Z}$-equivalence that is very quickly calculated.

LEMMA 14.12

Given a square integral matrix A, there exists a unique integral diagonal matrix D that is $\mathbb{Z}$-equivalent to A, say

$$D = diag(d_1, d_2, \ldots, d_r, 0, 0, \ldots, 0),$$

where r is the rank of A. The d_i are positive integers, and d_i divides d_{i+1} when $1 \leq i < r$. The greatest common divisor of the $k \times k$ subdeterminants of A is $d_1 d_2 d_3 \ldots d_k$. If A is $\mathbb{Z}$-equivalent to

$$\begin{bmatrix} D_1 & O \\ O & E \end{bmatrix}$$

where $D_1 = \mathrm{diag}\,(d_1, d_2, \ldots, d_k)$, then the greatest common divisor of the nonzero elements of E is d_{k+1}.

The proof may be found in books on linear algebra such as [25]. The matrix D is called the *Smith normal form* of A.

In particular, consider the Smith normal form of a Hadamard matrix H of order $4m$. Since H is invertible, none of the diagonal elements is zero. Write $D = D(H) = \text{diag}(d_1, d_2, \ldots, d_{4m})$ for the Smith form of H.

LEMMA 14.13

For any Hadamard matrix H,

$$d_i d_{4m-i} = 4m.$$

Proof. Suppose P and Q are integer matrices with integer inverses such that $PHQ = D$. Then

$$(PHQ)(Q^{-1}H^T P^{-1}) = PHH^T P^{-1} = 4mPP^{-1} = 4mI,$$

so

$$Q^{-1}H^T P^{-1} = \text{diag}\left(\frac{4m}{d_1}, \frac{4m}{d_2}, \ldots, \frac{4m}{d_{4m}}\right),$$

and consequently H^T has Smith normal form

$$\text{diag}\left(\frac{4m}{d_{4m}}, \frac{4m}{d_{4m-1}}, \ldots, \frac{4m}{d_1}\right).$$

But obviously every matrix is $\mathbb{Z}$-equivalent to its transpose, so

$$\text{diag}\left(\frac{4m}{d_{4m}}, \frac{4m}{d_{4m-1}}, \ldots, \frac{4m}{d_1}\right) = \text{diag}\,(d_1, d_2, \ldots, d_{4m}).$$

Equating corresponding entries we get

$$d_i = \frac{4m}{d_{4m-i+1}},$$

so $d_i d_{4m-i+1} = 4m$. □

It is clear that $d_1 = 1$ and easy to see that $d_2 = 2$. So $d_{4m-1} = 2m$ and $d_{4m} = 4m$. Suppose m is square-free; if $i \leq 2m$ then d_i divides d_{4m-i+1}; say $d_{4m-i+1} = \alpha d_i$. Then $\alpha d_i^2 = 4m$. Consequently $d_i = 1$ or 2. Unless $i = 1$, d_2 divides d_i, so $d_i = 2$. So:

THEOREM 14.14

If m is square-free, then H has Smith normal form

$$\text{diag}\left(1, 2^{(2m-1\ times)}, 2m^{(2m-1\ times)}, 4\right).$$

In general, the number of possible Smith normal forms of order $4m$ is limited by the sort of factors that m can have. For example, when $m = 9$, the facts that the d_i divide d_{4m} and that $d_{4m} = 4m$ mean that the invariants must be divisors of 36; as $d_2 = 2$, they must be chosen from $\{1, 2, 4, 6, 12, 18, 36\}$. Since $4m/d_i$ must also be an invariant, 4 and 12 are impossible. So the form is

$$\text{diag}\left(1, 2^{(\alpha \text{ times})}, 6^{(\beta \text{ times})}, 18^{(\alpha \text{ times})}, 36\right)$$

where $\alpha \geq 0, \beta \geq 0$ and $2\alpha + \beta = 34$. A similar analysis applies whenever m is a prime square.

There are further restrictions on Smith normal forms:

LEMMA 14.15 [161]

Any $n \times n$ $(0,1)$-matrix with nonzero determinant has at least $\lfloor \log_2 n \rfloor + 1$ entries 1 in its Smith normal form.

THEOREM 14.16 [161]

Any Hadamard matrix of order $4m$ has at least $\lfloor \log_2(4m-1) \rfloor + 1$ entries 2 in its Smith form.

Proof. Assume the Hadamard matrix is normalized. Subtract column 1 from every other column, then subtract row 1 from every other row. Negate all columns except the first. The resulting matrix is

$$\left[\begin{array}{c|c} 1 & 0 \ \dots \ 0 \\ \hline 0 & \\ \vdots & 2B \\ 0 & \end{array}\right]$$

for some nonsingular (0, 1)-matrix B. Now apply Lemma 14.15 to B. □

If $m = fg^2$, where f and g are coprime and square-free, then a Hadamard matrix of side $4m$ has Smith normal form

$$\text{diag}\left(1, 2^{(\alpha \text{ times})}, 2g^{(\beta \text{ times})}, 2fg^{(\beta \text{ times})}, 2m^{(\alpha \text{ times})}, 4m\right)$$

where $\alpha + \beta = 2m - 1$, so α completely determines the Smith normal form (and the $\mathbb{Z}$-equivalence class). For this reason, α is called the *Smith class* of H.

From Theorem 14.16 we see that the Smith class of 16×16 Hadamard matrices must be 4, 5, 6, or 7. All of these classes can be realized; the matrices H_1, H_2, H_3 and H_4 of Figure 14.4 are of classes 4, 5, 6 and 7, respectively.

For side 32 and for other sides divisible by 8, it is useful to observe the following fact. If H and K are Hadamard matrices of the same side, then

$$\begin{bmatrix} H & H \\ K & -K \end{bmatrix}$$

is also Hadamard. If $K = H$, then the Smith normal form of M is easy to calculate (see Exercise 14.5.4), but if K and H are different, then the situation is not clear. A range of inequivalent matrices can be obtained merely by taking K to be formed from H by column permutation, and in this way one can use the matrices of side 16 to obtain matrices of side 32 of all possible Smith classes from 5 to 15.

Exercises 14.5

14.5.1* Verify that $h(8) = 1$.

14.5.2 Prove that if H and K are equivalent Hadamard matrices, then H^T and K^T are equivalent.

14.5.3* The *weight* of a Hadamard matrix H is defined to be the number of entries +1 in the matrix; $w(H)$ is the maximum weight among all matrices equivalent to H.

(i) Prove that if H is of side n, then

$$w(H) \geq \tfrac{1}{2}(n+4)(n-1).$$

(ii) If, further, $n = 8t + 4$, prove that

$$w(H) \geq \tfrac{1}{2}(n+6)(n-2).$$

14.5.4* Suppose H is an integer matrix that is $\mathbb{Z}$-equivalent to the diagonal matrix D. Define

$$K = \begin{bmatrix} H & H \\ H & -H \end{bmatrix}.$$

(i) Show that K is $\mathbb{Z}$-equivalent to

$$\begin{bmatrix} D & O \\ 0 & 2D \end{bmatrix}.$$

(ii) In particular, if H is a Hadamard matrix of side $8m$, where m is odd and square-free, what are the invariants of K?

(iii) If H is any Hadamard matrix of side $8m$, prove that K has at least $12m - 1$ invariants divisible by 4.

14.5.5 Prove that if there is a conference matrix (see 14.4.10) of side n, then there is a Hadamard matrix of side $2n$ with Smith normal form

$$\operatorname{diag}\left(1, 2^{(n-1 \text{ times})}, n^{(n-1 \text{ times})}, 2n\right).$$

14.5.6 Suppose H is a skew-Hadamard matrix (see Exercise 14.4.8) of side $4m$. Show that

$$\begin{bmatrix} H & H \\ H^T & -H^T \end{bmatrix}$$

has invariants

$$\operatorname{diag}\left(1, 2^{(4m-1 \text{ times})}, 4m^{(4m-1 \text{ times})}, 8m\right).$$

Use this result and Exercise 14.5.4 to prove that whenever there is a skew-Hadamard matrix of side $8m$, there are integrally inequivalent Hadamard matrices of side $16m$.

Chapter 15

Room Squares

15.1 Definitions

A *Room square* of *side* r (or of *order* $r+1$) is a square array with r cells in each row and each column, such that each cell is either empty or contains an unordered pair of symbols chosen from a set of $r+1$ elements. Each row and each column contains each element precisely once, from which it follows that r must be odd; say $r = 2n-1$. Without loss of generality, we can take the elements as the numbers $1, 2, \ldots, 2n-1$ and the symbol ∞. So each row and each column of the design contains $n-1$ empty cells and n cells each containing a pair of symbols. Further, each of the $n(2n-1)$ possible distinct pairs of symbols is required to occur exactly once in a cell of the square. An example, of side 7, is shown in Figure 15.1.

The first mention of a Room square occurred in [89], in relation to Kirkman's schoolgirl problem. Consider the square of Figure 15.1, with its columns labeled a, b, c, d, e, f, g as shown. From it we construct a triple system as follows. From the column headed "a" we construct the four triples $a47$, $a\infty5$, $a12$ and $a36$, which are obtained from the pairs in the column by appending a to them; proceeding similarly with the other columns, we form 28 triples.

a	b	c	d	e	f	g
			35	17	$\infty2$	46
	26	$\infty4$			15	37
	13	57	$\infty6$	24		
47		16		$\infty3$		25
$\infty5$		23	14		67	
12	$\infty7$			56	34	
36	45		27			$\infty1$

FIGURE 15.1: A Room square of side 7.

To each row there corresponds one more triple, made up of the labels on the three columns that have no entries in that row. It will be observed that this is in fact a Kirkman triple system: Each parallel class consists of the triple corresponding to a row and the four triples that came from pairs in that row; the classes are

$$\begin{array}{lllll}
\{abc\} & \{d35\} & \{e17\} & \{f\infty 2\} & \{g46\} \\
\{ade\} & \{b26\} & \{c\infty 4\} & \{f15\} & \{g37\} \\
\{afg\} & \{b13\} & \{c57\} & \{d\infty 6\} & \{e24\} \\
\{bdf\} & \{a47\} & \{c16\} & \{e\infty 3\} & \{g25\} \\
\{bge\} & \{a\infty 5\} & \{c23\} & \{d14\} & \{f67\} \\
\{cdg\} & \{a12\} & \{b\ 7\} & \{e56\} & \{f34\} \\
\{cef\} & \{a36\} & \{b45\} & \{d27\} & \{g\infty 1\}.
\end{array}$$

A similar approach was taken by Anstice [5, 6]. These arrays have appeared sporadically in the literature and are used in duplicate bridge tournaments (see Section 15.5). In 1955 they were reintroduced in mathematics by Room [120], for whom they are named.

It is clear that the rows of a Room square are all one-factors, as are the columns. To construct one we must first find a one-factorization and allocate one of its factors to each row; then we must distribute the entries in each row into columns so that the columns are all factors as well. So a Room square can be interpreted as two related one-factorizations, a *row* factorization and a *column* factorization. The relationship is as follows: If R_i is any factor from the row factorization and C_j is any factor from the column factorization, then either $R_i \cap C_j$ is empty or it contains precisely one pair. Two such factorizations are called *orthogonal*. Conversely, given two orthogonal one-factorizations, one can construct a Room square—the (i, j) cell contains the pair in $R_i \cap C_j$.

The properties of Room squares are not dependent on the set of symbols, and the ordering of rows and columns is a matter of convenience. So we define two Room squares to be isomorphic if one can be obtained from the other by permuting rows and columns and relabeling the elements. In particular, it is possible in this way to produce a square with diagonal entries

$$(\infty 1, \infty 2, \ldots, \infty r)$$

from any square of side r. We say that such a Room square is *standardized*. For example, Figure 15.2(a) shows a standardized Room square obtained from the square of Figure 15.1 by permuting the rows and columns. There is more than one standardized square corresponding to a given square; Figure 15.2(b) was obtained from the square in (a) by interchanging the last two rows, interchanging the last two columns and exchanging symbols 6 and 7.

The side of a Room square is necessarily odd. It is natural to ask which

∞1				36	27	45
46	∞2	17			35	
25		∞3	16	47		
37	15		∞4			26
	67		23	∞5	14	
		24	57		∞6	13
	34	56		12		∞7

(a)

∞1				37	45	26
47	∞2	16				35
25		∞3	17	46		
36	15		∞4		27	
	67		23	∞5		14
	34	57		12	∞6	
		24	56		13	∞7

(b)

FIGURE 15.2: Standardized Room squares of side 7.

odd numbers occur as sides. The 1×1 array

∞ 1

is a Room square of side 1. There can be no Room square of side 3, for if such a square existed, it would have ∞1 as one of its entries. Assume (without loss of generality) that it lies in the $(1, 1)$ cell. The square has the form

∞ 1	?	?
?	?	?
?	?	?

.

Now each of ∞, 1, 2 and 3 must appear once in the first row. So 23 must occur in one of the cells of that row. Similarly, 23 must occur in the first column. But this is impossible, since the pair cannot be repeated and the common cell of row 1 and column 1 is already occupied.

A similar argument can be used to show that no Room square of side 5 exists. However, this fact also follows directly from the discussion of one-factorizations of K_6 in Section 10.2; since any two factorizations always have a common factor, they cannot be orthogonal.

Sides 3 and 5 are exceptional. We shall prove in Section 15.4 that there is a Room square of every odd side greater than 5.

The nonexistence of a Room square of side 5 leads to an easy proof of the nonexistence of a large set of triple systems on seven treatments [169]. For suppose a large set of triple systems of order 7 existed. We shall use the symbols $\{A, B, 1, 2, 3, 4, 5\}$. Without loss of generality the large set is $\{\mathcal{S}_1, \mathcal{S}_2, \mathcal{S}_3, \mathcal{S}_4, \mathcal{S}_5\}$ where

$$\mathcal{S}_i = \{ABi, Ax_iy_i, Az_it_i, Bx_iz_i, By_it_i, ix_it_i, iy_iz_i\}.$$

Now define

$$\begin{aligned}\mathcal{F} &= \{F_i = \{0i, x_iy_i, z_it_i\} : 1 \le i \le 5\}\\ \mathcal{G} &= \{G_i = \{0i, x_iz_i, y_it_i\} : 1 \le i \le 5\}.\end{aligned}$$

Since $\mathcal{S}_i$ is a triple system, F_i is a one-factor of the K_6 with vertices $\{0, 1, 2, 3, 4, 5\}$; from the fact that a large set contains all triples Aij for $0 \le i, j \le 5$, $\mathcal{F}$ is a one-factorization of the K_6; and similarly for $\mathcal{G}$. But it is easy to check that $F_i \cap G_i = \{0i\}$, and that if $i \ne j$ then $F_i \cap G_j$ can contain at most one edge (if $j = x_i$, then the intersection either consists of the edge z_it_i or it is empty). So the one-factorizations $\mathcal{F}$ and $\mathcal{G}$ are orthogonal, and are equivalent to a Room square of side 5. Therefore the large set cannot exist.

Exercises 15.1

15.1.1 Suppose there is a Room square of side 5. Without loss of generality, one may assume that its first row is

$\infty 1$	23	45		

Prove that this first row cannot be completed to a Room square.

15.1.2* Find another standardized Room square of side 7.
Hint: Take the square in Figure 15.1; apply a symbol permutation; then apply suitable row and column permutations.

15.2 Starter Constructions

Consider the Room square of Figure 15.3. It has been constructed as follows: The $(i+1, j+1)$ entry was obtained from the (i, j) entry by adding 1 to each member, subject to the rules of addition modulo 7 and "$\infty + 1 = \infty$." That is, both one-factorizations come from starters in $GF(7)$. Initially, the starter $\{16, 25, 34\}$ has been placed in the columns of the first row in such a way that the resulting first column is also a starter.

Can this be done more generally? Any starter in any Abelian group of order $2n-1$ will satisfy the conditions that every pair appears exactly once and every symbol appears exactly once per row. To satisfy the column condition, one must make some more restrictions on the starter. So we make the following definition.

Suppose $S = \{x_1y_1, x_2y_2, \ldots, x_{n-1}, y_{n-1}\}$ is a starter in an Abelian group G of order $2n - 1$. An *adder* A_S for S is an ordered set of $n - 1$ nonzero elements $a_1, a_2, \ldots, a_{n-1}$ of G, all different, with the property that the $2n-2$ sums $x_1 + a_1, y_1 + a_1, x_2 + a_2, y_2 + a_2, \ldots, y_{n-1} + a_{n-1}$ include each nonzero element of G once.

THEOREM 15.1

If there is a starter S in a group of order $2n-1$ with an adder A_S, then there is a Room square of side $2n-1$.

Proof. Label the elements of G as $g_0, g_1, \ldots g_{2n-2}$ in some order, where g_0 is the identity element. (In the case of a cyclic group, $g_i = i \pmod{2n-1}$.) We construct an array $\mathcal{H}$. We label the rows 0, 1, ..., $2n-2$, so that they correspond to the group elements. In the first row (row g_0), the first column (column g_0) contains $\{\infty, g_0\}$; column g_j contains an empty cell if $-g_j$ is *not* a member of the adder A_S, but if $-g_j \in A_S$ then column g_j contains the corresponding pair from the starter: If $-g_j = a_k$ then cell (g_0, g_j) contains $\{x_k, y_k\}$. The other rows are constructed from the first row. If the cell in the (g_0, g_j) position is empty then so is the $(g_h, g_j + g_h)$ cell; and if the (g_0, g_j) cell contains $\{x_k, y_k\}$ then cell $(g_h, g_j + g_h)$ contains $\{x_k + g_h, y_k + g_h\}$. The diagonal element in row g_j is $\{\infty, g_j\}$. Then $\mathcal{H}$ is easily seen to be a Room square. □

If we replace g_i by i for every i, throughout the foregoing construction, the resulting Room square is standardized.

Given the starter $\{x_1, y_1\}, \{x_2, y_2\}, \ldots, \{x_{n-1}, y_{n-1}\}$, consider the elements $a_i = -x_i - y_i$. We have

$$x_i + a_i = -y_i, \quad y_i + a_i = -x_i$$

for every i; the $2n-1$ sums $x_i + a_i$ and $y_i + a_i$ will be distinct and nonzero because the elements of pairs in a starter are distinct and nonzero. So, provided

∞0	34	16		25		
	∞1	45	20		36	
		∞2	56	31		40
51			∞3	60	42	
	62			∞4	01	53
64		03			∞5	12
23	05		14			∞6

FIGURE 15.3: Room square from a starter and adder.

the elements a_i (or equivalently, the elements $-a_i$) are themselves distinct and nonzero, we have an adder. We say a starter is *strong* if the sums $x_i + y_i$ are distinct and nonzero; what we have just said means that every strong starter has an adder.

THEOREM 15.2 [109]

Suppose p^n is an odd prime power of the form $p^n = 2^k t + 1$ where t is an odd integer greater than 1. Then there is a strong starter in $GF(p^n)$.

Proof. Suppose x is a primitive element in $GF(p^n)$ (see Section 5.2). For convenience write δ for $2k - 1$; then $x^{2\delta}t = 1$, but $x^\alpha \neq 1$ when $1 \leq \alpha < 2\delta t$. We shall verify that X,

$$X = \{X_{ij} : 0 \leq i \leq \delta - 1, 0 \leq j \leq t - 1\},$$

is a strong starter, where

$$X_{ij} = \{x^{i+2j\delta}, x^{i+(2j+1)\delta}\}.$$

It is easy to see that the totality of the entries in the X_{ij} constitute all nonzero elements of $GF(p^n)$; in fact, the left-hand members of $X_{00}, X_{10}, \ldots, X_{\delta-1,0}$ are $x^0, x^1, \ldots, x^{\delta-1}$, and the right-hand members are $x^\delta, x^{\delta+1}, \ldots, x^{2\delta-1}$; similarly, $\{X_{il}\}$ contains the members $x^{2\delta}, x^{2\delta+1}, \ldots, x^{4\delta-1}$, and so on. The differences between the elements of X_{ij} are $\pm x^{i+2j\delta}(x^\delta - 1)$. Since $x^\delta - 1$ is nonzero and $GF(p^n)$ is a field, the entries $x^{i+2j\delta}(x^\delta - 1)$ and $x^{k+2l\delta}(x^\delta - 1)$ will be equal if and only if $x^{i+2j\delta} = x^{k+2l\delta}$, and this can occur only when $i = k$ and $j = \delta$; so the δt differences with a + sign are all different. Similarly, the differences with a $-$ sign are all different. Moreover, if

$$x^{i+2j\delta}(x^\delta - 1) = -x^{k+2l\delta}(x^\delta - 1)$$

then

$$x^{i+2j\delta-k-2l\delta} = -1$$

and squaring yields

$$x^{2(i-k)+4\delta(j-l)} = 1.$$

This would imply that

$$2(i - k) + 4\delta(j - l) \equiv 0(\mathrm{mod}\ 2\delta t)$$

since $2\delta t$ is the order of x. In particular,

$$2(i - k) \equiv 0(\mathrm{mod}\ 2\delta)$$

and since $0 \leq i < \delta$ and $0 \leq k < \delta$ we must have $i - k = 0$. Therefore

$$4\delta(j - l) \equiv 0(\mathrm{mod}\ 2t)$$

from which

$$2(j-l) = 0(\bmod t);$$

but $-t < j - l < t$, so $j = l$ or $2(j-l) = \pm t$. The second case contradicts the fact that t is odd, and if $j = l$ then

$$2x^{i+2j\delta}(x^\delta - 1) = 0,$$

which is impossible since p^n is odd and consequently 2, x and $(x^\delta - 1)$ are all nonzero in $GF(p^n)$. So no two of the $2\delta t$ differences can be equal. Therefore the set of differences constitutes all nonzero elements of $GF(p^n)$.

This means that the set X satisfies the definition of a starter. To show that the starter is strong, we have to show that the δt elements formed by adding the two members of X_{ij} are different. The sum resulting from X_{ij} is

$$x^{i+2j\delta}(x^\delta + 1),$$

and these objects are all different because $x^\delta + 1$ is nonzero. □

THEOREM 15.3 **[34]**

If k is an integer greater than 1, then there is a strong starter in the additive group of integers modulo $2^{2k} + 1$.

Proof. Since $k \geq 2$, we can write $2^{2k} = 16t^2$, where $t = 2^{k-2}$ is a positive integer. The required starter consists of the following four pairs, for each i and j satisfying $1 \leq i \leq 2t$, $0 \leq j \leq t-1$:

$$\begin{array}{l}(i + 4jt, 4it - j)\\ (j - 4it, -8t^2 - 4jt - 2t - i)\\ (8t^2 + 4jt + 2t + i, 8t^2 + 4it - 2t - j)\\ (-8t^2 - 4it + 2t + j, -i - 4jt).\end{array}$$

The required verification is straightforward. □

Observe that Theorems 15.2 and 15.3 between them provide starters for Room squares of all prime orders except 3 and 5: Theorem 15.2 covers all primes that are not of the form $2^n + 1$; Theorem 15.3 covers all primes greater than 5 of the form $2^n + 1$ for n even; and if n is odd, $2^n \equiv 2 \pmod 3$, so 3 divides $2^n + 1$ and there is no prime greater than 3 of this type.

Exercises 15.2

15.2.1 Verify that the Room squares of side 7 given in Figures 15.1 and 15.3 are nonisomorphic.

15.2.2* Find adders for the three starters in the cyclic group of order 7.

15.2.3 Prove that there is no adder for the patterned starter in either group of order 9.

15.2.4* Prove that the patterned starter is never strong.

15.2.5 Verify that the pairs given in Theorem 15.3 do in fact form a strong starter.

15.2.6 A *skew Room square* $\mathcal{R}$ is a standardized Room square with the property that when $i \neq j$, either the (i, j) or the (j, i) cell of $\mathcal{R}$ is occupied, but not both.

(i) Suppose $\mathcal{R}$ is a Room square constructed from a starter and adder. Prove that in the notation used in Theorem 15.1, the $(i, 1)$ cell of $\mathcal{R}$ is empty if and only if g_i is not a member of A_S. Consequently, prove that the Room square constructed from a starter and adder is skew if and only if the adder never contains the negative of one of its elements.

(ii) Prove that the Room squares developed from strong starters in Theorems 15.2 and 15.3 are skew and standardized.

(iii) Suppose $\mathcal{R}$ is a skew Room square based on $\{\infty, 1, 2, \ldots, v\}$. Consider the set of all blocks $\{x, y, i, j\}$, where $1 \leq x < y \leq v$ and $\{x, y\}$ lies in the (i, j) cell of $\mathcal{R}$. Prove that these blocks form a $(2n + 1, 2n^2 + n, 4n, 4, 6)$-balanced incomplete block design.

15.2.7* A Room square of side $r = 2n - 1$ is called *special* if it has a set of $n - 1$ cells, occupied by $n - 1$ disjoint pairs of objects, which all lie in different rows and all lie in different columns.

(i) Show that a Room square constructed from a strong starter is special.

(ii) Exhibit a special Room square of side 9.

15.2.8 Suppose $\mathcal{R}$ is a skew, standardized Room square of side r (as defined in Exercise 6) based on $\{\infty, 1, 2, \ldots, r\}$. Define $\mathcal{P}$ to be the array formed by deleting the diagonal entries from $\mathcal{R}$, $\mathcal{Q}$ to be the array formed by adding r to every entry in $\mathcal{P}$, and $\mathcal{A}$ to be the result of superimposing $\mathcal{P}$ on the transpose of $\mathcal{Q}$. Select orthogonal idempotent Latin squares L and M of side r; denote by $\mathcal{B}$ the array whose diagonal cells are empty and whose (i, j) cell contains $\{l_{ij}, r + m_{ij}\}$ when $i \neq j$. Write $\mathcal{C}$ for $\text{diag}(\mathcal{A}, \mathcal{B})$.

(i) Show that the entries in the cells of $\mathcal{C}$ are all different.

(ii) Which unordered pairs on $\{1, 2, \ldots, 2r\}$ do not appear in $\mathcal{C}$?

(iii) Prove that it is possible to insert the "missing" pairs into $\mathcal{C}$ in such a way that the result is a Room square of side $2r - 1$.

15.3 Subsquare Constructions

In a Room square of side r we say a set of s rows and s columns forms a *subsquare of side* s if the $s \times s$ array formed by their intersection is itself precisely a Room square based on some $s + 1$ of the symbols. Trivially, a Room square is a subsquare of itself; if $s < r$ we say the subsquare is *proper*.

THEOREM 15.4 [72]

Suppose there exist a Room square of side r and a Room square of side s with a subsquare of side t. Suppose further that there exist orthogonal Latin squares of side $s - t$. Then there is a Room square of side $v = r(s - t) + t$ that has subsquares of sides r, s and t.

Proof. We construct a Room square based on the $v + 1$ symbols in the set

$$X = \{\infty, 1_1, 1_2, \ldots, 1_r, 2_1, 2_2, \ldots, n_r, n + 1, \ldots, s\}$$

where $n = s - t$. We assume the existence of standardized Room squares $\mathcal{R}$ of side r based on $\{\infty, 1, 2, \ldots, r\}$ and $\mathcal{S}$ of side s based on $\{\infty, 1, 2, \ldots, s\}$. We further assume $\mathcal{S}$ to have the form

$$\mathcal{S} = \begin{array}{|c|c|} \hline \mathcal{A} & \mathcal{B} \\ \hline \mathcal{C} & \mathcal{T} \\ \hline \end{array}$$

where the subarray $\mathcal{T}$ is a Room square of side t based on $\{\infty, n+1, n+2, \ldots, s\}$. We use orthogonal Latin squares L and M of side n on the symbols $\{1, 2, \ldots, n\}$. L_i denotes the array derived from L by giving every entry subscript i (if L has (p, q) entry x, then L_i has (p, q) entry x_i) and M_j is defined similarly. $\mathcal{N}_{ij}$ denotes the $n \times n$ array of unordered pairs whose (k, l) entry consists of the (k, l) entries of L_i and M_j. We subscript the arrays $\mathcal{A}$, $\mathcal{B}$ and $\mathcal{C}$ in a similar way; for example, in $\mathcal{A}_i$ the entry $\{x, y\}$ in $\mathcal{A}$ is replaced by $\{x_i, y_i\}$, except for entries ∞ or entries greater than n (if you prefer, we could define $x_i = x$ when $x = \infty$ or $x > n$) and the empty cells of $\mathcal{A}$ correspond to empty cells in $\mathcal{A}_i$.

We first convert $\mathcal{R}$ into an $nv_1 \times nv_1$ array $\mathcal{U}$ by replacing each cell of $\mathcal{R}$ by an $n \times n$ array. An empty cell is replaced by an empty $n \times n$ array; the entry $\{\infty, i\}$ is replaced by $\mathcal{A}_i$; and the entry $\{i, j\}, i \neq 0, j \neq 0$, is replaced by $\mathcal{N}_{ij}$.

Now define

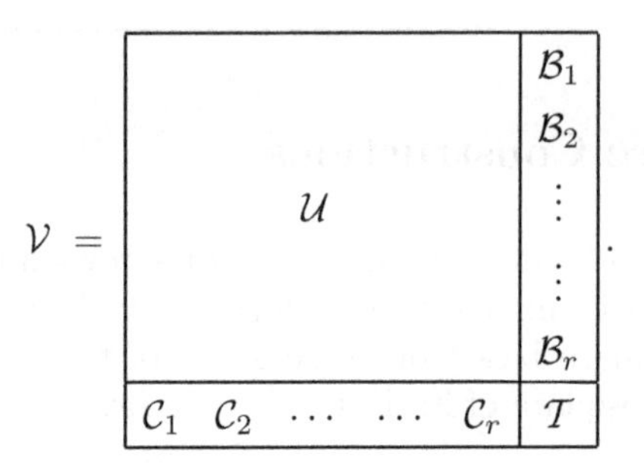

Row i of $\mathcal{R}$ contains $\{\infty, i\}$ and a set of entries $\{j, k\}$ whose elements collectively exhaust $\{1, 2, \ldots, i-1, i+1, \ldots, r\}$, So the i-th row of blocks of $\mathcal{V}$, $i \le i \le r$, contains $\mathcal{A}_i$, $\mathcal{B}_i$, and a set of $\mathcal{N}_{jk}$ such that the collection of all the j and k is $\{0, 1, \ldots, i-1, i+1, \ldots, r\}$. Each row of $\mathcal{N}_{jk}$ contains all the x_j and y_k, $1 \le x, y \le n$, once each. Each row of

$$\boxed{\mathcal{A}_i \mid \mathcal{B}_i}$$

contains all the objects

$$\{0_i, 1_i, \ldots, s_i\},$$

that is

$$\{0, 1_i, \ldots, n_i, n+1, \ldots, s\},$$

once each. So every row in the i-th row of blocks in $\mathcal{V}$ contains every element of X once, for $1 \le i \le r$. In the last row of blocks, since each row of $\mathcal{C}_i$ contains $\{1_i, 2_i, \ldots, n_i\}$ once each, $\{\mathcal{C}_1, \mathcal{C}_2, \ldots, \mathcal{C}_r\}$ will contain every element of X except $0, n+1, n+2, \ldots, s$ precisely once, and the missing elements are contained once in each row of $\mathcal{T}$. So every row of $\mathcal{V}$ contains every member of X once. A similar proof applies to columns.

If we take the whole of $\mathcal{V}$, we have every entry of $\mathcal{N}_{ij}$ once, for $1 \le i \le r$ and $i \le j \le r$. So every possible unordered pair of the form $\{x_i, y_j\}$, $1 \le x, y \le n$, $1 \le i, j \le r$, appears once. We also have entries $\mathcal{A}_i$, $\mathcal{B}_i$ and $\mathcal{C}_i$, $1 \le i \le r$; for fixed i, these three arrays will contain every pair of the form $\{x_i, y_i\}$, $1 \le x, y \le n$, and every entry $\{x_i, z\}$, $1 \le x \le n, z = 0$ or $n \le z \le s$. $\mathcal{T}$ will contain all the pairs $\{z, t\}$, where both z and t belong to $\{0, n+1, n+2, \ldots, s\}$. So $\mathcal{V}$ contains every unordered pair of members of X.

Finally, $\mathcal{A}_i$ and $\mathcal{B}_i$ together contain $\frac{1}{2}(s+1)$ pairs per row, each row of each $\mathcal{N}_{ij}$ contains n pairs, $\mathcal{C}_i$ contains $\frac{1}{2}n$ pairs in each row, and $\mathcal{T}$ has $\frac{1}{2}(t+1)$ pairs. So the number of pairs in $\mathcal{V}$ is

$$\begin{aligned}
&\left[\tfrac{1}{2}(r-1)n + \tfrac{1}{2}(s+1)\right] + t\left[\tfrac{1}{2}rn + \tfrac{1}{2}(t+1)\right] \\
&= \tfrac{1}{2}\left[rn(rn - n + n + t + 1) + t(rn + t + 1)\right] \\
&= \tfrac{1}{2}(rn + t)(rn + t + 1) \\
&= \tfrac{1}{2}v(v+1).
\end{aligned}$$

This is the number of unordered pairs that can be chosen from X, so each pair must appear precisely once. Therefore $\mathcal{V}$ is a Room square.

If we take the intersection of the first n and last t rows with the corresponding columns and delete everything else, we have

$\mathcal{A}_1$	$\mathcal{B}_1$
$\mathcal{C}_1$	$\mathcal{T}$

which is isomorphic to $\mathcal{S}$. The last t rows and columns form a subsquare isomorphic to $\mathcal{T}$.

To discover $\mathcal{R}$, take the intersection of rows $1, n+1, \ldots, n(r-1)+1$ and the corresponding columns. The array formed has entry $\{0, 1_i\}$ where $\mathcal{R}$ had $\{0, i\}$, and $\{1_i, 1_j\}$ where $\mathcal{R}$ had $\{i, j\}$. □

Although there cannot be a subsquare of side 0, the proof above works when $t = 0$. If $t = 1$, the subsquare size is 1 and any occupied cell is a subsquare of side 1. The only problem will arise when $n = 2$ or 6, because of the lack of orthogonal Latin squares. When $t = 0$, neither of these cases can occur, and when $t = 1$, only $n = 6$ (or $s = 7$) can occur. So we have:

COROLLARY 15.4.1

If there are Room squares of sides r and s, then there is a Room square of side rs with subsquares of sides r and s, and there is a Room square of side $r(s-1)+1$ with subsquares of sides r and s except possibly when $s = 7$.

The preceding theorem was originally proven as a device to construct Room squares of previously unknown orders. However, subsquares are of interest in their own right. There is a lower bound for the possible orders of proper subsquares.

THEOREM 15.5

If there is a Room square of side r with a subsquare of side s, where $s < r$, then $r \geq 3s + 2$.

Proof. Suppose there is a Room square $\mathcal{R}$ of side r based on $\{\infty, 1, 2, \ldots, r\}$, and suppose it has a subsquare $\mathcal{S}$ of side s based on $\{\infty, 1, 2, \ldots, s\}$, where $s < r$. Moreover, suppose $\mathcal{S}$ is located in the top left corner of $\mathcal{R}$: say

$$R = \begin{array}{|c|c|}\hline \mathcal{S} & \mathcal{A} \\ \hline \mathcal{B} & \mathcal{C} \\ \hline \end{array}.$$

Only symbols in $\{\infty, 1, 2, \ldots, s\}$ appear among entries in the subarray $\mathcal{S}$, and only symbols outside that set (greater than s) appear in $\mathcal{A}$ or $\mathcal{B}$.

Select any integer x with $s < x \leq r$. There must be a pair involving x in each of rows $1, 2, \ldots, s$; suppose $\{x, y_i\}$ is a pair in row i. Since $x > s$, this pair must lie in the subarray $\mathcal{A}$, so each y_i is greater than s. Similarly, there is a pair $\{x, z_j\}$ in column j for $j = 1, 2, \ldots, s$. These pairs lie in $\mathcal{B}$, so each z_j is greater than s. All of these elements must be different. So we know at least $3s + 2$ of the symbols in $\mathcal{R}$:

$$\infty, 1, 2, \ldots, s, x, y_1, y_2, \ldots, y_s, z_1, z_2, \ldots, z_s.$$

So $r \geq 3s + 1$. But both r and s must be odd, so $r \geq 3s + 2$. □

It has been shown [164] that every Room square of side s occurs as a subsquare of a Room square of side $3s + 2$, except for the trivial case $s = 1$.

Exercises 15.3

15.3.1 Prove that if there is a Room square of side r with a subsquare of side s, and there is a Room square of side s with a subsquare of side t, then there is a Room square of side r with a subsquare of side t.

15.3.2* Suppose $\mathcal{S}$ and $\mathcal{T}$ are Room squares of the same side, and suppose there is a Room square of side r with $\mathcal{S}$ as a subsquare. Show that there is a Room square of side r with $\mathcal{T}$ as a subsquare.

15.3.3 Suppose there is a pairwise balanced design $\mathcal{D}$ with parameters $(v; K; 1)$, and suppose there is a Room square $\mathcal{R}_k$ of side k for every $k \in K$. Without loss of generality, one may assume that each $\mathcal{R}_k$ is standardized. If $B = \{b_1, b_2, \ldots, b_k\}$ is a block of $\mathcal{D}$ of size k, then define a set $f(b_i, b_j)$ as follows:
- If the (i, j) cell of $\mathcal{R}_k$ is empty, $f(b_i, b_j) = \emptyset$.
- If the (i, j) cell of $\mathcal{R}_k$ contains $\{p, q\}$, then $f(b_i, b_j) = \{b_p, b_q\}$.

(i) Prove that if x and y are any distinct integers between 1 and v, then $f(x, y)$ is uniquely defined by the rules above.

(ii) Define a $v \times v$ array $\mathcal{S}$: if $x \neq y$, then $s_{xy} = f(x, y)$; and $s_{xx} = \{\infty, x\}$. Prove that $\mathcal{S}$ is a Room square of side v with subsquares of side k for every $k \in K$.

15.4 The Existence Theorem

From Section 15.3 we know that the existence of Room squares of sides r and s implies the existence of a square of side rs. In Section 15.2 we constructed squares of all prime sides except 3 and 5. So repeated multiplication will provide a Room square of every side m for every m prime to 15.

Theorem 15.2 provides squares of sides $25 = 5^2$ and $125 = 5^3$, so we can construct a square of side 5^k for every $k > 1$. There is a square of side 27, from Theorem 15.2 again, and there are Room squares of side 9; for example, one can be constructed by cyclically developing the first row

$\infty,1$	$6,9$	$4,8$					$5,7$	$2,3$

Similarly, a square of side 15 can be developed from

$\infty,1$	$3,8$				$11,14$			$7,13$	$5,6$		$4,15$	$10,12$		$2,9$

So

LEMMA 15.6

There is a Room square of every odd integer side r except possibly when $r = 3v$ or $r = 5v$, where $(v, 15) = 1$.

If $(v, 15) = 1$, we can construct a Room square of side v, so the cases missing in Lemma 15.6 are all three or five times the side of a Room square. Therefore, it will be sufficient for our purposes to provide a theorem that allows us to produce squares of sides $3r$ and $5r$ from a given square of side r.

Suppose n is an odd integer. Write $\mathcal{A}_n$ for the $n \times n$ array of ordered pairs chosen from the n-set $N = \{1, 2, \ldots, n\}$ of the first n natural numbers, whose (i, j) entry satisfies

$$a_{ij} \equiv (j - i + 1, i + j - 1) \pmod{n}.$$

For example,

$$A_3 = \begin{array}{|ccc|} \hline 11 & 22 & 33 \\ 32 & 13 & 21 \\ 23 & 31 & 12 \\ \hline \end{array}, \quad A_5 = \begin{array}{|ccccc|} \hline 11 & 22 & 33 & 44 & 55 \\ 52 & 13 & 24 & 35 & 41 \\ 43 & 54 & 15 & 21 & 32 \\ 34 & 45 & 51 & 12 & 23 \\ 25 & 31 & 42 & 53 & 14 \\ \hline \end{array}.$$

Some important properties of A_n are summarized in the following lemma.

LEMMA 15.7

The entries of $\mathcal{A}_n$ consist of the ordered pairs of members of N taken once each. The entries in a given row or column of $\mathcal{A}_n$ contain between them every member of N once as a left member and once as a right member. If the pair (x, y) occurs in a given column, then (y, x) also occurs in that column.

Proof. To see that any ordered pair of members of N appears precisely once in $\mathcal{A}_n$, observe that the pair (x, y) appears in position (i, j), where $1 \le i, j \le n$ and

$$\begin{aligned} i &\equiv \tfrac{1}{2}(n+1)(y-x)+1 \pmod{n} \\ j &\equiv \tfrac{1}{2}(n+1)(x+y) \pmod{n}. \end{aligned}$$

Now consider the rows and columns. In row i of A_n, the left-hand members of the entries are

$$\{2-i, 3-i, \ldots, -i, 1-i\}$$

(reduced modulo n), and the right-hand members are

$$\{i, i+1, \ldots, i-2, i-1\}.$$

In column j, the left-hand members are

$$\{j, j-1, \ldots, j+2, j+1\},$$

while the right-hand positions contain

$$\{j, j+1, \ldots, j-2, j-1\}.$$

In every row and every column, the set of entries is N.

Finally, suppose the pair (x, y) occurs in column j of $\mathcal{A}_n$. Then the pair (y, x) will also lie in column j, since $\frac{1}{2}(n+1)(y+x) = \frac{1}{2}(n+1)(x+y)$. $\square$

Now suppose $\mathcal{R}$ is a Room square of side v, based on the symbols $\{\infty, 1, \ldots, v\}$, and n is an odd integer less than v. We shall construct a Room square of side nv based on the symbol set

$$S = \{\infty, 1_1, 2_1, \ldots, v_1, 1_2, 2_2, \ldots, v_1, \ldots, v_n\}.$$

For convenience assume $\mathcal{R}$ to be standardized. We define an array $\mathcal{R}_{ij}$ for $1 \le i \le n$ and $1 \le j \le n$, as follows:

(i) delete all diagonal entries of $\mathcal{R}$;

(ii) if $x < y$, replace the entry $\{x, y\}$ of $\mathcal{R}$ by $\{x_i, y_j\}$.

The arrays $\mathcal{R}_{ij}$ will contain between them all unordered pairs of the symbols other than ∞, except for the pairs $\{x_i x_j\}$. If $\mathcal{R}_{ij}$ and $\mathcal{R}_{jk}$ are placed adjacent to one another, then row x of the resulting array will contain each of $1_j, 2_j, \ldots, v_j$ except for x_j precisely once; a similar remark applies to columns when $\mathcal{R}_{ij}$ is placed above $\mathcal{R}_{jk}$.

LEMMA 15.8

Given a Room square of $\mathcal{R}$ of side v, where $v = 2s + 1$, there are s permutations $\phi_1, \phi_2, \ldots, \phi_s$ of $\{1, 2, \ldots, v\}$ with the properties that $k\phi_i = k\phi_j$ never occurs unless $i = j$, and that cell $(k, k\phi_i)$ is empty whenever $1 \le k \le v$ and $1 \le i \le s$.

Proof. Consider the $r \times r$ matrix M whose (k, l) position contains 1 if $\mathcal{R}$ has its (k, l) cell empty, but is 0 otherwise. (M is called the *occupancy matrix* of $\mathcal{R}$.) M is a matrix of zeros and ones with every row and column sum equal to s. So, by Theorem 1.5, M is a sum of s permutation matrices, say

$$M = P_1 + P_2 + \cdots + P_s.$$

We define ϕ_i to be the permutation corresponding to P_i, as follows. If P_i has (k, l) entry 1, then $k\phi_i = l$. If $k\phi_i$ were to equal $k\phi_j$ when $i \ne j$, then P_i and P_j would both have 1 in position $(k, k\phi_i)$, so M would have an entry 2 or greater, which is impossible. That the $(k, k\phi_i)$ cell of $\mathcal{R}$ is empty follows from the definition of M. □

As an example, consider the square shown in Figure 15.2(a) of side 7. Three suitable permutations are as follows (where $(a, b, \ldots)$ represents the permutation ϕ for which $1\phi = a$, $2\phi = b$, $\ldots$):

$$\alpha = (2, 4, 6, 3, 7, 5, 1)\,,\ \beta = (3, 5, 7, 6, 1, 2, 4)\,,\ \gamma = (4, 7, 2, 5, 3, 1, 6)\,. \quad (15.1)$$

THEOREM 15.9 [163]

If v and n are odd integers, $v \ge n$, and if there is a Room square $\mathcal{R}$ of side v, then there is a Room square of side nv.

Proof. For convenience write $v = 2s + 1$ and $n = 2t + 1$. We define $\mathcal{A}_n$ and the arrays $\mathcal{R}_{ij}$ as before, and construct a Room square of side nv by replacing every entry of $\mathcal{A}_n$ by a $v \times v$ block. The square will be based on the symbol-set S, defined above. It is also convenient to index the rows and columns of the new square with the elements (other than ∞) of $\mathcal{S}$; for example, row i_j would be the $(i + v(j - 1))$-th row of the square.

For each j, select permutations $\phi_{j1}, \phi_{j2}, \ldots, \phi_{jn}$ satisfying:

(i) $\phi_{jk} = \phi_{jl}$ if and only if (k, l) and (l, k) appear in column j of $\mathcal{A}_n$;

(ii) if the entry in row j and column j of $\mathcal{A}_n$ is (x, y), then ϕ_{jx} and ϕ_{jy} equal the identity permutation;

(iii) all the ϕ_{jk} except for the identity permutation are selected from the set of s permutations associated with $\mathcal{R}$ according to Lemma 15.8.

(This will be possible since $v \geq n$ and consequently $s \geq t$.) Now replace the entry (k, l) in column j of $\mathcal{A}_n$ by the array $\mathcal{R}_{kl}\phi_{jk}$, constructed by performing the permutation ϕ_{jk} on the columns of $\mathcal{R}_{kl}$.

From Lemma 15.7 and the properties of the arrays $\mathcal{R}_{kl}$ it follows that the resulting array will contain every member of S other than ∞ in each row and each column, except for two problems. For each x, symbol x_k is missing from row x_j; and for those triples j, k, l such that (k, l) is an entry in column j of $\mathcal{A}_n$, symbols x_k and x_l are missing from column $(x\phi_{jk})_j$. Moreover, the array contains every unordered pair of members of $S \backslash \{\infty\}$ except for the pairs of the form $\{x_k, x_l\}$, and contains each precisely once.

We now insert some more entries into the j-th diagonal block of the new array. For each k, if (k, l) is an entry in column j of $\mathcal{A}_n$, we place $\{x_k x_l\}$ in the $(x, x\phi_{jk})$ position of this block (i.e., in the $(x_j, (x\phi_{jk})_j)$ position of the new square), except that $\{\infty, x_j\}$ is used instead of $\{x_j, x_j\}$ in the relevant position. Lemma 15.8 means that we shall not finish with two entries in the same cell, so this step is possible. The finished array now contains every entry from S once per row and once per column, and contains every unordered pair from that set precisely once. So it is the required Room square. $\square$

So there exist Room squares of all odd sides except 3 and 5.

The proof of Theorem 15.9 is somewhat opaque, so we shall walk through some of an example in the case $n = 3$, $r = 7$, constructing a Room square of side 21. We take $\mathcal{R}$ to be the Room square of side 7 shown in Figure 15.2(a). $\mathcal{R}$ is based on symbols $\{\infty, 1.2. \ldots, 7\}$, and we construct a Room square based on $\{\infty, 1_1, 2_1, \ldots, 7_1, 1_2, 2_2, \ldots, 7_2, \ldots, 1_3, 2_3, \ldots, 7_3\}$. We construct nine square arrays $\mathcal{R}_{ij}$; the array $\mathcal{R}_{23}$ is shown in Figure 15.4. These nine arrays contain all unordered pairs of the symbols other than the pairs $\{x_i, x_j\}$ for $x = 1, 2, \ldots, 7$, $i, j = 1, 2, 3$ and the pairs involving ∞.

The conditions on the permutations ϕ_{jk} imply that ϕ_{11}, ϕ_{21}, ϕ_{23}, ϕ_{31} and ϕ_{32} each equal the identity, and $\phi_{12} = \phi_{13}$. In terms of (15.1), let us take $\phi_{12} = \phi_{13} = \alpha$, $\phi_{22} = \beta$ and $\phi_{33} = \gamma$. The array constructed in the proof by

				3_26_3	2_27_3	4_25_3
4_26_3		1_27_3			3_25_3	
2_25_3			1_26_3	4_27_3		
3_27_3	1_25_3					2_26_3
	6_27_3		2_23_3		1_24_3	
		2_24_3	5_27_3			1_23_3
	3_24_3	5_26_3		1_22_3		

FIGURE 15.4: The array $\mathcal{R}_{23}$ of Theorem 15.9.

substituting subarrays into A_n will be, in our example,

$$A_3 = \begin{array}{|c|c|c|}\hline \mathcal{R}_{11} & \mathcal{R}_{22}\beta & \mathcal{R}_{33}\gamma \\ \hline \mathcal{R}_{32}\alpha & \mathcal{R}_{13} & \mathcal{R}_{21} \\ \hline \mathcal{R}_{23}\alpha & \mathcal{R}_{31} & \mathcal{R}_{12} \\ \hline \end{array}\,.$$

In this block array, the symbols x_i never appear in the x-th row of any block. Moreover, the choice and use of the permutations ensures that, if ij appears in column k of A_n, then x_i and x_j are missing from column $x\phi_{ij}$ of the blocks in column j of the block array. Then, when the last part of the proof requires that further entries be placed in the diagonal blocks of the block array, the construction ensures that each entry x_i will be placed in precisely those rows and columns where they are needed in order that every symbol should appear precisely once per row and per column, and that the cells to be used are currently empty. The construction reads: "If (k, l) is an entry in column j of $\mathcal{A}_n$, we place $\{x_kx_l\}$ in the $(x, x\phi_{jk})$ position of the j-th diagonal block." When $j = 1$, this means that entry $\{x_2, x_3\}$ must be placed in cell $(x, x\phi_{12})$ (that is, cell $(x, x\alpha)$) of the $(1, 1)$ block. (The rule would also require that $\{x_3, x_2\}$ should be placed in cell $(x, x\phi_{21})$, but this is exactly the same entry and the same cell.) So $\{1_2, 1_3\}$ goes in cell $(1, 1\alpha)$ of the top left block, $\{2_2, 2_3\}$ in cell $(2, 2\alpha)$, and so on. It is also required that $\{\infty, x_1\}$ be placed in the $(x, x\phi_{11})$ cell; since ϕ_{11} is the identity permutation, this means that $\{1_1, 1_1\}$, $\{2_1, 2_1\}$, ... go down the diagonal. Figure 15.5 shows the effect of inserting elements into the array $\mathcal{R}_{11}$ as required.

After similar insertions are made in the other two diagonal blocks, a square of side 21 is constructed.

Exercises 15.4

15.4.1 Suppose x and y are integers modulo n, where n is odd. Prove that the unique integers i and j modulo n that satisfy $j - i + 1 \equiv x$ and

$1_1 1_1$	$1_2 1_3$			$3_1 6_1$	$2_1 7_1$	$4_1 5_1$
$4_1 6_1$	$2_1 2_1$	$1_1 7_1$	$2_2 3_3$		$3_1 5_1$	
$2_1 5_1$		$3_1 3_1$	$1_1 6_1$	$4_1 7_1$	$3_2 3_3$	
$3_1 7_1$	$1_1 5_1$	$4_2 4_3$	$4_1 4_1$			$2_1 6_1$
	$6_1 7_1$		$2_1 3_1$	$5_1 5_1$	$1_1 4_1$	$5_2 5_3$
		$2_1 4_1$	$5_1 7_1$	$6_2 6_3$	$6_1 6_1$	$1_1 3_1$
$7_2 7_3$	$3_1 4_1$	$5_1 6_1$		$1_1 2_1$		$7_1 7_1$

FIGURE 15.5: Another array to illustrate Theorem 15.9.

$i + j - 1 \equiv y$ are given by

$$\begin{aligned} i &\equiv \tfrac{1}{2}(n+1)(y-x)+1, \\ j &\equiv \tfrac{1}{2}(n+1)(x+y). \end{aligned}$$

15.4.2 Construct the array A_7.

15.4.3 Construct a Room square of side 35.

15.5 Howell Rotations

As we said at the beginning of this chapter, Room squares arise in duplicate bridge.

In an ordinary bridge game, the relative scores of the two partnerships depend on the fall of the cards to a certain extent. Duplicate bridge attempts to remove this element of luck. In a duplicate tournament, a large number of partnerships[1] compete against each other. After the cards are dealt, into four hands labeled North, East, South and West, several pairs of partnerships play the deal independently, and then the *relative* performance of all the partnerships playing the North-South cards (NS) is compared, and similarly for the East-West (EW) partnerships. Each deal is called a *board* (this is also the term for the container in which they are stored; originally they were made of wood). If partnership i is NS on a particular board and partnership j is EW on that board, we say i *plays with* j on that board; if i and j both play NS or both play EW on a board, we say i *competes against* j on the board.

[1]These are usually called *pairs*, but we retain the word *partnerships* to avoid confusion.

The particular pairing that play against each other, and the board they play, will be called a *table.*

In the actual play of a tournament, various partnerships play various boards at the same time. The set of pairings and boards that play simultaneously is called a *round.* The following are desirable properties in a bridge tournament:

(a) for every i and j, the number of times i plays with j is constant, μ say;

(b) for every i and j, the number of times i competes against j is constant, k say;

(c) no board is played twice in the same round;

(d) every partnership plays every board once;

(e) every partnership plays in every round.

(Properties (a), (b) and (d) are for reliability of comparisons; (c) simplifies the running of the tournament and (e) is for the enjoyment of the players.)

Tournaments with these properties were developed by Edwin C. Howell, an MIT mathematics professor, in 1897. A *balanced Howell rotation* is a design for a tournament with $\mu = 1$ in which (a), (b), (c) and (d) all hold; if (e) is also true it is a *complete* balanced Howell rotation (or CBHR). Necessarily the number of partnerships will be even when (e) is true. We shall assume there are at least four partnerships ("duplicate" bridge for two partnerships is meaningless).

THEOREM 15.10 [127]

If there is a complete balanced Howell rotation for $2n$ *partnerships, then there is a Room square of side* $2n - 1$.

Proof. Suppose the partnerships have been numbered $0, 1, \ldots, 2n - 1$. We construct a square array $\mathcal{R}$ of side $2n - 1$ that contains $\{x, y\}$ in its (i, j) cell when partnerships x and y meet in round j and play board i there. Then R contains every unordered pair once, by (a), and every pair occurs once per row and once per column because of (c) and (d). So $\mathcal{R}$ is the required Room square. □

Every Room square can be used as a bridge tournament. In fact the term "complete Howell rotation" has been used in bridge to describe such a system, one that satisfies (a), (c) and (d) with $\mu = 1$. However, in a Howell rotation, an ordering is imposed on each pairing, resulting in an *ordered Room square*. This tells us which partnership competes against which. Then property (b) brings further restrictions.

THEOREM 15.11 [15, 112, 127]

If there is a complete balanced Howell rotation for $2n$ *partnerships, then* n *is even.*

Proof. Write B_i for the set of partnerships that compete against partnership x on board i. Then $|B_i| = n - 1$. There are $2n - 1$ boards, and $2n - 1$ partnerships other than x. By (b), each of these partnerships belong to k of the B_i, so $k = n - 1$.

Suppose there are λ boards on which partnerships y and z both compete against x; that is, $\{y, z\} \subseteq B_i$ for λ values of i. Then

$$\begin{array}{lll} y \in B_i, & z \notin B_i & \text{for } n-1-\lambda \text{ values of } i, \\ y \notin B_i, & z \in B_i & \text{for } n-1-\lambda \text{ values of } i, \end{array}$$

so $y \notin B_i$ and $z \notin B_i$ is true for $2n - 1 - \lambda - 2(n - 1 - \lambda) = \lambda + 1$ values of i. The cases where y competes against z are those where they are both sitting in the same direction as x (both in B_i or both in the opposite direction (neither in B_i), so they compete $\lambda + (\lambda + 1)$ times. Therefore $k = 2\lambda + 1$. But $k = n - 1$, so $n = 2(\lambda + 1)$. □

In the above proof, it is in fact clear that the sets B_i must form a symmetric balanced incomplete block design with parameters $(2n - 1, n - 1, \frac{1}{2}n - 1)$.

COROLLARY 15.11.1

If there exists a complete balanced Howell rotation for $2n$ *partnerships, then there exists a* $(2n-1, n-1, \frac{1}{2}n-1)$*-BIBD, or equivalently a Hadamard matrix of order* $2n$.

For this reason, the Room squares that correspond to complete balanced Howell rotations are sometimes called *Hadamard Room squares.*

Small examples of Howell rotations appear in bridge directors' manuals, such as [58]. Some small CBHRs are listed in [112].

THEOREM 15.12 [15]

There is a complete balanced Howell rotation for $4s$ *partnerships whenever* $4s - 1$ *is a prime power.*

Proof. We verify that the Room square of Theorem 15.2 provides a CBHR when the order is congruent to 3 (mod 4). Suppose x is a primitive element in $GF(4s - 1)$. Then Theorem 15.2 shows that

$$\{x, x^2\}, \{x^3, x^4\}, \ldots, \{x^{4s-3}, x^{4s-2}\}$$

is a strong starter. Denote the corresponding Room square by $\mathcal{R}$.

Assume the elements of $GF(4s-1)$ are $g_1, g_2, \ldots g_{2n-1}$ in some order, where g_1 is the identity element. In the i-th row of $\mathcal{R}$, the pairings are ordered (∞, g_i) and $(x^{2k} + g_i, x^{2k-1} + g_i)$. The set of pairs that play NS on the i-th board is $\{\infty\} \cup \{x^{2k} + g_i\}$, and the EW set is the complement. The collection of these sets as i varies is precisely the set of blocks of the Hadamard 3-design (see Exercise 14.2.2) constructed from the Paley Hadamard matrix. □

As an example, consider the case $4s - 1 = 7$ with $x = 3$ and the elements in order 0,1,2,3,4,5,6. The (ordered) difference set is $23, 46, 15$. Figure 15.6 shows the ordered Room square of side 7 and the corresponding Howell rotation. The notation x, y means x is NS, y is EW.

$\infty 0$			46		23	15
26	$\infty 1$			50		34
45	30	$\infty 2$			61	
	56	41	$\infty 3$			02
13		60	52	$\infty 4$		
	24		01	63	$\infty 5$	
		35		12	04	$\infty 6$

Round 1	$\infty, 0$	$4, 6$	$2, 3$	$1, 5$
Round 2	$\infty, 1$	$5, 0$	$3, 4$	$2, 6$
Round 3	$\infty, 2$	$6, 1$	$4, 5$	$3, 0$
Round 4	$\infty, 3$	$0, 2$	$5, 6$	$4, 1$
Round 5	$\infty, 4$	$1, 3$	$6, 0$	$5, 2$
Round 6	$\infty, 5$	$2, 4$	$0, 1$	$6, 3$
Round 7	$\infty, 6$	$3, 5$	$1, 2$	$0, 4$

FIGURE 15.6: Room square and Howell rotation.

In the above example, all the pairs listed in a column would play at the same (physical) table.

Exercises 15.5

15.5.1 Construct a CBHR for 12 partnerships.

15.5.2* A *Howell design* $\mathcal{H}(s, 2n)$ is an $s \times s$ array based on a $2n$-set V whose entries are empty or unordered pairs from V, such that no pair is repeated and such that every row and column contains every element of V precisely once.

(i) What design corresponds to an $\mathcal{H}(n, 2n)$?

(ii) What design corresponds to an $\mathcal{H}(2n - 1, 2n)$?

(iii) Find an $\mathcal{H}(4, 6)$.

15.5.3 In the notation of Exercise 15.5.2, find an $\mathcal{H}(s, 2n)$.

Chapter 16

Further Applications of Design Theory

16.1 Statistical Applications

We have pointed out that several parts of combinatorial design theory were studied because of statistical applications. In this section we shall sketch some of these applications. We give some detail in the first case (designs for weighing experiments). For further details and a description of the statistical methods used in analyzing experiments, see classical books on experimental design such as [42] or more recent volumes like [8]. A good book on statistical applications that makes frequent reference to the combinatorial background is [146].

Weighing Designs

A *weighing design* is a scheme for weighing a number of objects by weighing them in groups rather than one at a time. If the individual weights are small, as would often be true in chemical experiments, the error in weighing (either due to components of the scale or observational error) could well be large enough to be significant.

The following instructive example is from [68]. Assume four objects R_1, R_2, R_3, R_4 are to be weighed on a balance. There will be a small error in each weighing; suppose the errors are random and their average is 0. (If not, the balance could presumably be adjusted.) We can treat the errors as readings from a (normally distributed) random variable ϵ with mean 0 and variance σ^2. This means that if e is a reading of the error, its expected value is $E(e) = 0$, and $E(e^2) = \sigma^2$. If e_1 and e_2 are two separate error readings then they are independent, so $E(e_1e_2) = E(e_1)E(e_2) = 0$ and $E(e_1+e_2) = E(e_1)+E(e_2) = 0$.

Suppose the actual weight of R_i is μ_i, and the reading (weight recorded) for that object is m_i. Then $m_i = \mu_i + e_i$, where e_i is a reading of the error variable ϵ. Then the best estimate of μ_i is clearly m_i, and this is an unbiased estimate (the expected value of $\mu_i - m_i$ is 0). The *square error* is the expected value of $(\mu_i - m_i)^2$, and this is a measure of the reliability of the weighings. The *mean square error*, the expected value of the squared error, is σ^2.

Yates [187] observed that more reliable results can be obtained if the objects are weighed in groups. Say we are using a spring balance, and the following four weighings are carried out:

W1: objects R_2, R_3 and R_4, resulting weight w_1;

W2: objects R_1 and R_2, resulting weight w_2;

W3: objects R_1 and R_3, resulting weight w_3;

W4: objects R_1 and R_4, resulting weight w_4.

The results are

$$\begin{aligned} w_1 &= \phantom{\mu_1 + {}} \mu_2 + \mu_3 + \mu_4 + e_1, \\ w_2 &= \mu_1 + \mu_2 \phantom{{} + \mu_3 + \mu_4} + e_2, \\ w_3 &= \mu_1 \phantom{{} + \mu_2} + \mu_3 \phantom{{} + \mu_4} + e_3, \\ w_4 &= \mu_1 \phantom{{} + \mu_2 + \mu_3} + \mu_4 + e_4, \end{aligned} \tag{16.1}$$

where the e_i are again errors. Solving the equations (16.1) we get

$$\begin{aligned} 3\mu_1 &= -w_1 + w_2 + w_3 + w_4 + e_1 - e_2 - e_3 - e_4, \\ 3\mu_2 &= w_1 + 2w_2 - w_3 - w_4 - e_1 - 2e_2 + e_3 + e_4, \\ 3\mu_3 &= w_1 - w_2 + 2w_3 - w_4 - e_1 + e_2 - 2e_3 + e_4, \\ 3\mu_4 &= w_1 - w_2 - w_3 + 2w_4 - e_1 + e_2 + e_3 - 2e_4. \end{aligned}$$

The unbiased estimate of μ_1 is $\frac{1}{3}(-w_1 + w_2 + w_3 + w_4)$. The mean square error is

$$\begin{aligned} &\frac{1}{9}E((e_1 - e_2 - e_3 - e_4)^2) \\ &= \frac{1}{9}E(e_1^2 + e_2^2 + e_3^2 + e_4^2 - 2e_1e_2 - 2e_1e_3 - 2e_1e_4 + 2e_2e_3 + 2e_2e_4 + 2e_3e_4) \\ &= \frac{1}{9}\left(E(e_1^2) + E(e_2^2) + E(e_3^2) + E(e_4^2) - 2E(2e_1e_2) - 2E(2e_1e_3) - 2E(e_1e_4)\right. \\ &\qquad \left. +2E(e_2e_3) + 2E(e_2e_4) + 2E(e_3e_4)\right) \\ &= \frac{4}{9}\sigma^2. \end{aligned}$$

Similarly, μ_i has unbiased estimate $w_i - \frac{1}{3}(w_1 + w_2 + w_3 + w_4)$ and mean square error $\frac{7}{9}\sigma^2$ for $i = 2, 3, 4$.

If a beam balance is used, even better results can be obtained. Some objects can be placed in one pan, others in the other pan, and weights added until the pans balance. For example, say R_1 and R_2 are placed in the left-hand pan, while R_3 and R_4 are in the right-hand pan. The weights added to the right-hand pan to make this balance will form an estimate for $\mu_1 + \mu_2 - \mu_3 - \mu_4$. (If the right-hand pan is heavier, the weights added to the left estimate the negative of this.) Consider four weighings. In the first, all four objects are in the left pan; then $\{R_1, R_2\}$, $\{R_1, R_3\}$ and $\{R_1, R_4\}$ respectively are in the left. We get

$$\begin{aligned} w_1 &= \mu_1 + \mu_2 + \mu_3 + \mu_4 + e_1, \\ w_2 &= \mu_1 + \mu_2 - \mu_3 - \mu_4 + e_2, \\ w_3 &= \mu_1 - \mu_2 + \mu_3 - \mu_4 + e_3, \\ w_4 &= \mu_1 - \mu_2 - \mu_3 + \mu_4 + e_4. \end{aligned} \tag{16.2}$$

In this example it is found that the mean square error for each weight estimate is $\frac{1}{4}\sigma^2$.

All these experiments can be described by a *weighing matrix* A: $a_{i,j} = 1$ if R_j is in the left pan in weighing i, -1 if it is in the right and 0 if it is not weighed. (Treat a spring balance design as if all objects are in the left pan.) The three weighing methods we have described have matrices

$$\begin{bmatrix} 1&0&0&0 \\ 0&1&0&0 \\ 0&0&1&0 \\ 0&0&0&1 \end{bmatrix}, \quad \begin{bmatrix} 0&1&1&1 \\ 1&1&0&0 \\ 1&0&1&0 \\ 1&0&0&1 \end{bmatrix}, \quad \begin{bmatrix} 1&1&-1&-1 \\ 1&1&1&1 \\ 1&-1&1&-1 \\ 1&-1&-1&1 \end{bmatrix}.$$

Suppose a weighing matrix $A = (a_{ij})$ is used to estimate the weights of n objects. To ensure unique estimates we require that A should be square and invertible; write $A^{-1} = (b_{ij})$. Say the result of weighing i is w_i and the weight of object R_j is μ_j; write $\boldsymbol{w}$ and $\boldsymbol{m}$ for the vectors of w's and μ's. Then $\boldsymbol{w} = A\boldsymbol{m} + \boldsymbol{e}$, where $\boldsymbol{e}$ is the vector of errors, so $m = A^{-1}(\boldsymbol{w} - \boldsymbol{e})$. The error is $A^{-1}\boldsymbol{e}$, so the mean square error in the estimate of μ_i is $\sum_j b_{ij}^2\sigma^2$.

Write $\boldsymbol{a}$ and $\boldsymbol{b}$ for column i of A and row i of A^{-1} respectively and denote the length of the vector $\boldsymbol{v}$ by $\|v\|$ as usual. Then the above mean square error is $\|\boldsymbol{b}\|^2\sigma^2$.

Cauchy's inequality tells us that for any two vectors $\boldsymbol{u}$ and $\boldsymbol{v}$,

$$\boldsymbol{u} \bullet \boldsymbol{v} \leq \|\boldsymbol{u}\| \times \|\boldsymbol{v}\|.$$

In particular

$$\boldsymbol{a} \bullet \boldsymbol{b} \leq \|\boldsymbol{a}\| \times \|\boldsymbol{b}\|.$$

Now $\boldsymbol{a} \bullet \boldsymbol{b} = (A^{-1}A)_{ii} = 1$; as every element of any weighing matrix has square 0 or 1, $\|\boldsymbol{a}\| \leq \sqrt{n}$; so

$$\|\boldsymbol{b}\|^2\sigma^2 \geq \frac{\sigma^2}{n}.$$

We have:

THEOREM 16.1 [74]

The mean square error associated with an $n \times n$ weighing matrix is at least σ^2/n.

To attain this bound we require an invertible matrix with every entry either 1 or -1. In fact it may be shown that the rows must be orthogonal. That is, the weighing matrix is a Hadamard matrix.

COROLLARY 16.1.1

There is an $n \times n$ weighing matrix with associated mean square error σ^2/n if and only if $n = 1, 2$ or $n \equiv 0 \pmod 4$ and there exists a Hadamard matrix of order n.

When there is no Hadamard matrix, there are some very good weighing designs that approximate optimality, especially for large orders. For orders $n \equiv 3 (\bmod 4)$, if there is a skew-Hadamard matrix of order $n+1$, use the *core* as defined in Exercise 14.4.9 (delete the first row and column of a normalized skew-Hadamard matrix and set the diagonal equal to zero). When $n \equiv 2 (\bmod 4)$ a *conference matrix* (see Exercise 14.4.10) can be used if one is available. If $n \equiv 1 \pmod 4$ and there is a conference matrix C of order $n + 1$, we can define the core of C as follows: Multiply rows and columns of C to obtain a conference matrix with first row and column all 1's; delete that row and column. This is a very good weighing matrix.

Comparative Experiments and BIBDs

Consider an agricultural experiment. One might plant equally sized plots of land with wheat, and weigh the yield from each plot. In the simplest case, suppose we want to test the effects of different fertilizers. Assume all the plots are of equal fertility, all the wheat is the same, and all plots are treated in the same way (watering and so on). The yields will not all be the same; there will be some random variation. This variation is sometimes called an *error*, by analogy with the case of weighings. The model is linear:

$$y_i = \mu + e_i,$$

where $\mu = E(y)$ is the expected (or mean) yield, and y_i and e_i are the yield and error, respectively, for plot i. We assume there are variables y, the yield, and e, the error; y_i and e_i are the readings of y and e observed at plot i. The mean of y is best estimated by the average $\overline{y}$ of the y_i; in this model, the mean of the e_i is zero. In many (but not all) cases it is reasonable to assume that the e_i are random values of some error variable e (the same variable for all plots) that is distributed normally with mean 0.

The usual indicator of variability is the *variance* of the yields, the expected value of $y_i - \overline{y}$:

$$var(y) = E(y - \mu)^2.$$

In this simple case the variance equals the variance (in the usual statistical sense) of the error variable. In fact, one of the reasons for carrying out such a simple experiment is to estimate the variability, and therefore the distribution, of e. The variance of e is often called the *mean square error.*

Many experiments are *comparative.* For example, in the agricultural experiment, there might be several different types of wheat, or several different fertilizers. Say there are v treatments (such as fertilizers) and r plots are treated with each, giving vr plots. The mean yield for a plot with treatment i will be $\mu + \alpha_i$, where α_i shows the effect of applying treatment i (called a *treatment effect*) and the yield y_{ik} of the k-th plot receiving treatment i satisfies

$$y_{ik} = \mu + \alpha_i + e_{ik},$$

where e_{ik} is a reading from e. The object of the experiments is to estimate the effects α_i.

The assumption here is that there are at least vr plots of uniform fertility where there are v treatments (numbered $1, 2, \ldots, v$) to be tested and we wish to use each treatment on r plots. This may not be true. To overcome the problem, *blocking* is used. A *block* is a set of plots that are believed to be uniform. Suppose there are s blocks numbered $1, 2, \ldots b$, and each block contains $k = vt$ plots. Then we could use each treatment on t plots per block, giving $r = st$. This is called a *complete block design*; the treatments are allocated randomly within the blocks, and the result is a *randomized complete block experiment.* The model now is as follows: in block j, y_{ijk} is the yield of the k-th plot receiving treatment i, and

$$y_{ijk} = \mu + \alpha_i + \beta_j + e_{ijk},$$

where α_i is the treatment effect for treatment i, β_j is an effect for block j and e_{ijk} is the error reading.

Problems arise when the blocks are so small that fewer than v plots are available per block. You might try making the plots smaller, but then the error becomes too large compared to the values α_i, and it is difficult to obtain significant results. This is why Yates [187] introduced balanced incomplete block designs. The v objects in the definition of a BIBD correspond to the v treatments, and each (physical) block is divided into k plots; in them are applied the k treatments corresponding to the elements of the (set-theoretic) block. In this case there will be at most one plot in a block that receives a given treatment. If treatment i is applied to block j then the yield is $y_{ij} = \mu + \alpha_i + \beta_j + e_{ij}$, where α_i and β_j are defined as before, and $E(y_{ij}) =$

$\mu+\alpha_i+\beta_j$; if treatment i is not applied to block j then the yield is $y_{ij}\mu+e_{ij}$, and $E(y_{ij})=\mu$.

The existence of two models, one when a treatment is applied in a block and the other when it is not, is confusing. To simplify matters, we define a new vector $\boldsymbol{z}$ of length bk whose entries are the yields of the plots (those y_{ij} where treatment i is used on block j, in some order). The first k entries of $\boldsymbol{z}$ are the yields of block 1, ordered internally by treatment (if k were 3 and block 1 were $\{2,6,7\}$ then the first three entries of $\boldsymbol{z}$ would be y_{12}, y_{16}, y_{17}); then come the yields of block 2, and so on. The *plot-treatment matrix* $T=(t_{xy})$ of dimensions $bk\times v$ has $t_{ij}=1$ when treatment j is in plot i and 0 elsewhere, and the *plot-block matrix* $B=(b_{xy})$ is a $bk\times v$ $(0,1)$-matrix with $b_{ij}=1$ if and only if block j contains plot i. (Column j of B will have 1 in positions $(j-1)k+1, (j-1)k+2, \ldots, jk$, and 0 elsewhere.) Then the usual (treatment-block) incidence matrix of the design will be A, where

$$A = T^T B. \tag{16.3}$$

(Verification is left as an exercise.) Defining $\boldsymbol{\alpha}$ for the vector with i-th entry α_i, and similarly for $\boldsymbol{\beta}$, and $\boldsymbol{e}$ for a vector whose i-th value is the error for plot i, the model is

$$\boldsymbol{z} = \mu\boldsymbol{j} + \boldsymbol{\alpha} + \boldsymbol{\beta} + \boldsymbol{e}$$

($\boldsymbol{j}$ is a vector of all 1's).

Suppose $\boldsymbol{t}$ is the vector whose entry is the total yield of plots with treatment i, and $\boldsymbol{b}$ is defined similarly for blocks. If $\hat{\boldsymbol{\alpha}}$ is the vector of estimates of treatment effects, then

$$(rI - k^{-1}AA^T)\hat{\boldsymbol{\alpha}} = \boldsymbol{t} - k^{-1}A\boldsymbol{b}.$$

If $M = rI - k^{-1}AA^T$ were invertible, with inverse G, then we would have $\hat{\boldsymbol{\alpha}} = G(\boldsymbol{t} - k^{-1}A\boldsymbol{b})$. However, M is not invertible. In the case of a balanced incomplete block design,

$$M = \frac{\lambda v}{k}(I - \frac{1}{v}J). \tag{16.4}$$

This matrix is singular. The best estimate is found if G is a *generalized inverse* of M—a matrix such that $MGM = M$. It is easy to check that $(I - \frac{1}{v}J)$ is idempotent, so for a BIBD, we can put $G = \frac{k}{\lambda v}(I - \frac{1}{v}J)$.

A design is more efficient if the estimates of the treatment effects are more reliable—it is desirable to minimize the variances of these estimates. Among designs with v treatments and block size k, a design is called *A-optimal* if $tr(G)$ is minimal, *D-optimal* if $det(G)$ is minimal and *E-optimal* if the maximum eigenvalue of G is minimal (see [85]). A BIBD, if one exists, is optimal for all three of these criteria.

The above analysis of comparative experiments is largely from [8], [45] and [148].

Room Squares as Experimental Designs

A Room square can be interpreted as a design for a three-factor experiment (see [7, 130]). To continue the agricultural example, one might have different varieties, different treatments, and block effects. For example, one might have several varieties of wheat and wish to study the effects of several fertilizers on them. Suppose there are r blocks, r varieties and $r+1$ treatments. We shall show how to conduct an efficient experiment in the case where $r \equiv 1 \pmod 4$, and the Room square is a Hadamard Room square.

Suppose $\mathcal{R}$ is a given Hadamard Room square of side $r = 4m - 1$. We first interpret $\mathcal{R}$ as a balanced incomplete block design for r varieties in r blocks. This is done by ignoring the contents of the cells of $\mathcal{R}$, and only noting whether or not they are occupied. There are $2m$ varieties used in each block; variety i is used in block j if and only if the (i, j) cell of $\mathcal{R}$ is occupied. So there are $2m$ plots in each block.

Now each plot is divided into two subplots, so there will be $r + 1 = 4m$ subplots in each block. If $\{x, y\}$ is the entry in the (i, j) cell of $\mathcal{R}$, then the two subplots of block j that contain variety i receive treatments x and y respectively.

The analysis of this design is given by Shah [130]. It is found to be very efficient.

Designing Experiments Using Latin Squares

Latin squares provide another model for experiments where there are several different factors. As an example, consider a motor oil manufacturer who claims that using its product improves gasoline mileage in cars. Suppose you wish to test this claim by comparing the effect of four different oils on gas mileage. However, because the car model and the driver's habits greatly influence the mileage obtained, the effect of an oil may vary from car to car and from driver to driver. To obtain results of general interest, you must compare the oils in several different cars and with several different drivers. If you choose four car models and four drivers, there are 16 car-driver combinations. The type of car and driving habits are so influential that you must consider each of these 16 combinations as a separate experimental unit. If each of the four oils is used in each unit, 4×16, or 64, test drives are needed.

In this case there are three factors; the oil to be used in a test run is one, and so are the driver and the type of car.

An experiment is designed using a 4×4 Latin square. Suppose the cars are denoted X_1, X_2, X_3, X_4, and the drivers are Y_1, Y_2, Y_3, Y_4. Call the oils Z_1, Z_2, Z_3, Z_4. We call these the four *levels* of the factors. We assign one oil to each of the 16 car-driver combinations in the following arrangement:

	Y_1	Y_2	Y_3	Y_4
X_1	Z_1	Z_2	Z_3	Z_4
X_2	Z_2	Z_1	Z_4	Z_3
X_3	Z_3	Z_4	Z_2	Z_1
X_4	Z_4	Z_3	Z_1	Z_2

For example, the cell in the row marked X_2 and the column marked Y_3 contains entry Z_4. This means, in the test drive when car X_2 is driven by Y_3, oil Z_4 is to be used. The diagram is a plan for 16 test drives, one for each combination of car and driver. It tells us which oil to use in each case.

The array underlying the experiment is a Latin square, so each oil appears four times, exactly once in each row (for each car) and also exactly once in each column (for each driver). This setup can show how each oil performs with each car and with each driver, while still requiring only 16 test-drives, rather than 64.

This technique can be used with any size Latin square. Suppose there are three types of treatment, X, Y and Z, and each is to be applied at n levels (X at levels $X_1, X_2, \ldots, X_n$, and so on). An $n \times n$ Latin square provides n^3 comparisons with n^2 tests; if the (i, j) element is k, then treatment levels X_i, Y_j and Z_k are applied in one test.

In experiments where interaction between the different factors is unlikely, orthogonal Latin squares allow a large amount of information to be obtained with relatively little effort. This idea has been applied in a number of areas other than experimentation. One interesting application is the use of orthogonal Latin squares in compiler testing; for details, see [106].

Exercises 16.1

16.1.1 In the weighing design described by (16.2), verify that the mean square error for each weight estimate is $\frac{1}{4}\sigma^2$.

16.1.2 What is the mean square error for weighing experiments with the following matrices, if the error variable has variance σ^2?

(i) $\begin{bmatrix} 1 & 1 & 0 \\ 1 & 0 & 1 \\ 0 & 1 & 1 \end{bmatrix}$ (ii) $\begin{bmatrix} 1 & -1 & 1 \\ 1 & 1 & -1 \\ -1 & 1 & 1 \end{bmatrix}$

16.1.3 Assume the error variable has variance σ^2. What is the mean square error for weighing experiments with the following matrices?

(i) The core of a skew-Hadamard matrix of order $n \equiv 3 \pmod 4$.

(ii) A symmetric conference matrix of order $n \equiv 2 \pmod 4$.

(iii) The core of a symmetric conference matrix of order $n \equiv 1 \pmod 4$.

16.1.4 Verify the relation (16.3).

16.1.5 Find the matrices T and B for the $(7, 3, 1)$-BIBD constructed in Section 1.1.

16.1.6* Verify (16.4) and show that $(I - \frac{1}{v}J)$ has zero determinant.

16.1.7 Prove that if the matrices involved are $v \times v$ then $(I - \frac{1}{v}J)$ is idempotent.

16.2 Information and Cryptography

There are a number of relationships between combinatorial objects and information structures. In this section we point to some areas for further study. For further information on error-correcting codes, one of the best sources is [105]. Applications of design theory in information science and cryptography are surveyed in [38].

Network Design

Balanced incomplete block designs have been used in the construction of switching networks. The technique was introduced in [189]. The design provides a packet switched network with three desirable properties: It is *scalable*, which means that under arbitrary pattern the throughput increases by adding links and switches; it is *congestion-free*; and it guarantees that packets will reach their destination (this is called *convergence.*)

Previous approaches had not guaranteed all these properties. Those that are based on a simple linear topology like a bus or a ring (examples include ethernets and token rings) are congestion-free but not scalable. Network design based on an arbitrary topology is scalable but not congestion-free.

Yener et al. synthesized a network from linear structures called *virtual rings*, which were sets of nodes (that is, switches). The nodes correspond to the elements, and the virtual rings to the blocks, of a balanced incomplete block design. The block size provides a bound on the length of routings within the network.

For details of this correspondence between designs and packet switched networks, see [189].

Error-Correcting Codes

Information is commonly stored or transmitted by electronic means, typically as computer-readable files. For this reason it will usually be represented by a vector of binary digits (0s and 1s), or *bits*. This is called *encoding* the information, and the set of all representatives is called a (binary) *code*. So a *binary code* is a collection of binary vectors of some fixed length n. The vectors are called *codewords*, and the number of 1s in the word is called its *weight*. For example, the ASCII code and its descendants give 128 8-bit codewords representing all the upper and lower case letters and a number of other symbols; in this case $n = 8$.

Often an error will occur when a codeword is transmitted. Suppose one error occurs in sending a codeword. If two words in a code differ in only one place, it is possible that one of these words will be sent and the other received, and the error will not be noticed. However, if no two words have distance smaller than d, any error involving fewer than d symbols will be detected. This means that the minimum distance between codewords is important. If every two distinct codewords differ in at least d places, and d is the smallest such integer, the code is said to have *Hamming distance* d.

Nearest-neighbor decoding is the following process. Find the codeword whose distance from the received message is smallest; decode to that codeword. If fewer than $d/2$ errors are made in transmitting a word, and the code has Hamming distance d, any received word will be decoded correctly. We say the code will *detect any* $(d-1)$ *errors and correct any* $\lfloor(d-1)/2\rfloor$ *errors*; it is a $(d-1)$*-error correcting,* $\lfloor(d-1)/2\rfloor$*-error detecting* code.

The *binary sum* of two codewords is simply their sum when considered as vectors over $GF(2)$. A *linear code* is one that is closed under binary sum: The binary sum of any two of the codewords is also a codeword. Among the important properties of linear codes are the fact that the zero word $0000\ldots$ is always one of the codewords, and the Hamming distance of the code equals the minimum of all weights of nonzero codewords.

Several codes derived from Hadamard matrices and block designs are defined in [68]. Suppose H is a normalized Hadamard matrix of order $4m$. Write H^* for the matrix derived from H by deleting the first row and column. The *S-matrix* associated with H is the matrix $S = \frac{1}{2}(J - H^*)$. S will be the incidence matrix of a $(4m-1, 2m, m)$-SBIBD. A *simplex code* is the code whose words are the rows of S, along with the all-zero word. This code has $4m$ words of length $4m-1$, and Hamming distance $2m$.

In particular, suppose $4m-1$ is a prime power, and suppose S was derived from the quadratic elements of $GF(4m-1)$. Consider the rows of S_n as vectors over $GF(2)$. The elements of the subspace generated by them form a linear code containing $2^{(n+l)/2}$*codewords.* When $4m-1$ is prime, this is called a *quadratic residue code.*

Finally, consider the Hadamard matrices of order 2^d constructed by the Kronecker product (Sylvester matrices). Assuming we start with a normalized 2×2 matrix, the resulting matrix H is normalized. Take the rows of this matrix and their negatives, and replace each -1 by 0 (alternatively, take the rows of $\frac{1}{2}(J+H)$ and $\frac{1}{2}(J-H)$. The result is a *first-order Reed-Muller code* $RM(d,1)$ with 2^{d+1} words and Hamming distance 2^{d-1}.

Threshold Schemes

Suppose a bank has three trustees, T_1, T_2 and T_3. It is desired that the vault can be opened when any two of the trustees are present, but no lone trustee can open the vault.

One solution is to require a six-digit password, say $ABCDEF$. Say T_1 knows A, B, C, D, T_2 knows A, B, E, F, and T_3 knows C, D, E, F. Any two of them can reconstruct the password.

This is an example of a *threshold scheme*, a way of sharing a secret among several people so that approved combinations of them can together reconstruct the secret but other groups of them cannot. Typical secrets include passwords and cryptographic keys. Formally, a (v, t) threshold scheme consists of a v-set X of objects called *shares* or *shadows* (representing the trustees, or more precisely their pieces of information) and a set $\mathcal{B}$ of subsets of X, called *keys*; any t-subset of X belongs to precisely one key. In the example, $t = 2$; we could have taken the set of all trustees to be one key (of size 3), or the pairs to be three keys (of size 2).

This model can be used in more general situations. It is possible for several of the keys to correspond to the same password. There can also be several passwords, so that some groups have access to one secret and other groups to others; some of the keys could be worthless. For example, we could ask that some pairs of trustees can open the vault, but others cannot. In our earlier example, if we change T_3's share to A, C, D, F, then T_1 and T_2 can open the vault, as can T_2 and T_3, but T_1 and T_3 cannot.

One way to realize a (v, t)-threshold scheme is to use a t-wise balanced design with $\lambda = 1$. The keys correspond to the blocks, and every t-set of shares determines precisely one block. A set of fewer than t shares will determine some set of blocks, but not uniquely determine one.

A (v, t)-threshold scheme is called *perfect* if no set of fewer than t shares gives *any* information about the key. Stinson and Vanstone [141] showed that Kirkman triple systems provide perfect threshold schemes, and Cao and Du [28] do the same for Kirkman packings.

Another interesting application of design theory is described by Chung et al.

[36], who utilize a symmetric balanced incomplete block design to construct a protocol for distributing a cryptographic key; see also [95].

Exercises 16.2

16.2.1 Write down the codewords in $RM(2,1)$ and $RM(3,1)$.

16.2.2 Verify that $RM(2,1)$ has Hamming distance 2^{d+1}.

16.2.3* Find a scheme for four trustees T_1, T_2, T_3, T_4 to share a six-digit password in such a way that pairs T_1T_2, T_2T_3, T_3T_4 or T_4T_1 can open a vault but pairs T_1T_3 and T_2T_4 cannot.

16.3 Golf Designs

We conclude with another application to tournament scheduling. The teams representing $2n+1$ golf associations wish to play a round robin tournament. There are $2n+1$ rounds; each team serves as "host" for the one round in which they do not play, and the n matches in that round are played (simultaneously) on the course of the host team.

The teams wish to play $2n-1$ complete tournaments in such a way that every pair of teams play exactly once on each of the $2n-1$ available courses. (We shall call such an arrangement a *golf design*.) When is this possible?

This question arose in planning an actual golf tournament in New Zealand. It was reported by Robinson [119], who pointed out that the problem of constructing one round robin is equivalent to finding an idempotent symmetric Latin square; the square has (i,j) entry k, when $i \neq j$, if and only if i plays j at k. So the whole problem is equivalent to finding $2n-1$ symmetric idempotent Latin squares such that the $2n-1$ (i,j) entries cover $\{1,2,\ldots,2n+1\}$ with i and j deleted, for all $i \neq j$. He proved that case $n=5$ has no solution and exhibited a solution for case $n=7$ (see Figure 16.1).

We can use a Steiner triple system of order v to construct a $v \times v$ Latin square $L=(l_{ij})$. We set $l_{ii}=i$ for all i. If $i \neq j$, find the unique triple that contains both i and j. If it is $\{i,j,k\}$, then set $l_{ij}=k$. The resultant square L is idempotent and symmetric.

The Latin squares from two Steiner triple systems will coincide in an off-diagonal position if and only if the two systems have a common block. So we can construct a golf design in this way if and only if the set of all triples on

$$
\begin{array}{ccccccc}
0&2&4&5&1&6&3\\
2&1&6&4&5&3&0\\
4&6&2&0&3&1&5\\
5&4&0&3&6&2&1\\
1&5&3&6&4&0&2\\
6&3&1&2&0&5&4\\
3&0&5&1&2&4&6
\end{array}
\qquad
\begin{array}{ccccccc}
0&3&6&1&5&4&2\\
3&1&5&0&6&2&4\\
6&5&2&4&1&0&3\\
1&0&4&3&2&6&5\\
5&6&1&2&4&3&0\\
4&2&0&6&3&5&1\\
2&4&3&5&0&1&6
\end{array}
\qquad
\begin{array}{ccccccc}
0&4&5&6&2&3&1\\
4&1&3&2&0&6&5\\
5&3&2&1&6&4&0\\
6&2&1&3&5&0&4\\
2&0&6&5&4&1&3\\
3&6&4&0&1&5&2\\
1&5&0&4&3&2&6
\end{array}
$$

$$
\begin{array}{ccccccc}
0&5&3&2&6&1&4\\
5&1&4&6&3&0&2\\
3&4&2&5&0&6&1\\
2&6&5&3&1&4&0\\
6&3&0&1&4&2&5\\
1&0&6&4&2&5&3\\
4&2&1&0&5&3&6
\end{array}
\qquad
\begin{array}{ccccccc}
0&6&1&4&3&2&5\\
6&1&0&5&2&4&3\\
1&0&2&6&5&3&4\\
4&5&6&3&0&1&2\\
3&2&5&0&4&6&1\\
2&4&3&1&6&5&0\\
5&3&4&2&1&0&6
\end{array}
$$

FIGURE 16.1: A golf design for seven clubs.

$2n+1$ objects can be partitioned into $2n-1$ pairwise-disjoint Steiner triple systems—that is, a *large set* of systems. We have:

THEOREM 16.2 **[165]**

If there is a large set of Steiner triple systems of order $2n+1$*, then there is a golf design for* $2n+1$ *teams.*

So from Section 12.5 and Figure 16.1 it follows that there is a golf design of every order congruent to 1 or 3 (mod 6).

Tierlinck [151, 153] proved that if a golf design exists for order 11, then it exists for all odd orders other than 5 and possibly 41. Subsequently Colbourn and Nonay [40] found an example of order 11, and recently Chang [33] has constructed one of order 41. So the problem is completely solved.

To solve particular cases, one could try to construct symmetric idempotent Latin squares cyclically. If the first row of such a square is $0, a_1, a_2, \ldots, a_{2n}$ then the square would be

$$
\begin{array}{cccccc}
0 & a_1 & a_2 & a_3 & \ldots & a_{2n}\\
a_{2n}+1 & 1 & a_1+1 & a_2+1 & \ldots & a_{2n-1}+1\\
a_{2n-1}+2 & a_{2n}+2 & 2 & a_1+2 & \ldots & a_{2n-2}+2\\
 & \ldots & & & \ldots &
\end{array}.
$$

For symmetry we would require $a_{2n+1-j} = a_j - j \pmod{2n+1}$. In order that

2	5	9	14	16	13	15	12
3	1	10	12	11	15	4	13
4	7	12	2	13	16	1	14
5	8	2	13	15	1	14	11
6	9	15	3	1	14	11	10
7	16	13	8	3	11	2	9
8	14	16	9	6	5	10	2
9	4	14	16	10	7	3	6
10	13	4	7	2	12	5	16
11	15	7	5	14	8	6	3
12	10	5	1	4	9	13	15
13	6	8	15	7	3	16	1
14	11	1	6	8	10	12	7
15	3	11	10	12	5	9	4
16	12	6	4	9	2	8	5

FIGURE 16.2: Array to generate a golf design for 17 teams.

the rows be Latin we need $0, a_1, a_2, \ldots, a_n, a_n - n, a_{n-1} - (n-1), \ldots, a_1 - 1$ to be distinct modulo $2n+1$; a similar condition on columns exists. For a golf design we need $2n-1$ such sequences with different entries in all nondiagonal positions; this property can be derived from the starter rows. We have the following:

THEOREM 16.3 [165]

Suppose there exists a $(2n-1)\times(2n+1)$ array A with the following properties:

(a) *when $1 \le k \le n$, $a_{i,n+k} = a_{i,n+1-k} + n + k$;*

(b) *$a_{i0} = 0$;*

(c) *$a_{ij} + k \ne a_{ik} + j$ when $j \ne k$;*

(d) *each row of A is a permutation of $\{0, 1, \ldots, 2n\}$;*

(e) *column j of A is a permutation of $\{1, 2, \ldots, 2n\}\backslash\{j\}$.*

(All arithmetic is carried out modulo $2n+1$.) Then a golf design for $2n+1$ teams exists.

The $2n-1$ Latin squares are derived by developing the rows of A modulo $2n+1$.

The point of this theorem is that it yields a relatively easy computer construction. To construct row i of A it is only necessary to select the n entries $a_{i1}, a_{i2}, \ldots, a_{in}$. Then the row is

$$0, a_{i1}, a_{i2}, \ldots, a_{in}, (a_{in} + n + 1), \ldots, (a_{i2} + 2n - 1), (a_{i1} + 2n)$$

(reduced $\bmod 2n+1$). Moroever, you may as well fix a_{i1} to be $i+2$. Then properties (c) through (e) restrict the search considerably.

No solutions exist for $2n+1 = 5$, 7 or 11; it is found that there is a unique solution for $2n+1 = 9$, and exactly four solutions exist for $2n+1 = 13$.

A solution for $2n+1 = 17$ is illustrated in Figure 16.2. Only columns 1 through 8 of A are shown. Column 1 consists of all zeros; columns 9 through 16 are derived from columns 8 through 1 by adding $(9, 10, 11, 12, 13, 14, 15, 16)$ modulo 17.

Exercises 16.3

16.3.1 Prove that there is no golf design of order 5.

16.3.2 Verify that the array in Figure 16.2 yields a golf design for 17 teams.

References

[1] A. A. Albert and R. F. Sandler, *An Introduction to Finite Projective Planes* (Holt, Rinehart and Winston, New York, 1968).

[2] B. A. Anderson, M. M. Barge and D. Morse, "A recursive construction of asymmetric 1-factorizations," *Aeq. Math.* **15** (1977), 201–211.

[3] I. Anderson, *Combinatorial Designs and Tournaments* (Oxford U. P., Oxford, England, 1997).

[4] I. Anderson. "Balancing carry-over effects in tournaments," *Combinatorial Designs and their Applications* (F. C. Holroyd and C. Rowley, editors), (Chapman and Hall/CRC Press, Boca Raton, FL, 1999).

[5] R. R. Anstice, "On a problem in combinatorics," *Cambridge and Dublin Math. J.* **7** (1852), 279–292.

[6] R. R. Anstice, "On a problem in combinatorics (continued)," *Cambridge and Dublin Math. J.* **8** (1853), 149–154.

[7] J. W. Archbold and N. L. Johnson, "A construction for Room's squares and an application to experimental design," *Ann. Math. Statist.* **29** (1958), 219–225.

[8] A. C. Atkinson and A. N. Donev, *Optimum Experimental Designs* (Oxford U. P., Oxford, 1992).

[9] K. Balasubramanian, "On transversals of Latin squares," *Lin. Alg. Appl.* **131** (1990), 125–129.

[10] J. A. Barrau, "On the combinatory problem of Steiner," *K. Akad. Wet. Amsterdam Proc. Sect. Sci.* **11** (1908), 352–360.

[11] S. Bays, "Une question de Cayley relative au problème des triades de Steiner," *Enseignement Math.* **19** (1917), 57–67.

[12] L. D. Baumert, *Cyclic Difference Sets* (Springer-Verlag, Heidelberg, West Germany, 1971).

[13] A. F. Beecham and A. C. Hurley, "A scheduling problem with a simple graphical solution," *J. Austral. Math. Soc.* **21B** (1980), 486–495.

[14] M. B. Beintema, J. T. Bonn, R. W. Fitzgerald and J. L. Yucas, "Orderings of finite fields and balanced tournaments," *Ars Combin.* **49** (1998), 41–48.

[15] E. R. Berlekamp and F. K. Hwang, "Constructions for balanced Howell rotations for bridge tournaments," *J. Combinatorial Theory* **13A** (1972), 159–168.

[16] T. Beth, D. Jungnickel and H. Lenz, *Design Theory* (Cambridge U. P., Cambridge, England, 1986).

[17] K. N. Bhattacharya, "A new balanced incomplete block design," *Science and Culture* **11** (1944), 508.

[18] R. Bilous, C. W. H. Lam, L. H. Thiel, B. P. C. Li, G. H. J. van Rees, S. P. Radziszowski, W. H. Holzmann and H. Kharaghani, "There is no 2-$(22, 8, 4)$ block design," *J. Combin. Des.* (to appear).

[19] G. Birkhoff and T. C. Bartee, *Modern Applied Algebra* (McGraw-Hill, New York, 1970).

[20] J. A. Bondy and U. S. R. Murty, *Graph Theory with Applications* (Elsevier North-Holland, New York, 1976).

[21] J. T. Bonn, *Combinatorial Objects from Ordering the Elements of a Finite Field.* (Ph. D., Southern Illinois University, 1996.)

[22] R. C. Bose, "On the construction of balanced incomplete block designs," *Ann. Eugenics* **9** (1939), 353–399.

[23] R. C. Bose, "On the application of the properties of Galois fields to the problem of construction of hyper-Graeco-Latin squares," *Sankyha* **3** (1938), 323–338.

[24] R. C. Bose, S. S. Shrikhande, and E. T. Parker, "Further results on the construction of mutually orthogonal Latin squares and the falsity of Euler's conjecture," *Canad. J. Math.* **12** (1960), 189–203.

[25] R. A. Brualdi and H. J. Ryser, *Combinatorial Matrix Theory* (Cambridge U. P., Cambridge, 1991).

[26] R. H. Bruck and H. J. Ryser, "The non-existence of certain finite projective planes," *Canad. J. Math.* **1** (1949), 88–93.

[27] N. G. de Bruijn and P. Erdös, "On a combinatorial problem," *Indag. Math* **10** (1948), 1277–1279.

[28] H. Cao and B. Du, "Kirkman packing designs $KPD(\{3, s^*\}, v)$ and related threshold schemes," *Discrete Math.* **281** (2004), 83–95.

[29] H. Cao and Y. Tang, "On Kirkman packing designs $KPD(\{3, 4\}, v)$," *Discrete Math.* **279** (2004), 121–133.

[30] H. Cao and L. Zhu, "Kirkman packing designs $KPD(3, 5^*)$," *Des. Codes. Crypt.* **26** (2002), 127–138.

[31] A. Cayley, "On a tactical theorem relating to the triads of fifteen things," *Edinburgh and Dublin Phil. Mag. J. Sci.* (4)**25** (1863), 59–61.

[32] A. Černý, P. Horák and W. D. Wallis, "Kirkman's school projects," *Discrete Math.* **167/168** (1997), 189–196.

[33] Y. Chang, "The existence spectrum of golf designs," *J. Combin. Des.* (to appear).

[34] B. C. Chong and K. M. Chan, "On the existence of normalized Room squares," *Nanta Math.* **2** (1974), 8–17.

[35] S. Chowla and H. J. Ryser, "Combinatorial problems," *Canad. J. Math.* **2** (1950), 93–99.

[36] I. Chung, W. Choi, Y, Kim and M. Lee, "The design of conference key distribution system employing a symmetric balanced incomplete block design," *Inf. Proc. Lett.* **81** (2002), 313–318.

[37] C. J. Colbourn and J. H. Dinitz (editors), *The CRC Handbook of Combinatorial Designs* (2nd Ed.)(Taylor and Francis, 2006).

[38] C. J. Colbourn, J. H. Dinitz and D. R. Stinson, "Applications of combinatorial designs to communications, cryptography and networking," *Surveys in Combinatorics, 1999* (J. D. Lamb and D. A. Preece, editors), (London Math. Lecture Series **267**) (Cambridge U. P., 1999), 37–100.

[39] C. J. Colbourn and A. C. H. Ling, "Kirkman school project designs," *Discrete Math.* **203** (1999), 49–60.

[40] C. J. Colbourn and G. Nonay, "A golf design of order 11," *J. Statist. Plann. Inf.* **58** (1997), 29–31.

[41] R. J. Collens and R. C. Mullin, "Some properties of Room squares—A computer search," *Congressus Num.* **1** (1970), 87–111.

[42] W. G. Cochran and G. M. Cox, *Experimental Designs* (Wiley, New York, 1950).

[43] A. Cruse, "On embedding incomplete symmetric Latin squares," *J. Combinatorial Theory* **16A** (1974), 18–22.

[44] P. Dembowski, *Finite Geometries* (Springer-Verlag, New York, 1968).

[45] A. Dey, *Theory of Block Designs* (Wiley Eastern, New Delhi, 1986).

[46] L. E. Dickson and F. H. Safford, "Solution to problem 8 (group theory)," *Amer. Math. Monthly* **13** (1906), 150–151.

[47] J. H. Dinitz, "Room n-cubes of low order," *J. Austral. Math. Soc.* **36A** (1984), 237–252.

[48] J. H. Dinitz, D. K. Garnick and B. D. McKay, "There are 526,915,620 non-isomorphic one-factorizations of K_{12}," *J. Combin. Des.* **2** (1994), 273–285.

[49] H. L. Dorwart, *The Geometry of Incidence* (Prentice-Hall, Englewood Cliffs, N.J., 1966).

[50] D. A. Drake, G. H. J. van Rees and W. D. Wallis, "Maximal sets of mutually orthogonal Latin squares," *Discrete Math.* **194** (1999), 87–94.

[51] L. Euler, "Recherches sur une nouvelle espece de quarrées magiques," *Verhand. Zeeuwsch Gen. Wet. Vlissingen* **9** (1782), 85–239.

[52] T. Evans, "Embedding incomplete Latin squares," *Amer. Math. Monthly* **67** (1960), 958–961.

[53] R. A. Fisher, "An examination of the different possible solutions of a problem in incomplete blocks," *Ann. Eugenics* **10** (1940), 52–75.

[54] M. Gardner, "Euler's spoilers," *Martin Gardner's New Mathematical Diversions from Scientific American* (Allen & Unwin, London, 1969), 162–172.

[55] E. N. Gelling, *On one-factorizations of a complete graph and the relationship to round-robin schedules.* (MA Thesis, University of Victoria, Canada, 1973.)

[56] E. N. Gelling and R. E. Odeh, "On 1-factorizations of the complete graph and the relationship to round-robin schedules," *Congressus Num.* **9** (1974), 213–221.

[57] T. S. Griggs and A. Rosa, "Large sets of Steiner triple systems," *Surveys in Combinatorics* (London Math. Lecture Series **218**) (P. Rowlinson, editor), (Cambridge U. P., 1995), 25–39.

[58] A. Groner, *Duplicate Bridge Direction* (Barclay, New York, 1967).

[59] J. Hadamard, "Résolution d'une question relative aux déterminants," *Bull. Sci. Math.* (2) **17** (1893), 240–246.

[60] M. Hall, "Hadamard matrices of order 16," *JPL Research Summary 36-10,*.**1** (1961), 21–26.

[61] M. Hall, *Hadamard Matrices of Order* 20 (JPL Technical Report **32-761**, Pasadena, 1965).

[62] M. Hall, *Combinatorial Theory*, 2nd ed. (Wiley, New York, 1986).

[63] M. Hall and W. S. Connor, "An embedding theorem for balanced incomplete block designs," *Canad. J. Math.* **6** (1954), 35–41.

[64] P. Hall, "On representations of subsets," *J. London Math. Soc.* **10** (1935), 26–30.

[65] H. Hanani, "On quadruple systems," *Canad. J. Math.* **12** (1960), 145–157.

[66] G. H. Hardy and E. M. Wright, *An Introduction to the Theory of Numbers* (Oxford U. P., Oxford, England, 1938).

[67] F. Harary, *Graph Theory* (Addison-Wesley, Reading, MA, 1969).

[68] M. Harwit and N. J. A. Sloane, *Hadamard Transform Optics* (Academic Press, New York, 1979).

[69] A. Hedayat and W. D. Wallis, "Hadamard matrices and their applications," *Ann. Statist.* **6** (1978), 1184–1238.

[70] L. Heffter, "Uber Tripelsysteme," *Math. Ann.* **49** (1897), 101–112.

[71] J. Hirschfeld, *Projective Geometries over Finite Fields* (Oxford U. P., Oxford, 1983).

[72] J. D. Horton, "Variations on a theme by Moore," *Congressus Num.* **1** (1970), 146–166.

[73] J. D. Horton, "Quintuplication of Room squares," *Aeq. Math.* **7** (1971), 243–245.

[74] H. Hotelling, "Some improvements in weighing and other experimental techniques," *Ann. Math. Statist.* **15** (1944), 297–306.

[75] S. K. Houghten, L. H. Thiel, J. Janssen and C. W. H. Lam, "There is no $(46, 6, 1)$ block design," *J. Combin. Des.* **9** (2001), 60–71.

[76] D. R. Hughes and F. C. Piper, *Projective Planes* (Springer-Verlag, Heidelberg, West Germany, 1973).

[77] D. R. Hughes and F. C. Piper, *Design Theory* (Cambridge U. P., Cambridge, England, 1985).

[78] F. K. Hwang, "How to design round robin schedules," in *Combinatorics, Computing and Complexity (Tianjing and Beijing, 1988)* (D. Du and G. Hu, editors), Kluwer, Dordrecht, 1989, 142–160.

[79] H. Ibrahim and W. D. Wallis, "An enumeration of triad designs," *J. Combin., Inf., Syst. Sci.* (to appear).

[80] R. W. Irving, "Generalized Ramsey numbers for small graphs," *Discrete Math.* **3** (1974), 251–264.

[81] N. Ito, J. S. Leon and J. Q. Longyear, "Classification of 3-(24,12,5) designs and 24-dimensional Hadamard matrices," *J. Combinatorial Theory* **31A** (1981), 66–93.

[82] L. Ji, "A new existence proof for large sets of disjoint Steiner triple systems," *J. Combinatorial Theory* **112**A (2005), 308–327.

[83] D. M. Johnson, A. L. Dulmage, and N. S. Mendelsohn, "Orthomorphisms of groups and orthogonal Latin squares I," *Canad. J. Math.* **13** (1961), 356–372.

[84] H. Kharaghani and B. Tayfeh-Rezaie, "A Hadamard matrix of order 428," *J. Combin. Des.* **13** (2005), 435–440.

[85] J. Kiefer, "Optimum experimental designs," *J. R. Statist. Soc.* **21B** (1959), 272–319.

[86] H. Kimura, "New Hadamard matrix of order 24," *Graphs & Combin.* **2** (1986), 247–257.

[87] H. Kimura, "Classification of Hadamard matrices of order 28," *Discrete Math.* **133** (1994), 171–180.

[88] T. P. Kirkman, "On a problem in combinations," *Cambridge and Dublin Math. J.* **2** (1847), 191–204.

[89] T. P. Kirkman, "Note on an unanswered prize question," *Cambridge and Dublin Math. J.* **5** (1850), 255–262.

[90] T. P. Kirkman, "Query VI," *Lady's and Gentleman's Diary* (1850), 48.

[91] T. P. Kirkman, "Solution to Query VI," *Lady's and Gentleman's Diary* (1851), 48.

[92] K. K. Koksma, "A lower bound for the order of a partial transversal in a Latin square," *J. Combinatorial Theory* **7** (1969), 94–95.

[93] A. Kotzig and A. Rosa, "Nearly Kirkman systems," *Congressus Num.* **10** (1974), 607–614.

[94] C. W. H. Lam, L. Thiel and S. Swiercz, "The non-existence of finite projective planes of order 10," *Canad. J. Math.* **41** (1989), 1117–1123.

[95] T.-F. Lee and T. Hwang, "Improved conference key distribution protocol based on a symmetric balanced incomplete block design," *ACM SIGOPS Op. Sys. Rev.* **38** (2004), 58–64.

[96] H. Lenz, "A few simplified proofs in design theory," *Expositiones Math.* **1** (1983), 77–80.

[97] F. W. Levi, *Finite Geometrical Systems* (University of Calcutta, Calcutta, India, 1942).

[98] C. Lin, W. D. Wallis and L. Zhu, "Equivalence classes of Hadamard matrices of order 32," *Congressus Num.* **95** (1993), 179–182.

[99] C. Lin, W. D. Wallis and L. Zhu, "Equivalence classes of Hadamard matrices of order 32, II" Preprint **93-05** (Department of Mathematical Sciences, UNLV, 1993).

[100] C. C. Lindner, "A survey of embedding theorems for Steiner systems," *Ann. Discrete Math.* **7** (1980),175–202.

[101] C. C. Lindner, E. Mendelsohn and A. Rosa, "On the number of 1-factorizations of the complete graph," *J. Combinatorial Theory* **20A** (1976), 265–282.

[102] C. C. Lindner and A. Rosa, "Construction of large sets of almost disjoint Steiner triple systems," *Canad. J. Math.* **27** (1975), 256–260.

[103] J. X. Lu, "On large sets of disjoint Steiner triple systems I-VI," *J. Combinatorial Theory* **34**A (1983), 140–182, **37**A (1984), 136–192.

[104] H. L. MacNeish, "Euler squares," *Ann. Math.* **23** (1922), 221–227.

[105] F. J. MacWilliams and N. J. A. Sloane, *The Theory of Error-Correcting Codes* (2 Vols) (Elsevier/North-Holland, Amsterdam/New York, 1977).

[106] R. Mandl, "Orthogonal Latin squares: an application of experiment design to compiler testing," *Comm. ACM* **28** (1985), 1054–1058.

[107] H. B. Mann, *Addition Theorems* (Wiley, New York, 1965).

[108] H. B. Mann, "The construction of orthogonal Latin squares," *Ann. Math. Statist.* **13** (1942), 418–423.

[109] R. C. Mullin and E. Nemeth, "An existence theorem for Room squares," *Canad. Math. Bull.* **12** (1969), 493–497.

[110] H. W. Norton, "The 7×7 squares," *Ann. Eugenics* **9** (1939), 269–307.

[111] R. E. A. C. Paley, "On orthogonal matrices," *J. Math. and Phys.* **12** (1933), 311–320.

[112] E. T. Parker and A. M. Mood, "Some balanced Howell rotations for duplicate bridge sessions," *Amer. Math. Monthly* **62** (1955), 714–716.

[113] R. Peltesohn, "Eine Losung der beiden Heffterschen Differenzeprobleme," *Comp. Math.* **6** (1939), 251–267.

[114] N. C. K. Phillips, W. D. Wallis and R. S. Rees, "Kirkman packing and covering designs," *J. Combin. Math. Combin. Comput.* **28** (1998), 299–325.

[115] R. S. Rees and D. R. Stinson, "On resolvable group-divisible designs with block size 3," *Ars Combin.* **23** (1987), 107–120.

[116] R. S. Rees and W. D. Wallis, "Kirkman triple systems and their generalizations: a survey," *Designs 2002: Further Computational and Constructive Design Theory* (W. D. Wallis, editor), (Kluwer, Boston, 2003), 317–368.

[117] D. Raghavarao, *Constructions and Combinatorial Problems in Design of Experiments* (Wiley, New York, 1971).

[118] D. K. Ray-Chaudhury and R. M. Wilson, "Solution of Kirkman's schoolgirl problem," *Amer. Math. Soc. Symp. Pure Math.* **19** (1971), 187–204.

[119] D. F. Robinson, "Constructing an annual round-robin tournament played on neutral grounds," *Math. Chronicle* **10** (1981), 73–82.

[120] T. G. Room, "A new type of magic square," *Math. Gaz.* **39** (1955), 307.

[121] K. G. Russell, "Balancing carry-over effects in round robin tournaments," *Biometrika* 67 (198), 127–131.

[122] H. J. Ryser, "Matrices with integer elements in combinatorial investigations," *Amer. J. Math* **74** (1952), 769–773.

[123] H. J. Ryser, *Combinatorial Mathematics* (Wiley, New York, 1965).

[124] H. J. Ryser, "Neuere Problem in der Kombinatorik," *Vortraheuber Kombinatorik* (Oberwolfach, 1967), 69–91.

[125] H. J. Ryser, "The existence of symmetric block designs," *J. Combinatorial Theory* **32**A (1982), 103–105.

[126] A. Sade, "An omission in Norton's list of 7 × 7 squares," *Ann. Math. Stat.* **22** (1951), 306–307.

[127] P. J. Schellenberg, "On balanced Room squares and complete balanced Howell rotations," *Aeq. Math.* **9** (1973), 75–90.

[128] P. J. Schellenberg, G. H. J. van Rees, and S. A. Vanstone, "Four pairwise orthogonal Latin squares of order 15," *Ars Combin.* **6** (1978), 141–150.

[129] M. P. Schutzenberger, "A non-existence theorem for an infinite family of symmetrical block designs," *Ann. Eugenics* **14** (1949), 286–287.

[130] K. R. Shah, "Analysis of Room's square design," *Ann. Math. Statist.* **41** (1970), 743–745.

[131] J. Singer, "A theorem in finite projective geometry and some applications to number theory," *Trans. Amer. Math. Soc.* **43** (1938), 377–385.

[132] N. M. Singhi and S. S. Shrikhande, "Embedding of quasi-residual designs with $\lambda = 3$," *Utilitas Math.* **4** (1973), 35–53.

[133] N. M. Singhi and S. S. Shrikhande, "Embedding of quasi-residual designs," *Geom. Ded.* **2** (1974), 509–517.

[134] T. Skolem, "Some remarks on the triple systems of Steiner," *Math. Scand.* **6** (1958), 273–280.

[135] B. Smetaniuk, "A new construction on Latin squares I: a proof of the Evans conjecture," *Ars Combin.* **11** (1960), 155–172.

[136] R. G. Stanton, "The appropriateness of standard BIBD presentation," *Ars Combin.* **16**A (1983), 289–296.

[137] R. G. Stanton and I. P. Goulden,"Graph factorisation, general triple systems, and cyclic triple systems," *Aeq. Math.* **22** (1981), 1–28.

[138] R. G. Stanton and R. C. Mullin, "Techniques for Room squares," *Congressus Num.* **1** (1970), 445–464.

[139] R. G. Stanton and D. A. Sprott, "A family of difference sets," *Canad. J. Math.* **10** (1958), 73–77.

[140] J. Steiner, "Combinatorische Aufgabe," *J. Reine Angew. Math.* **45** (1853), 181–182.

[141] D. R. Stinson and S. A. Vanstone, "A combinatorial approach to threshold schemes," *SIAM J. Discrete Math.* **10** (1983), 84–88.

[142] D. R. Stinson and W. D. Wallis, "Snappy constructions for triple systems," *Austral. Math. Soc. Gaz.* **1** (1988), 230–236.

[143] D. R. Stinson and W. D. Wallis, "Two-fold triple systems without repeated blocks," *Discrete Math.* **47** (1983), 125–128.

[144] T. Storer, *Cyclotomy and Difference Sets* (Markham, Chicago, 1967).

[145] T. H. Straley, "Scheduling designs for a league tournament," *Ars Combin.* **15** (1983), 193–200.

[146] A. P. Street and D. J. Street, *Combinatorics of Experimental Design* (Oxford U. P., Oxford, 1987).

[147] A. P. Street and W. D. Wallis, *Combinatorics: A First Course* (Charles Babbage Research Centre, Winnipeg, Manitoba, Canada, 1982).

[148] D. J. Street. "Optimality and efficiency: comparing block designs," *The CRC Handbook of Combinatorial Designs* (C. J. Colbourn and J. H. Dinitz, editors), (CRC Press, 1996), 561–564.

[149] J. J. Sylvester, "Thoughts on inverse orthogonal matrices, simultaneous sign successions, and tessellated pavements in two or more colors, with applications to Newton's rule, ornamental tilework, and the theory of numbers," *Phil. Mag.* (4) **34** (1867), 461–475.

[150] G. Tarry, "Le problème des 36 officiers," *Comptes Rend. Assoc. Fr.* **1** (1900), 122–123; **2** (1901), 170–203.

[151] L. Teirlinck, "On the use of pairwise balanced designs and closure spaces in the construction of structures of degree at least 3," *Le Matematiche* **65** (1990), 197–218.

[152] L. Teirlinck, "A completion of Lu's determination of the spectrum for large sets of disjoint Steiner triple systems," *J. Combinatorial Theory* **57**A (1991), 302–305.

[153] L. Teirlinck, "Large sets of disjoint designs and related structures," *Contemporary Design Theory: A Collection of Surveys* (J. H. Dinitz and D. R. Stinson, editors), (Wiley, 1992), 561–592.

[154] V. Todorov, "Three mutually orthogonal Latin squares of order fourteen," *Ars Combin.* **20** (1985), 45–47.

[155] S. Vajda, *Patterns and Configurations in Finite Spaces* (Griffin, London, 1978).

[156] J. H. van Lint, "Non-embeddable quasi-residual designs," *Indag. Math.* **40** (1978), 269–275.

[157] G. H. J. van Rees, "Subsquares and transversals in Latin squares," *Ars Combin.* **29B** (1990), 193–204.

[158] O. Veblen and W. H. Bussey, "Finite projective geometries," *Trans. Amer. Math. Soc.* **7** (1906), 242–259.

[159] I. M. Vinogradov, *Elements of Number Theory* (Dover, New York, 1954).

[160] J. S. Wallis, "Two new block designs," *J. Combinatorial Theory* **7** (1969), 369–370.

[161] W. D. Wallis, "Integral equivalence of Hadamard matrices," *Israel J. Math.* **10** (1971), 359–368.

[162] W. D. Wallis, "Construction of strongly regular graphs using affine designs," *Bull. Austral. Math. Soc* **4** (1971), 41–49.

[163] W. D. Wallis, "On the existence of Room squares," *Aeq. Math.* **27** (1973), 260–266.

[164] W. D. Wallis, "All Room squares have minimal supersquares," *Congressus Num.* **36** (1982), 3–14.

[165] W. D. Wallis, "The problem of the hospitable golfers," *Ars Combin.* **15** (1983), 149–152.

[166] W. D. Wallis, "A tournament problem," *J. Austral. Math. Soc.* **24B** (1983), 289–291.

[167] W. D. Wallis, "Three orthogonal Latin squares," *Congressus Num.* **42** (1984), 69–86.

[168] W. D. Wallis *Combinatorial Designs* (Dekker, New York, 1988).

[169] W. D. Wallis, "Large sets and Room squares," *Bull. Inst. Combin. Appl.* **12** (1994), 93–94.

[170] W. D. Wallis, *One-factorizations* (Kluwer, Dortrecht, Netherlands, 1997).

[171] W. D. Wallis, "Subfactorizations and asymptotic numbers of one-factorizations," *Bull. Inst. Combin. Appl.* **27** (1997), 27–32.

[172] W. D. Wallis, *A Beginner's Guide to Graph Theory* (Birkhauser, Boston, 2000).

[173] W. D. Wallis, A. P. Street, and J. S. Wallis, *Combinatorics: Room Squares, Sum-Free Sets, Hadamard Matrices* (Lecture Notes in Math. **292**) (Springer-Verlag, Heidelberg, West Germany, 1972).

[174] S. P. Wang, *On Self-Orthogonal Latin Squares and Partial Transversals in Latin Squares* (Ph.D. thesis, Ohio State University, 1978).

[175] I. M. Wanless and B. S. Webb, "The existence of Latin squares without orthogonal mates," *Des. Codes. Crypt.* **40** (2006), 131–135.

[176] D. de Werra, "Scheduling in sports," In *Studies on Graphs and Discrete Programming* (North-Holland, Amsterdam, 1981), 381–395.

[177] D. de Werra, "On the multiplication of divisions: The use of graphs for sports scheduling," *Networks* 15 (1985), 125–136.

[178] D. de Werra, L. Jacot-Descombes and P. Masson, "A constrained sports scheduling problem," *Discrete Appl. Math.* **26** (1990), 41–49.

[179] J. Williamson, "Hadamard's determinant theorem and the sum of four squares," *Duke Math. J.* **11** (1944), 61–81.

[180] J. Williamson, "Note on Hadamard's determinant theorem," *Bull. Amer. Math. Soc.* **53** (1947), 608–613.

[181] W. S. B. Woolhouse, "Prize question 1733," *Lady's and Gentleman's Diary* (1844), 84.

[182] W. S. B. Woolhouse, *Lady's and Gentleman's Diary* (1845), 63–64.

[183] W. S. B. Woolhouse, "List of mathematical answers," *Lady's and Gentleman's Diary* (1846), 76.

[184] W. S. B. Woolhouse, *Lady's and Gentleman's Diary* (1847), 62–67.

[185] W. S. B. Woolhouse, "On triadic combinations of 15 symbols," *Lady's and Gentleman's Diary* (1862), 84–88.

[186] W. S. B. Woolhouse, "On triadic combinations," *Lady's and Gentleman's Diary* (1863), 79–90.

[187] F. Yates, "Complex experiments," *J. Roy. Stat. Soc. Supp.* **2** (1935), 181–247.

[188] F. Yates, "Incomplete randomized blocks," *Ann. Eugenics* **7** (1936), 121–140.

[189] B. Yener, Y. Ofek and M. Yung, "Combinatorial design of congestion-free networks," *IEEE/ACM Trans. Networking* **5** (1997), 989–1000.

Answers and Solutions

Chapter 1

Section 1.1

1.1.1 Without loss of generality, one can assume blocks 012, 034, 056, 078, 135. As 146 is impossible (since it would imply 178, i.e. 78 occurs twice), take 147 and 168. This is all unique up to isomorphism. The only possible extra blocks containing 3 are 238, 367. Then 246, 257, 248 are forced.

1.1.2 If 012,345 were 3-sets, there could be no other. Possibilities are 012, 034, and one or both of 135, 245 (up to isomorphism). Then the 2-sets are completely determined.

Section 1.2

1.2.2 The only possible factor including 12 is $\{12\ \ 34\}$. Similar thinking leads to the (unique) factorization

$$\{12\ \ 34\}\quad\{13\ \ 24\}\quad\{14\ \ 23\}.$$

1.2.7 8; possible edge-lists are $\emptyset$; 12; 13; 23; 12 13; 12 23; 13 23; 12 13 23. There are four isomorphism classes (all the graphs with the same number of vertices are isomorphic in this example).

Section 1.3

1.3.1 $n = 3$: Say $12, 12, 12$. You must also have $34, 34, 34$.
$n = 2$: Say $12, 12$. Without loss of generality add blocks $13, 24$. You must also have $34, 34$.
$n = 1$: No repeats. The only possibility is $12, 13, 14, 23, 24, 34$.

1.3.4 You could simply take two copies of each block in a $(7, 7, 3, 3, 1)$ design. Can you find another solution?

1.3.8 (i) $(16, 16, 6, 6, 2)$. (ii) Consider two elements in same row of t x t array. Necessarily $\lambda = t-2$. But for two elements not in same row or column, $\lambda = 2$. So $t - 2 = 2$, $t = 4$, is necessary for balance.

1.3.9 Parameters are as follows: Asterisk means "noninteger, cannot complete."
$(15, 35, 7, 3, 1)$, $(9, *, 6, 4, 2)$, $(14, 7, 4, 8, *)$, $(66, 143, 13, 6, 1)$, $(21, 28, 4, 3, *)$, $(17, *, 8, 5, 2)$, $(21, 30, 10, 7, 3)$, $(17,\ , *, 7, 1)$.

Section 1.4

1.4.2 123456, 124365, 143265, 143526, 143652, 213546, 214365, 241365, 243561, 314265, 314526, 314562, 321546, 324561, 341265, 341526, 341562.

1.4.3 (i) 13; (ii) 12; (iii) 13.

Chapter 2

Section 2.1

2.1.2 No. For suppose 0123 is a block. The only way to place 0 in further blocks size 3 and 4 is to have a block 0456. No completion is possible.

2.1.3 Suppose a $PB(8;\{4,3\};1)$ exists, with f_4 4-blocks and f_3 3-blocks. Then $6f_4 + 3f_3 = 28$. But 3 does not divide 28.

Section 2.2

2.2.2 $\left(v, \binom{v}{k}, \binom{v-1}{k-1}, k, \binom{v-2}{k-2}\right)$.

2.2.3 No.

2.2.4 $\left(v, \binom{v}{k} - b, \binom{v}{k} - r, k, \frac{v}{k}\binom{v-2}{k-2}\right)$.

2.2.9 $(v, nb, nr, k, n\lambda)$.

Section 2.4

2.4.3 t-$\left(v, k, \binom{v-t}{k-t}\right)$.

2.4.5 (i) No (try $s = 2$); (ii) yes; (iii) yes; (iv) no.

2.4.9 (ii) $(t-1)$-$(v-1, k-1, \lambda)$.

Chapter 3

Section 3.2

3.2.1 (i) $1 + 0 = 0$, so $a = a1 = a(1+0) = a1 + a0 = a + a0$. So $0 = (-a)+a = (-a)+(a+a0) = ((-a)+a)+a0 = 0+a0 = a0$. (ii) $a+(-a) = 0$ and $a + (-1)a = 1a + (-1)a = (1 + (-1))a = 0a = 0$. So $-a$ and $(-1)a$ are both inverses of a; by uniqueness, $(-1)a = -a$. (iii) If $a \neq 0$, then $a \in F^*$, so a^{-1} exists, so $ab = 0 \Rightarrow b = a^{-1}ab = a^{-1}0 = 0$.

3.2.4 The multiplicative identity is 6. Verification of the properties is now straightforward.

3.2.7 The polynomials are x^3+2x+1, x^3+2x+2, x^3+x^2+2, x^3+x^2+x+2, x^3+x^2+2x+1, x^3+2x^2+1, x^3+2x^2+x+1, x^3+2x^2+2x+2.

3.2.9 (i) $(0,0)$, $(1,1)$; (ii) any $(x,0)$ and $(0,y)$, $x \neq 0$, $y \neq 0$.

Section 3.3

3.3.1 The line joining $(0, 0)$ to $(1, 0)$ consists precisely of the points $x(0, 0) + y(1, 0)$ (i.e., points like $(y, 0)$). Neither $(0, 1)$ nor $(1, 1)$ has this form. So neither of the triples $\{(0, 0), (1, 0), (0, 1)\}$, $\{(0, 0), (1, 0), (1, 1)\}$ is collinear. Similarly for the other two triples.

Section 3.4

3.4.4 (i) $013, 103, 111, 124, 132, 140$; (ii) $x_1 + 4x_3 = 0$; (iii) 103.

3.4.7 An example is $\{001, 010, 100, 111\}$.

Chapter 4

Section 4.1

4.1.4 The tangents through the points are $x_2 + x_3 = 0$, $x_1 + 4x_2 + 2x_3 = 0$, $x_1 + x_2 + 2x_3 = 0$, $x_1 + x_2 + x_3 = 0$, $x_1 + 2x_2 + x_3 = 0$ and $x_1 + 2x_2 + 4x_3 = 0$ respectively. They are not concurrent: For example, the first two intersect in $(1, 2, 3)$ while the first and third meet in $(1, 1, 4)$.

Section 4.2

4.2.1 $ad \cap be \cap cf = (1, 1, 1)$; $ab \cap de$, $ac \cap df$ and $bc \cap ef$ lie on $x + 2y + 2z = 0$.

Chapter 5

Section 5.1

5.1.3 A $(13, 4, 1)$-difference set must contain two members, s and $s + 1$ say, with differences ± 1. By subtracting $s \pmod{13}$ from all elements, one obtains an example of the form $\{0, 1, x, y\}$. So assume that form. Then $\{x, y, x - 1, y - 1, y - x\} = \{\pm 2, \pm 3, \pm 4, \pm 5, \pm 6\}$. $x = 2$ is impossible (since $x - 1$ would be outside the permissible range). Try $x = 3$: by trial and error, the only possibility is $y = 9$. Other solutions are $\{x, y\} = \{5, 11\}, \{8, 10\}$.

5.1.4 Similar to Exercise 5.1.3: for example, $\{0, 1, 2, 4\}$.

5.1.7 From Corollary 5.2.1, $\lambda(v - 1) = k(k - 1)$. Put $v = 2k$. Then $\lambda \equiv 0 \pmod{k}$. But $k(k - 1) < k(2k - 1)$, obviously, so $\lambda < k$. So $\lambda = 0$, which is clearly impossible.

Section 5.2

5.2.5 $\{0, 1, 2, 4, 5, 8, 10\}$; $\{0, 1, 2, 3, 4, 6, 7, 8, 9, 12, 13, 14, 16, 18, 19, 21, 23, 24, 25, 26, 27, 28, 32, 36, 38, 39, 41, 42, 46, 48, 49, 50, 52, 53, 54, 56, 57, 63, 64, 65, 69, 72, 73, 75, 76, 78, 81, 82, 83, 84, 85, 91, 92, 96, 98, 100, 103, 104, 106, 108, 109, 112, 113, 114, 117, 123, 126, 128, 130, 133, 138\}$.

5.2.6 Theorem 5.7 gives $t = 1, 2, 3, 5, 6, 7, 8, 11, 12, 15, 17, 18, 20, 21, 26, 27, 30, 32, 33, 35, 38, 40, 41, 42, 45, 48, 50$. Theorem 5.8 gives $t = 4, 9, 16, 25, 36$. Eighteen values remain.

Section 5.3

5.3.1 As $\{x, y\}$ ranges through all pairs in D^*, $\{-x, -y\}$ ranges through all pairs in D. So the collection of values $(-x) - (-y)$ contains every nonzero member of D λ times. But $x - y = y - x = -((-x) - (-y))$, so the collection of values $x - y$ contains (the negative of) every nonzero member of D λ times.

Section 5.4

5.4.2 One can assume that the initial blocks are $\{0, 1, x\}$ and $\{0, y, z\}$. Difference ± 2 must arise; if it is from the first block, then $x = \pm 2$ or $x - 1 = \pm 2$, so $x = 2$ or 3 or 11 or 12. If $x = 2$ or 12 difference ± 1 is repeated; trial and error shows that $x = 3$ and $x = 11$ are also impossible. So try the form $\{0, 1, x\}$ and $\{0, 2, z\}$ for some x and z. An example that works is $\{0, 1, 4\}$, $\{0, 2, 8\}$.

5.4.3 $\{00, 01, 12, 22\}$ is a second block.

5.4.5 $\{\infty, 0, 1\}$, $\{0, 1, 3\}$, for example.

Section 5.5

5.5.2 If a cubic polynomial is reducible, it must have a linear factor. If $y-a$ is a factor of $f(y)$, then $f(a) = 0$. If $f(y) = 1-y+y^3$, then $f(0) = f(1) = f(2) = 1$.

5.5.3 Using $g(y) = 1+y+y^3$, the line $x_2 = 0$ yields set $\{0, 1, 3\}$ for $PG(2, 2)$. The same polynomial and $x_3 = 0$ yields $\{0, 1, 3, 8, 12, 18\}$ for $PG(2, 5)$.

5.5.4 (ii) {0, 1, 2, 3, 4, 8, 9, 10, 11, 13, 14, 16, 17, 18, 21, 24, 25, 27, 30, 32, 36, 41, 42, 45, 46, 47, 49, 51, 53, 54, 61}.

Chapter 6

Section 6.1

6.1.1 (i) $(10, 15, 6, 4, 3)$, $(6, 15, 5, 2, 1)$;
(ii) $(15, 21, 7, 5, 2)$, $(7, 21, 6, 2, 1)$; (iii) $(33, 44, 12, 9, 3)$, $(12, 44, 11, 3, 2)$;
(iv) $(4, 12, 9, 3, 6)$, $(9, 12, 8, 6, 5)$.

6.1.3 (i) $(n^2 + n + 1, n^2, n^2 - n)$; (ii) $(n + 1, n^2 + n, n^2, n, n^2 - n)$; (iii) it is a multiple of a known design.

Section 6.2

6.2.2 $\{0, 1, 3, 7\}$, $\{2, 4, 9, 10\}$, $\{\infty, 5, 6, 8\}$.

6.2.4 (i) No.

6.2.5 Every object belongs to $n(3n - 1)(3n - 2)/2$ triples, so this must equal the number of rows and the number of columns in the array. No example exists when $n = 2$; one may be found for $n = 3$.

Chapter 7

Section 7.1

7.1.2 (i) no (ii) $0^2 + 7^2$ (iii) $15^2 + 6^2$ (iv) $212 + 14^2$ (v) no (vi) no (vii) no (viii) $10^2 + 5^2$ (ix) no.

7.1.4 (i) If $x^2 = p$ then p is not prime (x is an integer). If $x^2 > p$ then y^2 is negative (impossible). So $x^2 < p$. Similarly for y, a, b.
(ii) $y^4 - x^4 = p(y^2 - x^2) \equiv 0$, same for $b^4 - a^4$. So $x^2 + y^2 = a^2 + b^2 \Rightarrow y^2 - a^2 = b^2 - x^2$. Squaring both sides, $y^4 - 2y^2a^2 + a^4 = b^4 - 2b^2x^2 + x^4$, so $y^4 - x^4 + 2b^2x^2 = b^4 - a^4 + 2y^2a^2$, and reducing mod p $2b^2x^2 \equiv 2y^2a^2$. We can cancel the 2.
(iii) $(xb - ya)(xb + ya)$ is a multiple of p; since p is prime, one must be a multiple of p. Now $0 < xb + ya < \sqrt{p}\sqrt{p} + \sqrt{p}\sqrt{p} = 2p$, so if p divides $xb + ya$, then $xb + ya = p$. Also $0 < xb < p$ and $0 < ya < p$, so $-p < xb - ya < p$, if this is the factor divisible by p, it must be 0.
(iv) By (7.1), $(x^2 + y^2)(a^2 + b^2) = (xb + ya)^2 + (yb - xa)^2$. The left hand equals p^2. If $xb + ya = p$ then the right hand equals $p^2 + (yb - xa)^2$, which is greater than p^2 because $yb > xa$. We have a contradiction. So $xb + ya \neq p$, and $xb - ya = 0$.
(v) b divides ya, but b and a are relatively prime, so b divides y. The equation is symmetric, so y divides b. So $y = b$. Similarly $x = a$.

7.1.6 (i) $1^2 + 1^2 + 1^2 + 2^2$; (ii) $0^2 + 0^2 + 1^2 + 3^2$; $1^2 + 1^2 + 2^2 + 2^2$; (iii) $0^2 + 1^2 + 1^2 + 9^2$; (iv) $0^2 + 1^2 + 1^2 + 4^2$; $0^2 + 0^2 + 3^2 + 3^2$; $1^2 + 2^2 + 2^2 + 3^2$; (v) $0^2 + 0^2 + 2^2 + 4^2$; $1^2 + 1^2 + 3^2 + 3^2$; (vi) $0^2 + 0^2 + 4^2 + 4^2$.

Section 7.2

7.2.2 (i) $28 + 369t + 243t^2$.

Chapter 8

Section 8.1

8.1.1 Up to equivalence, one may assume that the first row is 1 2 3 4 and so is the first column. There are only four ways to complete a Latin square:

$$A = \begin{matrix} 1\,2\,3\,4 \\ 2\,1\,4\,3 \\ 3\,4\,1\,2 \\ 4\,3\,2\,1 \end{matrix} \quad B = \begin{matrix} 1\,2\,3\,4 \\ 2\,1\,4\,3 \\ 3\,4\,2\,1 \\ 4\,3\,1\,2 \end{matrix} \quad C = \begin{matrix} 1\,2\,3\,4 \\ 2\,3\,4\,1 \\ 3\,4\,1\,2 \\ 4\,1\,2\,3 \end{matrix} \quad D = \begin{matrix} 1\,2\,3\,4 \\ 2\,4\,1\,3 \\ 3\,1\,4\,2 \\ 4\,3\,2\,1 \end{matrix}$$

Exchanging the symbols 3 and 4 in C gives D, and the permutation (243) changes B to C (after suitable row and column permutations in both cases).

8.1.4 (i, ii) Use row and column permutations. (iii) there are 1 at side 3 (obvious), 4 at side 4 (see solution to Exercise 8.1.1, above) and 56 at side 5:

the second column is one of

$$\begin{array}{ccccc|cccccc}
2&2&2&2&2&2&2&2&2&2&2\\
1&1&3&4&5&3&3&4&4&5&5\\
4&5&1&5&4&4&5&1&5&1&4\\
5&3&5&1&3&5&1&5&3&3&3\\
3&4&4&3&1&1&4&3&1&4&1
\end{array}$$

—the first five cases can be completed in four ways each, the others in six each.

8.1.7 Suppose a Latin square L has the form

$$\boxed{\begin{array}{cc} A & B \\ C & D \end{array}}$$

where A is an $s \times s$ Latin square. Select a symbol x of A. As x appears in every column of A, it can never appear in C, so it must lie in the first row of D. This is true for all s possible values of x. So row 1 of D has at least s entries. So D has at least s columns, and L has at least $2s$ columns.

Section 8.2

8.2.6 (iii)

$$\begin{bmatrix} 1&3&4&2\\ 4&2&1&3\\ 2&4&3&1\\ 3&1&2&4 \end{bmatrix} \qquad \begin{bmatrix} 1&3&2&5&4\\ 4&2&5&1&3\\ 5&4&3&2&1\\ 3&5&1&4&2\\ 2&1&4&3&5 \end{bmatrix}$$

Section 8.3

8.3.1 One example is:

$$\begin{bmatrix} 1&5&6&3&2&4\\ 3&2&1&6&4&5\\ 5&4&3&2&6&1\\ 6&1&5&4&3&2\\ 4&6&2&1&5&3\\ 2&3&4&5&1&6 \end{bmatrix}$$

8.3.2

$$\begin{bmatrix}
1 & 7 & 4 & 8 & 9 & 0 & 2 & 3 & 5 & 6 \\
7 & 2 & 0 & 9 & 6 & 8 & 1 & 5 & 4 & 3 \\
8 & 0 & 3 & 1 & 7 & 9 & 5 & 4 & 6 & 2 \\
3 & 9 & 8 & 4 & 0 & 7 & 6 & 1 & 2 & 5 \\
9 & 8 & 7 & 0 & 5 & 2 & 3 & 6 & 1 & 4 \\
0 & 5 & 9 & 7 & 8 & 6 & 4 & 2 & 3 & 1 \\
2 & 1 & 5 & 6 & 3 & 4 & 7 & 9 & 0 & 8 \\
4 & 6 & 1 & 3 & 2 & 5 & 0 & 8 & 7 & 9 \\
5 & 4 & 6 & 2 & 1 & 3 & 8 & 0 & 9 & 7 \\
6 & 3 & 2 & 5 & 4 & 1 & 9 & 7 & 8 & 0
\end{bmatrix}$$

8.3.3 (ii) Use the squares

$$\begin{bmatrix}
1 & 3 & 4 & 2 \\
4 & 2 & 1 & 3 \\
2 & 4 & 3 & 1 \\
3 & 1 & 2 & 4
\end{bmatrix}
\quad
\begin{bmatrix}
1 & 5 & 4 & 3 & 2 \\
3 & 2 & 1 & 5 & 4 \\
5 & 4 & 3 & 2 & 1 \\
2 & 1 & 5 & 4 & 3 \\
4 & 3 & 2 & 1 & 5
\end{bmatrix}$$

and the design

$$\left[\begin{array}{ccccc cccc cccc cccc}
0 & 1 & 2 & 3 & 4 & 1 & 6 & 11 & 16 & 2 & 7 & 9 & 16 & 3 & 8 & 9 & 14 \\
0 & 5 & 6 & 7 & 8 & 1 & 7 & 12 & 14 & 2 & 8 & 11 & 13 & 4 & 5 & 1l & 14 \\
0 & 9 & 10 & 11 & 12 & 1 & 8 & 10 & 15 & 3 & 5 & 10 & 16 & 4 & 6 & 9 & 15 \\
0 & 13 & 14 & 15 & 16 & 2 & 5 & 12 & 15 & 3 & 6 & 12 & 13 & 4 & 7 & 10 & 13 \\
 & 1 & 5 & 9 & 13 & 2 & 6 & 10 & 14 & 3 & 7 & 11 & 15 & 4 & 8 & 12 & 16
\end{array}\right]$$

to obtain

$$\begin{bmatrix}
0 & 4 & 3 & 2 & 1 & 8 & 7 & 6 & 5 & 12 & 11 & 10 & 9 & 16 & 15 & 14 & 13 \\
2 & 1 & 0 & 4 & 3 & 9 & 11 & 12 & 10 & 13 & 15 & 16 & 14 & 5 & 7 & 8 & 6 \\
4 & 3 & 2 & 1 & 0 & 12 & 10 & 9 & 11 & 16 & 14 & 13 & 15 & 8 & 6 & 5 & 7 \\
1 & 0 & 4 & 3 & 2 & 10 & 12 & 11 & 9 & 14 & 16 & 15 & 13 & 6 & 8 & 7 & 5 \\
3 & 2 & 1 & 0 & 4 & 11 & 9 & 10 & 12 & 15 & 13 & 14 & 15 & 7 & 5 & 6 & 8 \\
6 & 13 & 15 & 16 & 14 & 5 & 0 & 8 & 7 & 1 & 3 & 4 & 2 & 9 & 11 & 12 & 10 \\
8 & 16 & 14 & 13 & 15 & 7 & 6 & 5 & 0 & 4 & 2 & 1 & 3 & 12 & 10 & 9 & 11 \\
5 & 14 & 16 & 15 & 13 & 0 & 8 & 7 & 6 & 2 & 4 & 3 & 1 & 10 & 12 & 11 & 9 \\
7 & 15 & 13 & 14 & 16 & 6 & 5 & 0 & 8 & 3 & 1 & 2 & 4 & 11 & 9 & 10 & 12 \\
10 & 5 & 7 & 8 & 6 & 13 & 15 & 16 & 14 & 9 & 0 & 12 & 11 & 1 & 3 & 4 & 2 \\
12 & 8 & 6 & 5 & 7 & 16 & 14 & 13 & 15 & 11 & 10 & 9 & 0 & 4 & 2 & 1 & 3 \\
9 & 6 & 8 & 7 & 5 & 14 & 16 & 15 & 13 & 0 & 12 & 11 & 10 & 2 & 4 & 3 & 1 \\
11 & 7 & 5 & 6 & 8 & 15 & 13 & 14 & 16 & 10 & 9 & 0 & 12 & 3 & 1 & 2 & 4 \\
14 & 9 & 11 & 12 & 10 & 1 & 3 & 4 & 2 & 5 & 7 & 8 & 6 & 13 & 0 & 16 & 15 \\
16 & 12 & 10 & 9 & 11 & 4 & 2 & 1 & 3 & 8 & 6 & 5 & 7 & 15 & 14 & 13 & 0 \\
13 & 10 & 12 & 11 & 9 & 2 & 4 & 3 & 1 & 6 & 8 & 7 & 5 & 0 & 16 & 15 & 14 \\
15 & 11 & 9 & 10 & 12 & 3 & 1 & 2 & 4 & 7 & 5 & 6 & 8 & 14 & 13 & 0 & 16
\end{bmatrix}$$

Section 8.4

8.4.3 Yes.

8.4.4 (i) Block size $= k$, group size $= k-1$. (ii) There are r groups. So $r = k$ is necessary (and it is sufficient: the transversal designs are essentially affine planes, with one parallel class interpreted as groups).

Chapter 9

Section 9.1

9.1.2 (i)
$$\begin{matrix}[1,2 \mid 3,4] & [2,1 \mid 4,3] \\ [1,3 \mid 2,4] & [3,1 \mid 4.2] \\ [1,4 \mid 2,3] & [4,1 \mid 3,2]\end{matrix}$$

(ii) No. First show that up to isomorphism, the first three rounds are two symmetric $SR(4)$s glued together. Then completion is impossible.

Section 9.3

9.3.1 Consider the Latin squares of side 4:

$$L = \begin{array}{|cccc|}\hline 1&2&3&4\\ 2&1&4&3\\ 3&4&1&2\\ 4&3&2&1\\ \hline\end{array} \quad M_1 = \begin{array}{|cccc|}\hline 1&3&4&2\\ 4&2&1&3\\ 2&4&3&1\\ 3&1&2&4\\ \hline\end{array} \quad M_2 = \begin{array}{|cccc|}\hline 1&4&2&3\\ 3&2&4&1\\ 4&1&3&2\\ 2&3&1&4\\ \hline\end{array}$$

We construct 6×6 arrays from M_1 and M_2.

In M_1: replace by 5 all those entries where L has a 1; if the (i, j) entry contained x and was replaced, put x in the $(i, 5)$ and $(5, j)$ positions. (The cells where L has 1 form a transversal in M_1, and the operation is called *transporting* that transversal.) Do similarly for those cells with a 2 in L, replacing the cells of M_2 with 6. Finally, fill in the bottom 2×2 square with 5, 6, 6, 5.

In M_2: perform the same operations where L had 3 and 4 respectively.

We obtain

$$\begin{array}{|cccc|cc|}\hline 5&6&4&2&1&3\\ 6&5&1&3&2&4\\ 2&4&5&6&3&1\\ 3&1&6&5&4&2\\ \hline 1&2&3&4&5&6\\ 4&3&2&1&6&5\\ \hline\end{array} \qquad \begin{array}{|cccc|cc|}\hline 1&4&5&6&2&3\\ 3&2&6&5&1&4\\ 5&6&3&2&4&1\\ 6&5&1&4&3&2\\ \hline 4&3&2&1&5&6\\ 2&1&4&3&6&5\\ \hline\end{array}$$

Why does this work?

Chapter 10

Section 10.1

10.1.4 $(2n-1) \times (2n-1) \times \ldots \times 3 \times 3$.

10.1.7 The block size is constant, so it is sufficient to show that λ_3 is constant. So consider any 3-set xyz. The three pairs xy, xz, yz must all lie in different factors. Say xy is in F_1, and the pair zt is in F_1. Then $xyzt$ is the unique block derived from F_1 that contains xyz. Therefore this triple belongs to precisely 3 blocks. So $\lambda_3 = 3$. So we have a 3-design. The parameters are $v = 2n$, $b = \frac{1}{2}n(n-1)(2n-1)$, $r = (n-1)(2n-1)$, $k = 4$, $\lambda = 4(n-1)$, $\lambda_3 = 3$.

10.1.9 (ii) Consider the factors with isolate 0. In each case, it has the form $\{0\,1x\,\ldots\}$ and the entries x must all be different and nonzero. So at most $2n$ factorizations are possible.

Section 10.2

10.2.2 Since $(n-2)!2^{n-2} = (2n-4) \times (2n-6) \times \ldots \times 2$, the number given is actually $(2n-3) \times (2n-5) \times \ldots \times 3 \times 1$. This is the number of one-factors that contain a given pair (see Exercise 10.1.4). Using the notation of Theorem 10.2, suppose G is any one-factor containing $\{\infty, 0\}$. One can find a permutation ϕ of $\{1, 2, \ldots, 2n-2\}$ such that ϕ maps F_0 to G. Then ϕ maps the one-factorization of the theorem to one containing G. So the number of one-factorizations is at least equal to the number of factors containing $\{\infty, 0\}$.

10.2.8 $(15, 35, 7, 3, 1)$

Section 10.3

10.3.1 In Z_3: 12; in Z_5: 14 15; in Z_7: 16 25 34, 15 23 46, 13 26 45.

10.3.3 There are nine solutions:

01, 02	10, 20	11, 22	12, 21	01, 02	11, 21	12, 20	10, 22
01, 02	12, 22	10, 21	11, 20	10, 11	02, 22	01, 20	12, 21
10, 12	01, 21	11, 22	02, 20	11, 12	10, 20	02, 21	01, 22
20, 21	02, 12	11, 22	01, 10	20, 22	01, 11	02, 10	12, 21
21, 22	10, 20	01, 12	02, 11.				

Chapter 11

Section 11.2

11.2.5 The remaining two rounds must include matches 12, 13 and 23, and no two can be in the same round.

Section 11.3

11.3.1 Without loss of generality, we can assume $F_1 = \{01, 23, 45\}$ and $F_2 = \{01, 23, 45\}$ up to isomorphism. A short search shows we may assume F_3 is one of $\{03, 15, 24\}$, $\{04, 13, 25\}$, $\{05, 12, 34\}$ or $\{05, 13, 24\}$. In each case there is a pair $\{x, y\}$ such that x precedes y in round 1 and also in round 2: they are respectively $x, y = 04, 51, 30, 30$.

Chapter 12

Section 12.1

12.1.3 (i) $(n-k)/2$; (ii) the answer to part (i) is an integer; (iii) k must equal 1.

12.1.9 Say the $T(v-3, v-1)$ has treatment set S and block set $\mathcal{B}$. Let $\mathcal{P}$ be the set of all unordered pairs of elements of S; write $\mathcal{Q} = \{p \cup \{\infty\} : p \in P\}$. Then $\mathcal{Q} \cup \mathcal{B}$ is the block set of the required design.

Section 12.2

12.2.1 Putting $m_1 = 0, n_{i1} = i, n_{i2} = 9 + i$ we get blocks 0 1 10, 0 2 11, 0 3 12, ..., 0 9 18, together with 48 more blocks, 4 from each block of a $T(1, 9)$: For example, the block 1 2 3 gives rise to 1 2 12, 1 11 3, 10 2 3, 10 11 12. The other design is derived similarly from the theorem. It is easy enough to show that the design above contains no sub-$T(1, 7)$, while the other will do so; so they are not isomorphic.

12.2.5 1 2 3, 4 5 7, 5 6 8, ... ,10 11 13, 11 12 4, 12 13 5, 13 4 6, 4 8 1, 12 6 1, 10 5 1, 7 11 1, 9 13 1, 8 12 2, 6 10 2, 11 5 2, 13 7 2, 4 9 2, 13 8 3, 6 11 3, 7 12 3, 4 10 3, 5 9 3.

Section 12.4

12.4.1 $S = \{1, 2, 4\}$, which does not form a difference partition.

12.4.2 $v = 7$: $3 = 1 + 2$; $v = 13$: for example, $1 + 3 = 4$, $2 + 5 + 6 = 13$.

Section 12.5

12.5.3 Given x and y, there are $v - 2$ blocks containing both: the xyz where z takes each of the remaining $v - 2$ values. The design has constant block size 3, so it is a $T(v - 2, v)$. If there were four pairwise $T(1, 7)$s, their union—a $T(4, 7)$ would be a subdesign of this $T(5, 7)$, so its complement would be a $T(1, 7)$; together with the four systems, it would form a large set of order 7, which is impossible.

12.5.4 $(15, 35, 7, 3, 1)$.

Chapter 13

Section 13.1

13.1.2 Suppose a $PB(v; \{4, 3\}; 1)$ exists. In the notation of Section 2.1, $12f_4 + 6f_3 = v(v - 1)$; since f_4, f_3 and v are integers, 3 divides $v(v - 1)$, so $v = 0$ or $1 \pmod 3$. Case $v = 6$ is easily eliminated. If $v \equiv 1$ or $3 \pmod 6$, use the Steiner triple system on v points. If $v \equiv 4 \pmod 6$, take a Kirkman triple system on $v - 1$ points; add a new element A to every block in one parallel class. If $v \equiv 0 \pmod 6$, take a $KTS(v - 3)$ and use three new elements a, b, c; for each, select a parallel class, and add the new element to every block. Then add a block $\{a, b, c\}$. (In the latter case one needs at least three parallel classes; since a $KTS(v-3)$ contains $(v-4)/2$ parallel classes, the necessary condition is $v \geq 10$.)

13.1.3 See Exercise 13.1.2. Clearly, $v \equiv 0$ or $1 \pmod 3$ is still necessary. The constructions in the earlier exercise work for $v \equiv 0, 4 \pmod 6$, except trivially for $v = 4$. If $v \equiv 1 \pmod 6$, add four new elements to a $KTS(v-4)$, together with the block consisting of all four of them. This is possible provided the $KTS(v-4)$ contains more than four parallel classes (exactly four classes won't work; you get no 3-blocks), so we need $(v-5)/2 > 4$, $v > 13$. So $v = 7, 13$ are not covered. If $v \equiv 3 \pmod 6$, one can use a $KTS(v-12)$ with 12 parallel classes; the 12 extra elements are formed into the blocks of a $PB(12; \{4,3\}; 1)$ and added on. This works when $(v-13)/2 \geq 12$, or $v \geq 37$. (Case $v = 12$ does *not* have to be excluded this time, the $PB(12; \{4,3\}; 1)$ includes both a 4-block and a 3-block.) So cases 3, 9, 15, 21, 27 and 35 remain. Of the residue, 3, 4, 6, 7, and 9 are impossible. (For 7, see Exercise 2.1.2. For 9, say 0 belongs to a 4-block. Without loss of generality, blocks containing 0 are 0123, 0456, 078. Every other block will contain exactly one element from each of these three sets; there can be at most six (three containing each of 7 and 8). So $f_4 = 2$, $f_3 \leq 7$, $12f_4 = 6f_3 = 72$—impossible. For $v = 13$ or 21, delete three elements from one block of a $(v+3)$-treatment BIBD with $k = 4, \lambda = 1$. For $v = 15$ or 27, take one treatment from the $(v+1)$-treatment BIBD. For $v = 34$, use the $PB(34; \{7,4\}; 1)$ that was constructed in this section; replace the 7-block by a $(7,3,1)$-SBIBD, and delete one element not in the 7-block. So the values are $v \equiv 0$ or $1 \pmod 3$, $v \neq 3$, 4, 6, 7 or 9.

Section 13.2

13.2.3 Call the Latin squares A and B. Construct three copies $\mathcal{N}_1, \mathcal{N}_2, \mathcal{N}_3$ of an $NKTS(v)$; $\mathcal{N}_i$ has support set $\{1_i, 2_i, \ldots, v_i\}$. We use them to construct an $NKTS(3v)$. The one-factor is the union of the three one-factors in the three $\mathcal{N}_i$ Each of the next $\frac{1}{2}(v-2)$ rounds is the union of one triangle round from each $\mathcal{N}_i$. Finally append the v rounds R_j, where

$$R_i = \{i_1\,(a_{ij})_2\,(b_{ij})_3 : 1 \leq j \leq v\}.$$

13.2.5 $r(10)$: Say the first round is $\{0123, 456, 789\}$. The 4-set in the second round must contain at most 1 element from each of $\{0123\}$, $\{456\}$, $\{789\}$—impossible. To show $r(7) = 1$ is even easier.

13.2.11 Say $v = 3n+2$. Each round contains $n-2$ triples and two quadruples, so it covers $3(n-2) + 6 + 6 = 3n + 6 = v + 4$ unordered pairs. There are $\frac{1}{2}v(v-1)$ different pairs available. So

$$r(v+4) \leq \frac{v(v-1)}{2},$$

whence

$$r \leq \frac{v(v-1)}{2(v+4)} = \left(\frac{1}{2}(v-5)\right) + \frac{10}{v+4}.$$

As $v \geq 17$, $\frac{10}{v+4} < \frac{1}{2}$, and r is an integer, so $r \leq \lfloor \frac{1}{2}(v-5) \rfloor$.

Chapter 14

Section 14.1

14.1.1 There are four places, and each can hold two values, so the number of matrices is $2^4 = 16$. By inspection half are Hadamard.

14.1.2 One can construct $2^{16} = 65,536$ matrices. To count the Hadamard ones, first count the standardized ones: There are 6. So the number with first row 1111 is 48 (for each of the 2^3 possible first columns, there are 6 matrices, one for each standardized form). Each of the 48 gives rise to 8 matrices with $(1,1)$ entry 1. So there are $8 \times 48 = 384$ Hadamard matrices with $(1,1)$ element 1, and 384 more (their negatives) with $(1,1)$ element -1. So there are 768.

Section 14.2

14.2.1 No (you need $BC^T = O$).

Section 14.3

14.3.2 We show first rows. Side 3: one must be $+++$ or $---$ (two choices). The other must be chosen from $+--$ and $-++$ (four choices: 0, 1, 2, or 3 is $+--$). Total $2 \times 4 = 8$. Side 5: one must be $+----$ or $-++++$; one must be $+-++-$ or $-+--+$; the other two are chosen from $\{++--+, --++-\}$. So there are $2 \times 2 \times 3 = 12$ cases.

14.3.3 One example:

$$\begin{bmatrix} A & B & C & D & E & F & G & H \\ -B & A & D & -C & F & -E & -H & G \\ -C & -D & A & B & G & H & -E & -F \\ -D & C & -B & A & H & -G & F & -E \\ -E & -F & -G & -H & A & B & C & D \\ -F & E & -H & G & -B & A & -D & C \\ -G & H & E & -F & -C & D & A & -B \\ -H & -G & F & E & -D & -C & B & A \end{bmatrix}.$$

Section 14.4

14.4.4(iii) $$\begin{bmatrix} J & I & I & I & I \\ I & J & I & L & M \\ I & I & J & M & L \\ I & L & M & J & I \\ I & M & L & I & J \end{bmatrix}.$$

14.4.5 (i) The minimum polynomial must divide $x^2 - n$. (ii) Trace = sum of eigenvalues (since H is a real symmetric matrix). (iii) Trace $= n = (a-b)\sqrt{n}$, so $a - b = \sqrt{n}$. So $\sqrt{n}$ is integral.

14.4.6 (i) This follows from $\lambda = 1$. (ii) B_i has u elements and each belongs to $2u$ further blocks, which must all be different from part (i). So there are $2u^2$ values j such that $a_{ij} = 1$. One can also show that $a_{ij} = a_{ik} = 1$ for u^2 values i when $j \neq k$. So we have a $(4u^2-1, 2u^2, u^2)$-SBIBD. (iii) A is symmetric and has diagonal all zeros. So the corresponding Hadamard matrix is graphical.

14.4.7 (i) The matrix is necessary regular. (ii) For side 4, use first row $+++-$ (for example). For side 16, observe that the matrix is equivalent to a $(16, 6, 2)$-difference set in Z_{16}; a relatively short exhaustive search shows that none exists.

Section 14.5

14.5.1 This follows from the uniqueness of a $(7, 3, 1)$-SBIBD.

14.5.3 (i) Take H of side n. Normalize it, and then negate the result. One ends with all the first row and first column negative, and $n/2$ positive elements in each row and column except the first. Now negate rows 2 to n and negate columns 2 to n. The final matrix has $n-1$ positive entries in row 1 and $(n+1)/2$ positive entries in each other row. The weight is

$$n - 1 + (n-1)(\tfrac{1}{2}n + 1) = (n-1)(\tfrac{1}{2}n + 2) = \tfrac{1}{2}(n+4)(n-1).$$

(ii) Write $n = 4t$; t must be odd, say $t = 2s+1$. Again normalize H; reorder the columns so that the first t columns start $+++\ldots$, the next t start $++-\ldots$, next t start $+-+\ldots$, and the last t start $+++\ldots$. Say row i (where $i > 3$) contains q_i entries $+$ among the first t columns. It is easy to see that the second, third, and fourth sets each contain q_i +s also. Either $q_i \leq s$ or $q_i \geq s+1$. For each i such that $q_i \leq s$, negate row i. The result is a matrix equivalent to H in which three rows have weight $4s+2$ and the rest have weight at least $4s+4$. So the weight is at least $3(4s+2) + (8s+1)(4s+4) = 12s+6+32s^2+36s+4 = 32s^2+48s+10 = 2(4s+5)(4s+1) = (n+6)(n-2)/2$.

14.5.4 (i) First, show that the matrix is Z-equivalent to

$$\begin{bmatrix} D & D \\ D & -D \end{bmatrix}.$$

(ii) $1, 2^{(4m)}, 4^{(4m-1)}, 4m^{(4m-1)}, 8m^{(4m)}, 16m$.

Chapter 15

Section 15.2

15.2.2 $\{16, 25, 34\}$, adder $2, 4, 1$; $\{23, 46, 15\}$, adder $1, 2, 4$; $\{45, 13, 26\}$, adder $5, 3, 6$.

15.2.4 The sums $x_i + y_i$ are all equal, so the starter is not strong.

15.2.7
$$\begin{bmatrix} 01 & 69 & (48) & & & & & 57 & 23 \\ 34 & 02 & 17 & (59) & & & & & 68 \\ 79 & 45 & 03 & 28 & (16) & & & & \\ & 18 & 56 & 04 & 39 & (27) & & & \\ & & 29 & 67 & 05 & 14 & 38 & & \\ & & & 13 & 78 & 06 & 25 & 49 & \\ & & & & 24 & 89 & 07 & 36 & 15 \\ 26 & & & & & 35 & 19 & 08 & 47 \\ 58 & 37 & & & & & 46 & 12 & 09 \end{bmatrix}$$

Section 15.3

15.3.2 In the square of side r, simply delete S and put in T.

Section 15.5

15.5.2 (i) Orthogonal Latin squares of side n.
(ii) Room square of side $2n-1$.
(iii)
$$\begin{array}{|cccc|} \hline 01 & & 24 & 35 \\ 34 & 02 & 15 & \\ & 45 & 03 & 12 \\ 25 & 13 & & 04 \\ \hline \end{array}$$

Chapter 16

Section 16.1

16.1.6 The sum of the rows of $(I - \frac{1}{v}J)$ equals the zero vector.

Section 16.2

16.2.3 *ABCD*, *CDEF*, *ABCE*, *BDEF*.

Index